Military Adaptation in Afghanistan

Military Adaptation in Afghanistan

Edited by
Theo Farrell
Frans Osinga
and
James A. Russell

Stanford Security Studies
An Imprint of Stanford University Press
Stanford, California

Stanford University Press
Stanford, California

© 2013 by the Board of Trustees of the Leland Stanford Junior University.
All rights reserved.

No part of this book may be reproduced or transmitted in any form or by any means, electronic or mechanical, including photocopying and recording, or in any information storage or retrieval system without the prior written permission of Stanford University Press.

Printed in the United States of America on acid-free, archival-quality paper

Library of Congress Cataloging-in-Publication Data

Military adaptation in Afghanistan / edited by Theo Farrell, Frans Osinga, and James A. Russell.
    p. cm.
Includes bibliographical references and index.
ISBN 978-0-8047-8588-4 (cloth : alk. paper) —
ISBN 978-0-8047-8589-1 (pbk. : alk. paper)
    1. Afghan War, 2001– . 2. North Atlantic Treaty Organization—Armed Forces—Afghanistan. 3. Operational art (Military science)—Case studies. 4. Tactics—Case studies. 5. Military policy—Case studies. I. Farrell, Theo, editor of compilation. II. Osinga, Frans P. B., editor of compilation. III. Russell, James A. (James Avery), editor of compilation.

DS371.412.M54 2013
958.104'74—dc23   2012043940

Typeset at Stanford University Press in 10/14 Minion

Special discounts for bulk quantities of Stanford Security Studies are available to corporations, professional associations, and other organizations. For details and discount information, contact the special sales department of Stanford University Press. Tel: (650) 736-1782, Fax: (650) 736-1784

CONTENTS

   Acknowledgments   vii
   Contributors   ix
   Acronyms   xi

1. Introduction: Military Adaptation in War
   Theo Farrell   1
2. The Great Game and the Quagmire: Military Adaptation in the British and Soviet Wars in Afghanistan, 1839–1989
   Daniel Moran   24
3. Into the Great Wadi: The United States and the War in Afghanistan
   James A. Russell   51
4. ISAF and NATO: Campaign Innovation and Organizational Adaptation
   Sten Rynning   83
5. Back from the Brink: British Military Adaptation and the Struggle for Helmand, 2006–2011
   Theo Farrell   108
6. The Military Metier: Second Order Adaptation and the Danish Experience in Task Force Helmand
   Mikkel Vedby Rasmussen   136

7 Soft Power, the Hard Way: Adaptation by the Netherlands' Task Force Uruzgan
Martijn Kitzen, Sebastiaan Rietjens, and Frans Osinga   159

8 Mission Command Without a Mission: German Military Adaptation in Afghanistan
Thomas Rid and Martin Zapfe   192

9 Canadian Forces in Afghanistan: Minority Government and Generational Change while under Fire
Stephen M. Saideman   219

10 Military Adaptation by the Taliban, 2001–2011
Antonio Giustozzi   242

11 Shoulder-to-Shoulder Fighting Different Wars: NATO Advisors and Military Adaptation in the Afghan National Army, 2001–2011
Adam Grissom   263

12 Conclusion: Military Adaptation and the War in Afghanistan
Frans Osinga and James A. Russell   288

Index   329

Acknowledgments

This multinational project involved thirteen contributors from six different countries. For some, thirteen may be an unlucky number, but not for us! We were fortunate to work with a wonderful team of expert authors who enthusiastically engaged with the project, and responded with remarkably good cheer to editorial guidance. Two project workshops in delightful settings, at Kreuzlingen in Switzerland in December 2010, and the Cotswolds in England in September 2011, enabled us to run through iterative drafts of the project papers and, we think, produce a far stronger volume for it. We thank the Institute for Advanced Study at the University of Konstanz in Southern Germany, especially Fred Girod, for generously supporting our first workshop. We thank the US Navy International Programs Office and the Department of National Security Affairs at the US Naval Postgraduate School for funding our second workshop, and for providing very generous financial support to the project as a whole. Theo Farrell gratefully acknowledges funding from the UK Research Council's Global Uncertainties Programme (Grant RES-071-27-0069), which also supported the first workshop, production of the volume, and his own work on the project. We are most grateful to our ever-efficient research assistants, Martin Bayly and Mike Stevens, for their help on the project. It has been a huge pleasure to collaborate, once again, with the marvelous Geoffrey Burn and his team at Stanford University Press.

Contributors

THE EDITORS

Dr. Theo Farrell is Professor of War in the Modern World in the Department of War Studies at King's College London. His latest book is *Transforming Military Power since the Cold War: Britain, France and the United States, 1991-2011* (Cambridge University Press, forthcoming), coauthored with Sten Rynning and Terry Terriff. Over 2009–10 he acted as an advisor to the British military and ISAF command on the campaign in Afghanistan.

Dr. Frans Osinga is Air Commodore and Professor at the War Studies Department of the Netherlands Defence Academy. His recent publications include *Science, Strategy and War, the Strategic Theory of John Boyd* (Routledge, 2007) and *A Transformation Gap, American Innovations and European Military Change*, edited with Theo Farrell and Terry Terriff (Stanford University Press, 2010).

Dr. James A. Russell is an Associate Professor in the Department of National Security Affairs at the Naval Postgraduate School in Monterey, California. He is author of *Innovation, Transformation and War: U.S. Counterinsurgency Operations in Anbar and Ninewa, 2005–7* (Stanford University Press, 2011).

THE AUTHORS

Dr. Antonio Giustozzi is Visiting Senior Research Fellow in the Department of War Studies at King's College London. His publications include *The Art of Coercion* (Hurst, 2011), *Decoding the Taliban* (Hurst, 2009), and *Koran, Kalashnikov and Laptop* (Hurst, 2007).

Dr. Adam Grissom is a Senior Political Scientist at RAND and Senior Visiting Research Fellow at the Center for Security Studies in Georgetown University's Edmund A. Walsh School of Foreign Service. He has authored numerous articles and studies on military innovation, special operations, and foreign internal defense.

Dr. Martijn Kitzen is a Research Scientist at the War Studies Department of the Netherlands Defence Academy.

Dr. Daniel Moran is Professor of International and Military History in the Department of National Security Affairs at the Naval Postgraduate School in Monterey, California. Recent works include *The People in Arms* (Cambridge, 2006) and *Wars of National Liberation* (Harper-Collins, 2006).

Dr. Thomas Rid is a Reader in the Department of War Studies at King's College London. His most recent books are *War 2.0* (Praeger, 2009) and *Understanding Counterinsurgency* (Routledge, 2010).

Dr. Sebastiaan Rietjens is a Research Scientist the War Studies Department of the Netherlands Defence Academy.

Dr. Sten Rynning is Professor of International Relations at the Department of Political Science, the University of Southern Denmark. He is author of *NATO in Afghanistan: The Liberal Disconnect* (Stanford University Press, 2012) and *From the Hindu Kush to Lisbon: NATO, Afghanistan, and the Future of the Atlantic Alliance* (UNISCI Press, 2010), coedited with Antonio Marquina.

Dr. Stephen Saideman holds the Paterson Chair in International Affairs at the Norman Paterson School of International Affairs at Carleton University. He has authored two books on the international relations of ethnic conflict and coedited a volume on intrastate conflict. He spent one year on the US Joint Staff via a Council on Foreign Relations International Affairs Fellowship.

Dr. Mikkel Vedby Rasmussen is Professor of Military Studies at the Department of Political Science, University of Copenhagen. He is author of *The Risk Society at War* (Cambridge, 2006) and a book in (published in Danish) on Denmark's war in Helmand (2011).

Dr. Martin Zapfe heads the Global Security research group at the Center for Security Studies (CSS) at the ETH Zurich. He has published on foreign and security policy for the German Ministry of Defense, Stiftung Wissenschaft und Politik, and the NATO Defense College. Previously, he worked as a consultant on international and security affairs at Joschka Fischer & Company in Berlin.

Abbreviations

3D	Defense, Development, and Diplomacy
AAF	Afghan Air Force
ACO	Atlantic Command Operations
ACT	Allied Command Transformation
ADZs	Afghan Development Zones
ANA	Afghan National Army
ANASF	Afghan National Army Special Forces
ANSF	Afghan National Security Forces
AOR	Area of Responsibility
ASTOR	Airborne Stand-Off Radar
BG	Battle Group
C2	Command and Control
CAPTF	Combined Airpower Transition Force
CAS	Close Air Support
CBC	Canadian Broadcasting Corporation
CDS	Chief of Defence Staff (UK/Canada)
CENTCOM	Central Command
CEFCOM	Canadian Expeditionary Forces Command
CF	Canadian Forces
CFC-A	Combined Forces Command Afghanistan
CJSOTF-A	Combined Joint Special Operations Task Force Afghanistan
CJTF	Combined Joint Task Force
CIA	Central Intelligence Agency
C-IED	Counter-Improvised Explosive Device

CIMIC	Civil-Military Cooperation
CIVMIL	Civilian-Military
CLBs	Corps Logistics Battalions
CNAS	Center for a New American Security
COIN	Counterinsurgency
COMIJC	Commander ISAF Joint Command
COMISAF	Commander International Security Assistance Force
CONOPS	Concept of Operations
CS	Combat Support
CSPMP	Comprehensive Strategic Political-Military Plan
CSS	Combat Service Support
CSTC-A	Combined Security Transition Command—Afghanistan
CUAT	Commander's Unit Assessment Tool
DANCON	Danish Contingent
DCDC	Doctrine, Concepts, and Development Centre (UK)
DCDS	Deputy Chief of Defence Staff (UK/Canada)
DCU	Dutch Consortium for Uruzgan
DDR	Disarmament, Demobilization, and Reintegration
DND	Department of National Defence (Canada)
DoD	Department of Defense (US)
EBAO	Effects-Based Approach to Operations
EOD	Explosive Ordnance Disposal
ESC	Executive Steering Committee
ETT	Embedded Training Teams
EU	European Union
FAC	Forward Air Controller
FM 3-24	Field Manual 3-24 (US Marine Corps Counterinsurgency Doctrine Publication)
FOB	Forward Operating Base
G8	Group of Eight (largest economic powers)
GDP	Gross Domestic Product
GIRoA	Government of the Islamic Republic of Afghanistan
GoU	Government of Uruzgan
HMMMV	High Mobility Multipurpose Wheeled Vehicle, or "Humvee"
HQ	Headquarters
HTTs	Human Terrain Teams
HVT	High Value Target

IED	Improvised Explosive Device
IJC	ISAF Joint Command
IO	International Organization
ISAF	International Security Assistance Force
ISI	Directorate for Inter-Services Intelligence (Pakistan)
ISR	Intelligence, Surveillance, and Reconnaissance
JALLC	Joint Analysis and Lessons Learned Center
JFCOM	Joint Forces Command
JFHQ	Joint Force Headquarters
JFTC	Joint Force Training Center
JHP	Joint Helmand Plan
JIEDDO	Joint Improvised Explosive Device Defense Organization
JWC	Joint Warfare Center
KGB	Komitet Gosudarstvennoy Bezopasnosti (Committee for State Security)
KGv COIN	Konzeptionelle Grundvorstellungen zum militärischen Beitrag zur Herstellung von Sicherheit und staatlicher Ordnung in Krisengebieten (Conceptual Basic Thoughts on the Military Contribution to the Establishment of Security and Public Order in Crisis Regions, German Army Doctrine Publication)
KMTC	Kabul Military Training Center
KSK	Kommando Spezialkräfte
LoE	Lines of Effect
MC	Military Committee
MCWL	Marine Corps Warfighting Laboratory
MEB	Marine Expeditionary Brigade
Medevac	Medical Evacuation
MEF	Marine Expeditionary Force
MoD/MOD	Ministry of Defense
MOGs	Mobile Operations Groups
MOOTW	Military Operations Other Than War
MRAP	Mine-Resistant Ambush Protected
NAC	North Atlantic Council
NATO	North Atlantic Treaty Organization
NCO	Non-Commissioned Officer
NDHQ	National Defense Headquarters
NEC	Network Enabled Capability

NFS	NATO Force Structure
NGO	Non-Governmental Organization
N-KET	Non-Kinetic Effects Team
NTM-A	NATO Training Mission Afghanistan
NTM-I	NATO Training Mission Iraq
ODA	Operational Detachment Alpha
OEF	Operation Enduring Freedom
OGOs	Other Governmental Organizations
OIF	Operation Iraqi Freedom
OMF	Opposing Military Forces
OMLT	Operational Mentor and Liaison Team, pronounced "omelet"
OODA	Observe, Orient, Decide, and Act
Op	Operation
PJHQ	Permanent Joint Headquarters (UK)
PRT	Provincial Reconstruction Team
PSE	Psychological Element
PsyOps	Psychological Operations
QRF	Quick Reaction Force
R&D	Reconstruction and Development
RC	Regional Command
RC-E	Regional Command East
RC-N	Regional Command North
RC-S	Regional Command South
RCC	Route Clearance Company
RMA	Revolution in Military Affairs
ROEs	Rules of Engagement
RPG	Rocket Propelled Grenade
S&R	Stabilization and Reconstruction
SACEUR	Supreme Allied Commander Europe (NATO)
SCR	Senior Civilian Representative (NATO)
SFIR	Stabilization Force in Iraq
SHAPE	Supreme Headquarters Allied Powers Europe (NATO)
SMA	Sher Mohammed Akhonzada
SOCOM	Special Operations Command
SOF	Special Operations Forces
SOPs	Standard Operating Procedures
SSTR	Stability, Security, Transition, and Reconstruction

TCAF	Tactical Conflict Assessment Framework
TCAPF	Tactical Combat Advisory Planning Framework
TFU	Task Force Uruzgan
TICs	Troops in Contact
TLO	The Liaison Office
TTPs	Tactics, Techniques, and Procedures
UAVs	Unmanned Aerial Vehicles
UCP	Uruzgan Campaign Plan
UN	United Nations
UNAMA	United Nations Assistance Mission in Afghanistan
UNESCO	United Nations Educational, Scientific, and Cultural Organization
UOR	Urgent Operational Requirement
US	United States
USAID	United States Agency for International Development
USG	United States Government
USMC	United States Marine Corps
USSR	Union of Soviet Socialist Republics
VHF	Very High Frequency (radio)
VIP	Very Important Person
VSO	Village Stability Operations
VSP	Village Stabilization Program
WW1	World War I
WW2	World War II
ZOA	Zuidoost-Azië, meaning "Southeast Asia" in Dutch

Military Adaptation in Afghanistan

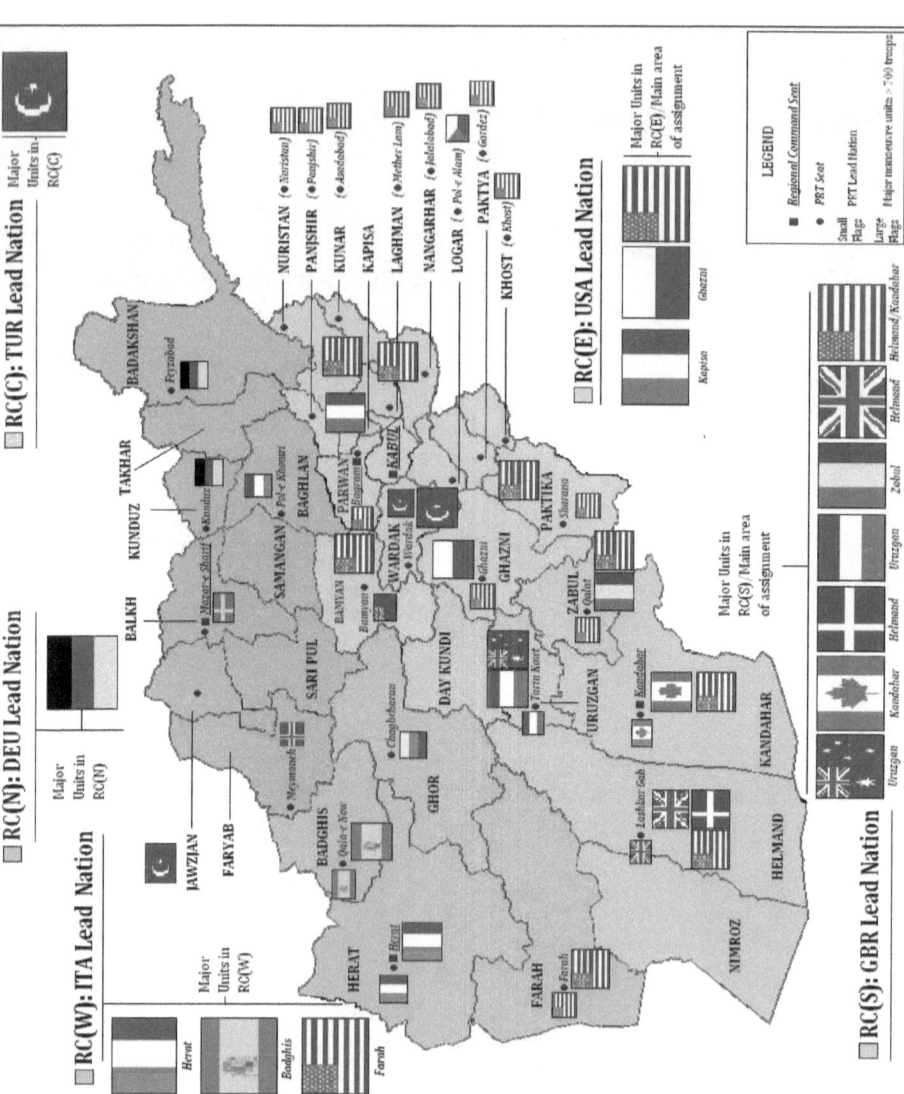

1

Introduction: Military Adaptation in War

Theo Farrell

The current war in Afghanistan has been ongoing now for almost a decade. How have Western states and militaries adapted to the challenges of this war? The North Atlantic Treaty Organization (NATO) took charge of the International Security Assistance Force (ISAF) for Afghanistan in 2003, and gradually expanded ISAF out from Kabul to the provinces from 2004 to 2006. Most of the European partners in ISAF conceptualized the mission their forces would conduct not as war at all, nor even counterinsurgency (COIN), but as a stabilization and reconstruction, only to find out that in Afghanistan this might actually require significant combat. How have their armed forces and the political leadership reacted? As ISAF expanded into the south and east of Afghanistan, it encountered a far more resistant and capable insurgency than had been anticipated. How did NATO and its member states respond? And as the campaign has evolved, key operational imperatives have clearly emerged, including military support to the civilian development effort, closer partnering with Afghan security forces, and greater military restraint. How have the different militaries in ISAF adapted in response to these imperatives?

History clearly shows that war forces states and their militaries to adapt. So it has been for the NATO partners in Afghanistan. All have adapted, in various ways and to varying extents, to campaign imperatives and pressures. This book explores how they have done so. We explore the *extent* and *processes* of military adaptation. Most of the cases we examine—Britain, Canada, Denmark, the Netherlands, and the United States—have been involved in the fighting in the south and east of Afghanistan. Because of this these states and militaries have faced the most intense operational and strategic pressures to adapt. We also examine the case of Germany, as a major troop contributing state facing the greatest political and cultural constraints, and whose military is deployed on a stabilization mission in the north. Afghanistan has been NATO's first true test as a global security organization, and so we examine how it has adapted to the task. Two other military organizations are also crucial to the story of war and adaptation in Afghanistan—namely, the Afghan National Army (ANA) and the Taliban. NATO's exit strategy for this war depends on the ANA's taking over responsibility for securing Afghanistan. We look at how well the ANA has adapted to the challenges it faces. To complete our picture, we must look at "the other side of the hill," to see how NATO's opponents are adapting. Of course, this is not the first time that foreign powers have intervened in Afghanistan. Thus, we also look at military adaptation in previous campaigns in Afghanistan by the British and Soviet militaries to see what specific lessons may be learned from history.

In this book, military adaptation is defined broadly as *change to strategy, force generation, and/or military plans and operations, undertaken in response to operational challenges and campaign pressures*. Force generation includes force levels, equipment, training, and doctrine. Thus our definition covers adaptation at the strategic level (strategy and the mobilization of resources) and operational level (preparations for and conduct of operations).

In most cases, states and militaries may be expected to adapt in order to improve military performance and campaign prospects. But we should recognize that, in some cases, states may adapt in response to political pressures in ways designed to minimize the costs of, or national commitment to, an operation. Equally, militaries may adapt in ways intended to reduce operational risks. In other words, military adaptation is not always for the betterment of the campaign. Such behavior has been included in our definition of adaptation because it reflects the reality of those states that contribute forces to a military coalition primarily for political reasons tangentially related to the outcome of the military campaign (that is, for reasons of domestic or alliance politics).

This book conceives of military adaptation as occurring at two levels. At the strategic level, states adapt when they change strategy, force levels, and/or resources (including acquiring new equipment) for the military campaign. Obviously, states may increase national effort to achieve strategic success, pouring more troops and resources into the campaign. But equally, as suggested above, states may adapt to a failing campaign by reducing resources or adopting a new strategy. At the operational level, military organizations adapt when they change how they prepare for, plan, and/or conduct operations. Thus we include adaptation in military tactics under the broader category of operational adaptation. Under this category adaptation may also cut both ways, and involve accepting more or less risk in operations, using more or less firepower, relaxing or tightening up rules of engagement, and so forth.

Here I introduce the common analytical framework for this book that is applied in the case studies. Our framework interrogates the mix of drivers and factors that has shaped military adaptation in each case. The most common and significant driver is operational challenges. Such challenges will be a powerful trigger of adaptation when they significantly increase risks to friendly forces or threaten to derail the mission. This should come as no surprise. Technological change is another significant driver, especially in opening up new opportunities to response to operational challenges. This too is to be expected. Perhaps more surprising is the extent to which four other factors—domestic politics, alliance politics, strategic culture, and civil-military relations—shape the process of military adaptation, and not always in ways that are helpful to the campaign. The case studies reveal how these factors often interact with operational challenges to shape how a state or military adapts.

At the same time, this book has sought to avoid weighing down our case studies with theory. All case study chapters follow the analytical framework. But our case study experts have been asked, first and foremost, to "tell the story" of military adaptation in their case. In some chapters the narrative focuses more on strategic than operational adaptation (such as the Canadian case) and vice versa (such as the US case).

In what follows, I begin by discussing the importance of adaptation in war. I then explore the distinction between military innovation and adaptation. Finally, I flesh out the common analytical framework.

WHY MILITARY ADAPTATION MATTERS

War invariably throws up challenges that require states and their militaries to adapt. Indeed, it is virtually impossible for states and militaries to anticipate all of the problems they will face in war, however much they try to do so.[1] Even when states have good intelligence data on enemy intentions and capabilities, and on the social and geographical environment of operations, mistakes are commonly made when it comes to analyzing this data, digesting the analysis, and devising appropriate responses.[2] Cognitive limitations, organizational politics, military culture, and civil-military relations, may operate individually or collectively, to pervert timely and accurate strategic assessments.[3] Hence Field

Marshall Helmuth von Moltke's famous principle that "no plan survives contact with the enemy."[4] States may underestimate the scale of the military challenge and accordingly the level of resources that must be committed for campaign success. Militaries may misunderstand the character of the conflict, or may be caught off guard by new technologies or tactics employed by opponents. Thus the imperative for adaptation is often a product of strategic, technological, or tactical surprise.[5] But it need not be; simple unfamiliarity with the political or battlefield terrain, one's allies, or one's own new technology, may require learning and adaptation on the job.

For example, Britain faced growing pressure for strategic and operational adaptation in World War I (WW1). Britain was slow to appreciate the need for general mobilization of the population and economy. It was two years into the war before the cabinet came around to the view of Field Marshall Lord Kitchener, the Minister for War, that Britain "must be prepared to put armies of millions in the field and to maintain them for several years."[6] It also took the British Army about two years to learn how to combine artillery barrages and infantry assaults effectively to breach the German defensive line.[7]

A more recent example is the British military campaign in Iraq from 2003 to 2008. Following the successful US-led invasion of Iraq, Britain was given responsibility for securing the country's second largest city, Basra, and the surrounding provinces in southeast Iraq. The British military mistook a growing Shia insurgency for general lawlessness, and failed to appreciate the determination and growing capabilities of Shia militias. Rather than increase its military effort in response to this growing threat, Britain continued to follow a phased drawdown of its forces from Iraq. By 2007, Shia militias had seized control of Basra from the British, and it was left to the Iraq Army to regain control a year later.[8] This was a clear case of failure to adapt at the operational and strategic level. In fairness, the US military was also slow to recognize and respond to the threat from the Shia insurgency.[9] However, the United States did finally adapt at the strategic level with a surge of forces, and the US military adapted at the operational level by developing new counterinsurgency tactics and capabilities.[10]

The above cases, and others from history, underline the importance of military adaptation. Simply put, states and militaries that fail to adapt risk defeat in war, as the British discovered in Basra in 2007–8. In contrast, the United States did adapt its strategy and approach to operations, and turned around a failing campaign, in 2007–8. Other prominent examples are the fall of France in 1940 and Soviet failure in Afghanistan from 1979 to 1989. In each case, defeat was suffered at the hands of a less powerful opponent (in material terms). The French lost to a less well equipped German army in 1940 because they were unable to adapt to German blitzkrieg tactics. Similarly, the Soviet Army that invaded Afghanistan was trained and equipped for conventional warfare. As Chapter 2 shows, the Soviets were slow in adapting to a far more dynamic guerrilla enemy, and remained wedded to firepower in their approach to counterinsurgency operations.[11]

The above cases also point to the difficulty that states and militaries can have in adapting when required. The point is that there is nothing natural or easy about military adaptation, be it at the strategic or operational level. A war strategy is often underpinned by considerable political and material investment, making it difficult for a state to contemplate changing it. Often, staying the course and "doubling-down" make better politics.[12] Furthermore, changing strategy can send the wrong signal to allies, enemies, and home publics (that of a failing campaign). In any case, it takes time for strategy to produce military and political effects, good or otherwise. This gives states added reason to delay, in order to give war strategy time to work.

Operational level adaptation should be easier because it usually involves smaller scale change. But the fact is that militaries find change difficult. Organizations are not designed to change, bound as they are by operating routines, bureaucratic interests, and cultural preferences. Organizations are understandably reluctant to abandon that which they have invested in and become good at: the concept of the "competency trap" nicely captures this problem.[13] Military organizations are especially disinclined to change, as closed and socially conservative communities that, especially in the West, exist apart from the rest of society.[14] Military routines, interests, and culture mutually reinforce an organization's preferred ways of war. Moreover, when at war, a military has strong incentives to stick with those ways of operating that have been tried and tested, and for which the organization has trained and is equipped: the opportunity costs of introducing new ways of operating in the midst of war are high, especially if the new way does turn out to be not so effective.[15] Where militaries deploy units on a rotation cycle through operations for a limited duration (as is common in Western forces), this can adversely impact operational adaptation insofar as it hinders the institutionalization of lessons learned.

This explains why learning is so difficult for militaries. And yet, states and militaries do adapt in war, raising questions of when and how. Our analytical framework is designed to address these questions.

MILITARY INNOVATION AND ADAPTATION

In the 1980s and 1990s a rich stream of scholarship by social scientists emerged on military change. Most of these studies focused on explaining military innovation. As Adam Grissom notes in his review of this literature, while there is no agreed upon definition in the field, a tacit definition may be discerned that encompasses three elements: military innovation involves major organizational change, "is significant in scope and impact," and is "equated with greater military effectiveness."[16] Basically, almost all works in the field look at historical cases of military change that have had a major impact on the conduct of warfare, and seek to unpack how that innovation came about.

This scholarly focus on military innovation is understandable for three reasons. First, it involves the puzzle of major organizational change by conservative militaries. If militaries are disinclined to change, this will be doubly true for innovation. Change on this scale is especially disruptive for organizations, often requiring "forceful abandonment of the old." For the military undertaking innovation, this process of "creative destruction" is often painful.[17] Second, military innovation involves change that by definition is important, and has consequences for national policy and international politics. Military innovation often involves major expenditure of national resources, and can have consequences for regional balances of power.[18] Finally, military innovation studies got caught up with the growing scholarly and policy interest in the 1990s in understanding revolutions in military affairs.[19] In contrast, military adaptation held little interest for social scientists. It remained a topic for military historians in the context of studies of specific battles and campaigns.

Those few studies that have explicitly considered military adaptation have taken it to mean less significant change with little or no impact on strategy and organizational structures. In an earlier study, Theo Farrell and Terry Terriff offered the following definitions: "Innovation involves developing new military technologies, tactics, strategies, and structures. Adaptation involves adjusting existing military means and methods." Farrell and Terriff suggested that adaptation can accumulate to innovation: "Adaptation can, and often does, lead to innovation when multiple adjustments over time gradually lead to the evolution of new means and methods."[20] Similarly, in his study of the US Army and US Marine Corps in Iraq in 2005–7, James Russell shows how "tactical adaptation can serve as a way station along the route toward more comprehensive innovation."[21]

However, there is a problem with the assumption that military adaptation is about less significant change. As this book will show, adaptation may occur at the strategic level. Indeed, the imperative for strategic adaptability is officially recognized in UK defense policy.[22] The point is that any change at this level is significant. Even adaptation at the lowest operational level, such as adapting tactics, techniques, and procedures (TTPs), can add up to significant change in a military's capabilities or approach to operations. Tighter TTPs to produce more restraint and great discrimination in the use of lethal force were central to improving the US approach to COIN in Iraq.[23] Conversely, failure to adapt at the tactical level may over time result in strategic failure.[24]

Overall, we do not consider it feasible or fruitful to draw too fine a distinction between adaptation and innovation. Indeed, it may be more helpful to think of the two as points on a sliding scale. In this book, military innovation implies a greater degree of novelty and disruptive organizational change than adaptation.[25] This is consistent with previous studies by Farrell and Russell. For Russell, wartime innovation involves the development of "new organizational structures and new organizational capacities built in war." In his study on the British COIN campaign in Afghanistan, Farrell argues that when military adaptation "involves doctrinal or structural change, or the acquisition of a brand new technology, it crosses the threshold into innovation."[26] Consistent with Farrell, Terriff, and Russell, we allow that adaptation may lead to innovation. But equally, it need not. Departing from the earlier scholarship, as outlined by Grissom, we do not consider "significance" of change to be a criterion for distinguishing adaptation and innovation. As discussed above, adaptation occurring at the strategic level can be very significant indeed.

Adaptation	Innovation
←——————————————————→	
Adjusting strategy	Switching strategy
Revising ROEs	Adjusting mandate
Supplementing forces	Surging forces
Retro-fitting equipment	Acquiring new equipment
Adjusting training	Developing new doctrine
Adjusting tactics	New approach to operations

FIG. 1. The sliding scale of military adaptation and innovation

THE COMMON ANALYTICAL FRAMEWORK

We use a common analytical framework to ensure a degree of commonality across our case study chapters—not so much to get in the way of "telling the story" but enough to facilitate comparisons across the case studies. All chapters adopt the definition of military adaptation given at the beginning of this chapter. All follow the book concept of adaptation at the strategic and operational levels. Each level may involve adaptation in a number of key areas that are identified in Table 1. As we noted before, the chapters do not give equal treatment to both levels in their analysis. For example, the German case is primarily about strategic adaptation, whereas the US case is more focused on operational adaptation. Overall the chapters provide a range of case studies of strategic and operational adaptation in Afghanistan.

In each case, we have asked authors, insofar as possible, to identify the key moments in their respective narratives when important adaptations occurred. Predictably, such "adaptive moments" are more evident in some case studies than in others. Our analytical framework recognizes two drivers and four shapers of military adaptation. These are also listed in Table 1 and elaborated below.

TABLE 1. Drivers and Shapers of Military Adaptation

Drivers	Shapers	Adaptations
Operational Challenges	1. Domestic politics	**Strategic:** Strategy, force levels and resources
	2. Alliance politics	
New Technologies	3. Strategic culture	**Operational:** Doctrine, training, plans and operations
	4. Civil-military relations	

ADAPTATION DRIVERS

In war, pressures from operations are the most important driver of military adaptation. Military history suggests as much.[27] These pressures can take the form of new operational challenges or intensification of existing operational challenges. Such challenges may require militaries to adapt their training, doctrine, plans, or conduct of operations. If the challenges are severe enough, they may require states to adapt their campaign strategy, force levels, or resources.

Operational Challenges

Our definition of military adaptation points to the importance of this driver. Operational challenges include intense combat over a protracted period, new enemy tactics, conducting operations at great strategic distance, operating in a demanding physical environment, having to depend on unreliable allies, and having to work with civilian partners to achieve campaign objectives.

Operational challenges may aggregate or reach such an intensity as to cause the military campaign to begin to fail. Military innovation theory suggests that such situations are highly likely to trigger change. In the seminal work in the field, Barry Posen noted the causal link between failure and innovation as a recurring theme in organization theory. "Events understood to be serious failures challenge the organization's basic existence."[28] Applied to the arena of war, this suggests that military defeat can spur innovation. Posen's study further suggests that the mere prospect of defeat is enough to trigger military innovation. One might reasonably suppose that there is a proportionate causal relationship between the scale of failure and the speed and extent of innovative change. However, bureaucratic routines, norms, and interests can prevent organizations from correctly diagnosing failure and taking remedial action.[29] Hence, failure is not enough; there must be recognition by military commanders and policy-makers that the campaign is in trouble. At the strategic level, this requires mechanisms to enable civilian and military leaders to track and assess military progress, identify underperformance, and adjust strategy, forces, or operations as required.[30]

New Technologies

A second driver of military adaptation is technological change. Unlike operational challenges, new technologies often provide imperatives *and* opportunities for military adaptation. Indeed it is best to think of these drivers as existing in symbiotic relationship. The arrival of new technologies on the battlefield, or adaptive use of old technologies by opponents, creates new operational challenges. These and other operational challenges generate requirements for new technologies and associated organizational capabilities. From the opponent's perspective, this in turn can create new challenges and requirements for new technological capabilities. The competition between Improvised Explosive Device (IED) tactics and technologies, and the counter-IED capabilities in Iraq and Afghanistan is a prime example of this.

Most military innovation studies emphasize the central role of technological change. This is well demonstrated in the revolutions in land, sea, and air war in the early- to mid-twentieth century. The thirty years stretching across the two world wars saw innovations in armored and mechanized warfare, carrier and subsurface warfare, and strategic bombing. New technology made all these innovations possible. However, technological change had to be accompanied by organizational and doctrinal change in order to realize these new ways of war.[31] The lesson is that technology alone does not produce military innovation. All the same, with the gathering pace of technological change in the twentieth and early twenty-first centuries, this macrodriver assumes ever greater importance.[32] The self-conscious attempt by the US military to harness the revolution in information technology and effect a revolution in military affairs (RMA) is a vivid illustration of this trend. The extent of emulation by allies and other states around the world of the US-led RMA suggests that this is more than a passing American fancy.[33]

SHAPING FACTORS

New operational challenges and new technologies do not automatically produce military adaptation. Domestic and alliance politics, strategic culture, and civil-military relations may each shape the process whereby states and militaries respond to imperatives and opportunities to adapt.

Domestic Politics

As an alliance of democracies, domestic politics looms large in how NATO and its member states wage war. Indeed, domestic politics makes it difficult for some states—Germany, the Netherlands, and Norway—even to recognize that NATO is at war in Afghanistan. This is not to suggest that democracies are pacific. The historical record is clear on this. Democracies are just as war prone as nondemocracies. Democracies rarely, if ever, fight each other. But they do have a habit of fighting nondemocratic regimes, as recently demonstrated in NATO's wars against Serbia and the Taliban in Afghanistan, the US-led war against Iraq, and the Anglo-French-led war against the Libyan regime of Colonel Gaddafi.[34]

Nonetheless, two historical trends are discernible with implications for the domestic politics of NATO states using military force. First, war has gone out of fashion in Western Europe. Over time, major war involving European societies has become unthinkable.[35] This trend is rooted the trauma of the two world wars that devastated European societies.[36] The civil wars that accompanied the bloody collapse of Yugoslavia suggest that organized violence still had a place in the politics of some European states, and this appears to remain the case in the eastern reaches of the continent. But with the end of the Cold War, Western Europe has steadily demilitarized, with states moving from large conscript forces to smaller, professional ones. European states still wage war, but these have been far removed from European societies, mostly smaller affairs, and handled by professional forces. As Colin McInnes put it, war has become a "spectator-sport" for European societies.[37]

This gets us to the second trend. The wars that the West has fought since WW2 are not wars of survival. Indeed, many of the military campaigns conducted by Western states have been humanitarian in purpose, intended to "save strangers," and not to advance national security.[38] These humanitarian wars combine a more assertive approach to peace operations by the United Nations, with a new "muscular liberalism" on the part of some Western states.[39] The distinction that has been drawn between the "wars of choice" of the post–Cold War era and earlier "wars of necessity" probably overstates the situation, but it does get to the essential point that recent wars have involved a much larger margin of discretion on the part of Western states, with regard to whether and how to get involved.[40] This, in turn, raises the significance of domestic politics.

The most obvious way that domestic politics may impact on national policy is by degree and trend in public support for the mission. However, the relationship between public support and policy is not as straightforward as one might think. Among the states we look at in this book, the level of public support for the Afghanistan war varies considerably. The average level of support among polled individuals from 2006 to 2009 was 55 percent in the United States, 49 percent in Denmark, 43 percent in the Netherlands, 40 percent in Canada, and 32 percent in Britain. The first thing to note is that among these countries, the level of support is lowest in Britain, and yet Britain is both the second largest troop contributor to ISAF, and tripled its force from 3,000 to 9,500 over this period. Sarah Kreps finds that poor or falling public support does not automatically lead to a policy decision to reduce the military commitment to Afghanistan. Indeed, average scores for public support are low in Spain (33%), France (33%), Poland (21%), and Turkey (18%), yet all of these states decide to *increase* their forces in Afghanistan over this period. Kreps argues that consensus among domestic political elites in favor of the campaign explains "how leaders are not running for the exits in Afghanistan when their publics would prefer that they do."[41] This finding points to the need to examine public opinion in the context of political debate.

Consistent with our case study approach, we have asked those authors that do examine strategic adaptation (that is, the British, Canadian, Danish, and German case studies) to consider how politics has affected decisions concerning strategy, force levels, and resources. Different aspects of domestic politics will have differing importance across these cases. Falling public opinion is not a significant factor in the British or Danish cases, but it is so in the Canadian and Dutch cases. Legislative oversight plays a minor role in the British case, but is very important in the German case. The lack of attention by political leaders on Afghanistan is a major problem in the British case; Afghanistan grabs the attention of political leaders in Canada, Germany, and the Netherlands. Financial pressures are significant in the British case but are less so for the other cases. Not only will the relative importance of public opinion, legislative oversight, political attention, and financial constraints vary across the cases, so will the extent and effect of their interaction.

Alliance Politics

Being a NATO-led war, alliance politics is also an important shaping factor in how individual states and their militaries adapt. Working within alliance invariably involves compromise and deliberation, at the cost of freedom and speed of action. Allies are also dependent on each other's militaries, and here too the ability of the most capable to adapt may be hindered by the shortcomings of the less capable militaries. Thus, in both world wars, the preferred strategy of the senior partner in the Western coalition (France in WW1 and the US in WW2) was frustrated by the inability of Britain to support it. The United States wanted to launch the liberation of France in 1943, in part to take pressure off the Eastern front and support the Soviet Union. But this was too early for the British Army, which was occupied in North Africa at the time, and the British government was content to see the Germans and Soviets fight it out; hence, the compromise was to delay by one year.[42] Similarly, in WW1, France wanted to launch a "big push" on the Western front at the Somme in 1916. However, once again, the British Army was not ready. However, this time Britain had no choice but to go along in order to demonstrate to France that it was fully committed to the coalition war effort. This case also demonstrates how alliance politics can frustrate the ability of the junior partner to adapt strategically. The original plan was thrown into disarray by the German offensive at Verdun in

February 1916. A large number of French divisions scheduled for the Somme offensive had to be sent instead to defend Verdun. Lacking the necessary forces to break through the German defensive line at the Somme, the coalition should have adapted its strategy. But the French High Command wanted the Somme offensive to proceed in order to relieve Verdun, and the British were committed to it for reasons of alliance politics.[43]

The 1999 Kosovo War also demonstrates the frictions of alliance warfare, and especially how alliance deliberation reduces adaptability. From the start, NATO planned for a limited and highly restrained war. A land invasion was ruled out because of US political sensitivities. Expecting Serbia to cave in easily, NATO planned a short air campaign against Serbian military targets in Kosovo. When Serb forces responded by increasing their attacks on Kosovo civilians, the Supreme Allied Commander of NATO sought to adapt the air campaign by escalating NATO bombing in tempo and the range of targets. Eventually, after nine weeks of deliberation, NATO escalated to really punishing bombing, hitting dual-use targets (that is, civilian objects of potential military use, such as transport, energy, and communications infrastructure) in and around the Serbian capital of Belgrade. The process of NATO adaptation was a torturous one. Some European states (especially Germany) were highly nervous about the expansion of the bombing campaign, and this held back those states (Britain and the United States) that favored supporting the Supreme Allied Commander Europe (SACEUR).[44] Each new target set had to be approved by the North Atlantic Council (NAC), NATO's senior decision-making body. Moreover, many sensitive targets had to be approved by all (then) twenty-eight NATO capitals.[45] This reduced the rate and effectiveness military operations. One official US government report found that the high-level approval process "led to approved targets being provided on a sporadic basis, which limited the military's ability to achieve planned effects and mass." Moreover, at the end of the war, "over 150 targets were still waiting approval."[46]

The literature on alliance politics points to another problem, or more accurately a dilemma typically facing partners in alliances between doing too much and being taken for a ride, and doing not enough and being sidetracked or even abandoned by the lead power.[47] NATO has struggled with this "burden-sharing" problem throughout its existence, with America regularly complaining that the Europeans do not pull their weight.[48] Arguably the desire to shore up the transatlantic relationship and keep America engaged was behind the European-led effort to have NATO take charge of ISAF in 2003 and expand out into Afghanistan.⁴⁹ Since 2010 European commitment has waned, and America is once again in the position of grumbling about the lackluster European effort in Afghanistan.⁵⁰

This burden-sharing dilemma manifests in alliance considerations that are specific to individual partners. Britain claims for itself a "special relationship" with the United States, and seeks whenever possible to be the number one partner in any military coalition undertaken with the United States. In contrast, Denmark sees Britain as its partner of choice, and is keen to integrate its military effort within a larger British force construct. For Germany, NATO is the essential framework for military action, and accordingly, its driving concern is to prove and sustain the viability of NATO as a global security organization. These more bespoke alliance concerns will affect the policy choices of these governments when it comes to adapting to the challenges in Afghanistan. Moreover, domestic politics and alliance politics may be expected to interact in shaping national policy. Naturally, governments routinely juggle international and domestic political imperatives. But governments may also engage in a two-level game, seeking to preserve policy options ironically by invoking domestic constraints in negotiations with allies, and invoking alliance obligations in domestic political discussions.⁵¹

For all the reasons elaborated above—strategic compromise, deliberation processes, multinational command challenges, and burden-sharing problems—when alliance politics is present as a shaping factor, generally we expect it to delay and limit military adaptation.

Strategic Culture

Culture also delimits and shapes military adaptation. Culture makes some options for military change preferable and some options simply inconceivable. Strategic culture is the sum of beliefs about the use of force that are shared by the military and policy communities of a state. Such beliefs, or norms, prescribe when and how military force may be used. Norms also proscribe certain modes and means of warfare. Some of these prohibitions are particular to national strategic cultures; others are codified in international law. Given the moral and political force of norms, it is very hard for states and militaries to act, or even imagine acting, in ways that are inconsistent with strategic culture.⁵²

For example, the rise of antimilitarist strategic cultures in German and Japan following WW2 has severely constrained the use of force by those states. Japan has not gone to war since 1945.⁵³ Germany likewise turned its back on military power. The experience of Afghanistan is slowly changing things, with the German polity beginning to accept that its military may have to use lethal force, under tight constraints, in order to support the NATO mission. Within Europe, it is possible to discern different strategic cultures producing differing national patterns of strategic behavior in the use of force. For instance, British and French both have strategic cultures that are both supportive of the use of force in a wide range of circumstances beyond strict self-defense. However, Britain prefers to fight alongside the United States, and is predisposed to joining US-led military campaigns, whereas France is far more reluctant in backing US military endeavors.[54]

Strategic culture also shapes the kind of militaries that states raise. For example, it has been argued that the United States has a technocentric strategic culture, and that this produces a style of warfare that relies intensively on technology. The impact of technological change on modern warfare has already been noted, and it would be true to say that all Western militaries became ever more enthusiastic consumers and users of technology as the twentieth century progressed. Moreover, navies and air forces literally depend on technology to stay afloat (or submerged but not sunk) and in the skies. With the most technologically advanced military in the world since WW2, there is plenty of evidence pointing to a US technocentric strategic culture.[55]

Military strategy and operations may be also shaped by strategic culture. For example, it has been argued that Germany's extremely risky and ultimately self-destructive strategy in WW1 was shaped by an extremist military culture that pursued military necessity to the brutal end without regard for strategic or moral considerations.[56] How militaries operate may be shaped by cultures that are particular to military organizations (as a subset of strategic culture). For example, the Royal Navy's great history and culture prizing ships of the line led to a focus on battleships in WW2, whereas the younger German Navy, with no such tradition, concentrated more on its U-boats. Accordingly, the German Navy was more prepared and better able than the Royal Navy to exploit unrestrained submarine warfare.[57]

Culture makes some options for military change possible, and others impossible. Culture frames how actors see the world, leading to a focus on some problems and a neglect of others. Culture also frames the search for solutions.[58] In this way, culture shapes what and how militaries learn. For example, John Nagl has argued that the British Army was an effective "learning organization" because it had an organizational culture that was suited to counterinsurgency operations. He contrasts this with the US Army in Vietnam, which was slow to learn and adapt to COIN because its organizational culture was focused on major conventional warfare.[59] Of course, the role of culture should not be overstated. Fast forward to 2004–7 in Iraq and, as we have noted, the British Army failed to adapt to the demands of the military campaign, whereas the US Army did so successfully.

If military change does come, it will usually be consistent with the core norms and identity of a community. Sometimes visionary political or military leaders perceive the need for more radical change, in strategy or military doctrine. In his study of the US Marines in the 1970s and 1990s, Terry Terriff has shown that where this challenges the core of strategic or military culture, it is less likely to succeed. Coming out of Vietnam, the US Marines had to reposture for the defense of Europe. The obvious option was to "heavy up" with armor in order to be able to fight and survive in a war against the Soviet Army. However, this option offended US Marine culture, which centered on its core mission of amphibious warfare. Accordingly, the US Marines ended up pursuing a compromise: some additional armor married with a new concept of maneuver warfare to permit US Marines to avoid head-to-head attritional battle against the Soviet armored and mechanized forces. The end of the Cold War presented the US Marines with yet another moment of possible change. This time, the commandant of the US Marine Corps, General Charles Krulak, attempted to change the very identity and mindset of the corps, from that of "warriors" to thinkers, to enable Marines to "out-innovate" their opponents and changing environment. Krulak's initiative centered on the Marines Corps Warfighting Laboratory (MCWL). However, Krulak was trying to push well beyond US Marine culture—indeed he was trying to effect cultural change. Krulak failed to change Marine identity, and his MCWL gained little traction.[60]

Generally, we may expect that when military adaptation does occur, it will be consistent with strategic (and military) culture. Should strategic or operational imperatives develop to challenge strategic culture, we may expect one of three possible outcomes. First, the strategic culture may be "bolstered" by pouring more resources behind the existing strategy or approach to operations. Second, strategic culture may be "stretched," when adaptation pushes to the very limits of what is accepted under that strategic culture. Third, strategic culture may be "broken," when military adaptation pushes beyond the boundaries of acceptable behavior.[61] Norm breaking involves situations of extreme necessity, often involving some great military failure or setback. Indeed, such traumatic events can result in old norms being ditched and strategic culture being remade. A stark example is the utter defeat experienced by Germany and Japan in WW2, which resulted in the highly militarized strategic cultures of both these states being replaced by antimilitarized ones.[62]

Civil-Military Relations

Civil-military relations occupy a central place in most studies on military innovation. There are two basic schools of thought. One sees militaries as inherently opposed to innovation (due to organizational politics and culture, as elaborated earlier). Accordingly, military innovation requires effective civilian leadership. Civilian policy-makers must see the need for innovation, in order to avoid defeat. They then must push militaries to change through effective oversight mechanisms or, if necessary, through direct intervention in military affairs. An alternative view is that military innovation must be internally led.[63] Only military leaders have the knowledge and legitimacy to lead programs of radical change. From this perspective, the role of civilian leaders is to support visionary military leaders when they seek to effect innovation within their organizations.[64]

There is a similar debate in the literature on civil-military relations concerning strategy in war. The traditional view is that, in democracies, civilian and military leaders do and should occupy different spheres of policy by virtue of their different areas of professional expertise. Lacking the necessary military expertise, civilians can do little more than dabble in strategy. Equally, military leaders lack the political knowledge to question policy; nor would it be advisable, for stable civilian control of the military, to encourage military leaders to develop political instincts. Accordingly, civilians should set national policy and leave it up to the military to apply this, in formulating and implementing an appropriate war strategy.[65] An alternative view is that war strategy invariably occupies the junction between civilian policy and military matters. Accordingly, civilians should take the lead in setting strategy, listening to but not slavishly following military advice. This, in turn, requires civilian leaders to cultivate an understanding of military matters.[66]

Of course, ideally the civil-military relationship should be one of partnership not competition. Risa Brooks's study on civil-military relations and strategy points to the benefits of partnership, in terms of sharing of information and coordination of strategy-making.[67] Moreover, Kimberly Zisk's study of Soviet military innovation shows how civilian intervention in military debate need not trigger competition. She shows how Soviet "expert policy communities" in defense, dominated by the military, expanded in the late 1980s to include new civilian members and their ideas.[68]

Ultimately, the literatures on military innovation and on civil-military relations and strategy point to the importance of leadership. Somebody must lead military change. That leadership may come from civilian policy-makers, or from military chiefs. Generally, we would expect military leadership to be more important than civilian leadership for adaptation at the operational level, given that this concerns how operations are planned, prepared for, and conducted. Obviously, operational adaptation may occur through a bottom-up process and involve military leadership at the lower command level. Russell's study on US Army and USMC adaptation in Iraq demonstrates this. So does the example of the US Army following D-Day in 1944, in how it adapted tactics and equipment as it fought its way out of Normandy.[69] Given that our cases concern democracies, we would expect civilian leadership to be more important than military leadership to adaptation at the strategic level. That said, in the absence of a strong civilian lead, it is entirely possible that the military may lead when it comes to adapting strategy and resources for the campaign. Moreover, it is also possible that adaptations at the operational level may shape a change in strategy. We see this in US strategy for Iraq in 2007. New COIN tactics developed by US units in the field in Iraq in 2005–7, which are swept up and codified in the new US Army and US Marine Corps COIN manual (FM 3-24), provided the framework for a new military strategy that combined a troop surge with closer engagement with local populations and security forces. Thus this case of strategic adaptation combined bottom-up and top-down processes.[70]

NOTES

1. Talbot C. Imlay and Monica Duffy Toft, eds., *The Fog of Peace and War Planning* (Abingdon, Oxon: Routledge, 2006). The classic study on what happens when militaries are surprised in war is Eliot A. Cohen and John Gooch, *Military Misfortunes: The Anatomy of Failure in War* (New York: Free Press, 1990).

2. Robert Jervis, *Why Intelligence Fails: Lessons from the Iranian Revolution and the Iraq War* (Ithaca, NY: Cornell University Press, 2010); Thomas G. Mahnken, *Uncovering Ways of War: U.S. Intelligence and Foreign Military Innovation, 1918–1941* (Ithaca, NY: Cornell University Press, 2002).

3. Risa A. Brooks, *Shaping Strategic: The Civil-Military Politics of Strategic Assessment* (Princeton: Princeton University Press, 2008); Robert Jervis, Richard Ned Lebow,

and Janice Stein, eds., *Psychology and Deterrence* (Baltimore, MD: Johns Hopkins University Press, 1985).

4. Field Marshall Helmuth von Moltke ("the elder") was chief of staff of the Prussian Army from 1857 to 1888.

5. Mier Finkel, *On Flexibility: Recovery from Technological and Doctrinal Surprise on the Battlefield* (Stanford: Stanford University Press, 2011).

6. Cited in William Philpott, *Bloody Victory: The Sacrifice on the Somme and the Making of the Twentieth Century* (London: Little, Brown, 2009), 133.

7. Paddy Griffith, *Battle Tactics of the Western Front: The British Army's Art of Attack, 1916–18* (New Haven: Yale University Press, 1994).

8. Frank Ledwidge, *Losing Small Wars: British Military Failure in Iraq and Afghanistan* (New Haven: Yale University Press, 2011); Richard North, *Ministry of Defeat: The British War in Iraq, 2003–2009* (London: Continuum, 2009).

9. Patrick Cockburn, *Muqtada al-Sadr and the Shia Insurgency in Iraq* (London: Faber and Faber, 2008).

10. Thomas E. Ricks, *The Gamble: General Petraeus and the Untold Story of the American Surge in Iraq, 2006–2008* (London: Allen Lane, 2009); James A. Russell, *Innovation, Transformation and War: Counterinsurgency Operations in Anbar and Ninewa Provinces, Iraq, 2005–2007* (Stanford: Stanford University Press, 2011); David H. Ucko, *The New Counterinsurgency Era: Transforming the U.S. Military for Modern Wars* (Washington, DC: Georgetown University Press, 2009).

11. For discussion on these cases, see Finkel, *On Flexibility*, chs. 11 and 12.

12. That was the case under President George W. Bush and the failing transition strategy for Iraq from 2004 to 2006. See Bob Woodward, *State of Denial: Bush at War, Part III* (New York: Simon and Schuster, 2006).

13. James G. March and Barbara Levitt, "Organizational Learning," in James G. March, ed., *The Pursuit of Organizational Intelligence* (Oxford: Blackwell, 1999), 78–79.

14. The classic study on this is Samuel Huntington, *The Soldier and the State: The Theory and Politics of Civil-Military Relations* (Cambridge: Harvard University Press, 1957).

15. Williamson Murray, *Military Adaptation in War: With Fear of Change* (Cambridge: Cambridge University Press, 2011).

16. Adam Grissom, "The Future of Military Innovation Studies," *Journal of Strategic Studies* 29, no. 5 (2006): 906–7.

17. Harvey M. Sapolsky, Brendan Rittenhouse Green, and Benjamin H. Friedman, "The Missing Transformation," in Sapolsky, Friedman, and Green, eds., *US Military Innovation since the Cold War: Creation without Destruction* (London: Routledge, 2009), 6.

18. Kier A. Lieber, *War and the Engineers: The Primacy of Politics over Technology* (Ithaca, NY: Cornell University Press, 2005).

19. Dima Adamsky, *The Culture of Military Innovation: The Impact of Cultural Factors on the Revolution in Military Affairs in Russia, the US, and Israel* (Stanford: Stanford University Press, 2010); Williamson Murray and Allan R. Millet, eds., *Military Innovation in the Interwar Period* (Cambridge: Cambridge University Press, 1996); Macgregor Knox and Williamson Murray, eds., *The Dynamics of Military Revolution, 1300–2050* (Cambridge: Cambridge University Press, 2001).

20. Theo Farrell and Terry Terriff, "The Sources of Military Change," in Farrell and Terriff, eds., *The Sources of Military Change: Culture, Politics, Technology* (Boulder, CO: Lynne Rienner, 2002), 6.

21. Russell, *Innovation, Transformation and War*, 8.

22. It is highlighted as a key theme in the 2010 Defence Green Paper. Moreover, "Adaptable Britain" was adopted as the UK's strategic posture following the 2011 Strategic Defence and Security Review. MOD, *Adaptability and Partnership: Issues for the Strategic Defence Review*, Cm 7794 (London: TSO, February 2010); HM Government, *Securing Britain in an Age of Uncertainty: The Strategic Defence and Security Review*, Cm 7948 (London: TSO, October 2010), 10.

23. Colin H. Kahl, "In the Crossfire or the Crosshairs? Norms, Civilian Casualties, and US Conduct in Iraq," *International Security* 32, no. 1 (2007); Chad C. Serena, *A Revolution in Military Adaptation: The US Army in the Iraq War* (Washington, DC: Georgetown University Press, 2011).

24. Arguably the failure of the US Army to adapt its counterinsurgency tactics in 2003–4 was a major contributing factor to strategic failure in Iraq over this period. Thomas E. Ricks, *Fiasco: The American Military Adventure in Iraq* (London: Allen Lane, 2006).

25. Russell, *Innovation, Transformation and War*, 8.

26. Theo Farrell, "Improving in War: Military Adaptation and the British in Helmand Province, Afghanistan, 2006–9," *Journal of Strategic Studies* 33, no. 4 (2010): 570.

27. Murray, *Military Adaptation in War*.

28. Barry R. Posen, *The Sources of Military Doctrine: France, Britain, and Germany between the Wars* (Ithaca, NY: Cornell University Press, 1984), 47.

29. Scott D. Sagan, *The Limits of Safety: Organizations, Accidents and Nuclear Weapons* (Princeton: Princeton University Press, 1993).

30. Scott Sigmund Gartner, *Strategic Assessment in War* (New Haven: Yale University Press, 1997).

31. Murray and Millet, *Military Innovation in the Interwar Period*.

32. Indeed, Barry Buzan and Eric Herring argue that the past one hundred years have witnessed a revolution in the pace of revolutionary technological change. Buzan and Herring, *The Arms Dynamic in World Politics* (Boulder, CO: Lynne Rienner, 1998).

33. Terry Terriff, Frans Osinga, and Theo Farrell, eds., *A Transformation Gap? American Innovations and European Military Change* (Stanford: Stanford University Press,

2010); Emily O. Goldman and Thomas G. Mahnken, eds., *The Information Revolution in Military Affairs in Asia* (Basingstoke: Palgrave, 2004); Chris C. Demchak, "Creating the Enemy: Global Diffusion of the Information Technology-Based Military Model," in Emily O. Goldman and Leslie C. Eliason, eds., *The Diffusion of Military Technology and Ideas* (Stanford: Stanford University Press, 2003), 307–47.

34. John M. Owen, *Liberal Peace, Liberal War: American Politics and International Security* (Ithaca, NY: Cornell University Press, 1997); Zeev Maoz and Bruce Russett, "Normative and Structural Causes of Democratic Peace," *American Political Science Review* 87 (1993): 624–38.

35. John Mueller has argued that this is a general postwar trend in the West, but America's wars in Korea, Vietnam, and twice against Iraq, and the level of mobilization that accompanied these wars, suggest that the United States polity is more militarized than most West European states. John Mueller, *Retreat from Doomsday: The Obsolescence of Major Wars* (New York: Basic Books, 1989); Michael Mandelbaum, "Is Major War Obsolete?" *Survival* 40, no. 4 (1998–99): 20–26.

36. James Sheehan, *The Monopoly of Violence: Why Europeans Hate Going to War* (London: Faber and Faber, 2007); John Mueller, *The Remnants of War* (Ithaca, NY: Cornell University Press, 2004).

37. Colin McInnes, *Spectator-Sport War: The West and Contemporary Conflict* (Boulder, CO: Lynne Rienner, 2002).

38. Nicholas J. Wheeler, *Saving Strangers: Humanitarian Intervention in International Society* (Oxford: Oxford University Press, 2000).

39. Trevor Findlay, *The Use of Force in UN Peace Operations* (Oxford: Oxford University Press, 2002); Michael Howard, *War and the Liberal Conscience* (London: Hurst, 2008).

40. Lawrence Freedman, "On War and Choice," *National Interest*, May–June 2010, at http://nationalinterest.org/article/on-war-and-choice-3440.

41. Sarah Kreps, "Elite Consensus as a Determinant of Alliance Cohesion: Why Public Opinion Hardly Matters for NATO-led Operations in Afghanistan," *Foreign Policy Analysis* 6 (2010): 209. Public opinion data from this article also.

42. Andrew Roberts, *Masters and Commanders: How Roosevelt, Churchill, Marshall and Alanbrooke Won the War in the West* (London: Allen Lane, 2008).

43. Philpott, *Bloody Victory*, 56–87.

44. This problem is not untypical in alliances. See Jeremy Pressman, *Warring Friends: Alliance Restraint in International Politics* (Ithaca, NY: Cornell University Press, 2008).

45. General Wesley Clarke, *Waging Modern War: Bosnia, Kosovo and the Future of Conflict* (New York: Public Affairs, 2001), 220–42; Ivo H. Daadler and Michael E. O'Hanlon, *Winning Ugly: NATO's War to Save Kosovo* (Washington, DC: Brookings Institution, 2000).

46. U.S. General Accounting Office, *Kosovo Air Operations: Need to Maintain Alli-*

ance Cohesion Resulted in Doctrinal Departures, GAO-01-784 (Washington, DC: GAO, July 2001), 8–9.

47. Glenn H. Snyder, *Alliance Politics* (Ithaca, NY: Cornell University Press, 1997).

48. John Duffield, *Power Rules: The Evolution of NATO's Conventional Force Posture* (Stanford: Stanford University Press, 1995).

49. Tim Bird and Alex Marshall, *Afghanistan: How the West Lost Its Way* (New Haven: Yale University Press, 2011), 114–18.

50. Ian Traynor, "US Defence Chief Blasts Europe over NATO," *Guardian*, June 10, 2011, at http://www.guardian.co.uk/world/2011/jun/10/nato-dismal-future-pentagon-chief.

51. Thomas Risse-Kappen, *Cooperation among Democracies: The European Influence on US Foreign Policy* (Princeton: Princeton University Press, 1995).

52. Theo Farrell, *The Norms of War: Cultural Beliefs and Modern Conflict* (Boulder, CO: Lynne Rienner, 2005); Ward Thomas, *The Ethics of Destruction: Norms and Force in International Relations* (Ithaca, NY: Cornell University Press, 2001); Peter J. Katzenstein, ed., *The Culture of National Security: Norms and Identity in World Politics* (New York: Columbia University Press, 1996). For comparative studies on strategic culture, see Jeannie L. Jonson, Kerry M. Kartchner, and Jeffrey A. Larsen, eds., *Strategic Culture and Weapons of Mass Destruction: Culturally Based Insights into Comparative National Security Policymaking* (New York: Palgrave, 2009); Lawrence Sondhuas, *Strategic Cultures and Ways of War* (Abingdon, Oxon: Routledge, 2006).

53. Thomas U. Berger, *Cultures of Antimilitarism: National Security in Germany and Japan* (Baltimore, MD: Johns Hopkins University Press, 1998); John S. Duffield, *World Power Foresaken: Political Culture, International Institutions, and German Security Policy after Unification* (Stanford: Stanford University Press, 1998); Peter J. Katzenstein, *Cultural Norms and National Security: Police and Military in Postwar Japan* (Ithaca, NY: Cornell University Press, 1996).

54. Christoph O. Meyer, *The Quest for a European Strategic Culture: Changing Norms on Security and Defence in the European Union* (Basingstoke: Palgrave, 2006).

55. Thomas G. Mahnken, *Technology and the American Way of War* (New York: Columbia University Press, 2008).

56. Isabel V. Hull, *Absolute Destruction: Military Culture and the Practices of War in Imperial Germany* (Ithaca, NY: Cornell University Press, 2005).

57. Jeffrey W. Legro, *Cooperation under Fire: Anglo-German Restraint during World War II* (Ithaca, NY: Cornell University Press, 1995).

58. Lynn Eden, *Whole World on Fire: Organizations, Knowledge and Nuclear Weapons Devastation* (Ithaca, NY: Cornell University Press, 2004); Hull, *Absolute Destruction*.

59. John A. Nagl, *Learning to Eat Soup with a Knife: Counterinsurgency Lessons from Malaya and Vietnam* (Chicago: University of Chicago Press, 2002).

60. Terry Terriff, "'Innovate or Die': Organizational Culture and the Origins of Ma-

noeuvre Warfare in the United States Marine Corps," *Journal of Strategic Studies* 29, no. 3 (2006): 475–503; Terriff, "Warriors and Innovators: Military Change and the Organizational Culture of the US Marine Corps," *Defence Studies* 6, no. 2 (2006): 215–47; Terriff, "Of Romans and Dragons: Preparing the US Marine Corps for Future Warfare," *Contemporary Security Policy* 28, no. 1 (2007): 143–62.

61. Theo Farrell, "World Culture and Military Power," *Security Studies* 14, no. 3 (2005): 448–88.

62. Berger, *Cultures of Antimilitarism*; Katzenstein, *Cultural Norms and National Security*.

63. Deborah D. Avant, *Political Institutions and Military Change: Lessons from Peripheral Wars* (Ithaca, NY: Cornell University Press, 1994); Posen, *The Sources of Military Doctrine*.

64. Stephen Peter Rosen, *Winning the Next War: Innovation and the Modern Military* (Ithaca, NY: Cornell University Press, 1991).

65. Huntington, *The Soldier and the State*.

66. Elliot A. Cohen, *Supreme Command: Soldiers, Statesmen, and Leadership in Wartime* (New York: Simon and Schuster, 2002).

67. Brooks, *Shaping Strategy*.

68. Kimberly Marten Zisk, *Engaging the Enemy: Organization Theory and Soviet Military Innovation, 1955–1991* (Princeton: Princeton University Press, 1993), 4.

69. James Jay Carafano, *GI Ingenuity: Improvisation, Technology, and Winning WWII* (Mechanicsburg, PA: Stockpole Books, 2006); Michael D. Doubler, *Closing with the Enemy: How GIs Fought the War in Europe, 1944–45* (Lawrence: University Press of Kansas, 1994).

70. Russell, *Innovation, Transformation and War*; Ricks, *The Gamble*; Philipp Rotmann, David Tohn, and Jaron Wharton, "Learning under Fire: Progress and Dissent in the US Military," *Survival* 51, no. 4 (2009): 31–48.

2
The Great Game and the Quagmire: Military Adaptation in the British and Soviet Wars in Afghanistan, 1839–1989

Daniel Moran

The struggle for mastery in Central Asia conducted by the Russians and the British in the nineteenth century was first called "the Great Game" by Arthur Conolly, an intelligence officer of the British East India Company who published an account of his exploits in the region in 1834;[1] though the phrase only seized the popular imagination when it was picked up by Rudyard Kipling in his novel *Kim* (1901), since which time it has attached itself to the enterprise of empire as a whole. The Russians, of darker sensibility, called it the "Tournament of Shadows," a more appropriately ominous name, perhaps, for a pastime that included massacre, murder, and deceit of every description; but nevertheless one in which the contestants accept that theirs is a struggle bounded, if not by rules exactly, then by mutually comprehensible interests and methods.

The term "quagmire" has an altogether different resonance. A quagmire is a natural phenomenon, created by water-soaked earth whose lubricious qualities may not be apparent, but whose suffocating effects can render otherwise normal terrain treacherous. It was the American journalist David Halberstam who first popularized the word as a metaphor for the war in Vietnam, in a prescient book published in 1965.[2] Since then the fear of getting stuck in a quagmire has cast its shadow over virtually every military intervention in the Third World, where even the best armies have sometimes found themselves trapped in conditions that have caught them by surprise, despite the ample precedents calling for caution that have accumulated since the 1960s. Such was the case of the Red Army in Afghanistan, whose invasion in December 1979 was supposed to mark something like a resumption of the Great Game, this time against the United States. In the end, though, it turned out that the playing field—Afghanistan and the people who lived there—was the real enemy, from whose suffocating grip the Soviets would finally extricate themselves ten years later.³

These contrasting metaphors provide a useful starting point for considering the efforts of two quite different nations, and four quite different armies (three British, one Soviet) to adapt to the military and political conditions encountered during—and created by—their efforts to have their way in Afghanistan. This is not to suggest that such popular catchphrases should be taken too seriously. The Great Game was a deadly business, which the rhetorical imperturbability of the Victorians cannot conceal. Yet the phrase still captures the sense of limited liability that shaped Britain's engagement in Central Asia. It is also a reminder that Afghanistan itself was neither the prize nor the opponent in the game, just part of the arena. Britain's first three Afghan wars were only incidentally about Afghanistan. They were fought, from the point of view of the men in Whitehall, to contain Russia and help secure Britain's interests in India and Persia. It is helpful to keep this in mind when judging the form that Britain's military commitments took, and the limited adaptive reactions that its sometimes appalling experiences inspired. There is no question that adaptability was necessary to play the Great Game. But the forces that reshaped the British Army over the game's course are not to be found in the Afghan wars, but in between them.

Similarly, the quagmire in which the Red Army lost its way was not a natural phenomenon at all, but a social, psychological, and political one, entirely man-made. Yet the feeling of being unaccountably stuck, of sinking inexorably into what was supposed to have been dry land, captures a crucial aspect of the Soviet experience in Afghanistan, and points toward a contrasting explanation for its adaptive difficulties. The British had limited incentives to adapt to the conditions they encountered in Afghanistan because they had bigger fish to fry elsewhere, and because the stakes were bounded by the terms of the game. For the Soviets, an enterprise that was envisioned as a routine piece of strategic coercion, in which a good outcome would be ensured by the disproportion of power between the belligerents, somehow turned into a test of moral and institutional strength so grinding that their thoughts turned more and more to escape, to never going back, to somehow putting the whole nightmarish business to rest. A quagmire is not something to which one adapts, except to the extent necessary to find a way out.

The Soviet experience highlights an aspect of military adaptation that is worth emphasizing: it always represents an unwelcome challenge, one that most people will seek to evade if they can. Adaptability is undoubtedly a military virtue, but it is not reckoned among the highest, ranking behind courage, loyalty, determination, and whatever complex of psychological characteristics are required to thrive in an environment dominated by chance and danger, which Clausewitz simply called "genius." Nevertheless, adaptability must surely be part of what Clausewitz had in mind. In his work, genius most commonly manifests itself in a capacity to look beyond doctrinally prescribed solutions, to weigh accurately the shifting and intangible forces at play in war, and to anticipate and forestall the actions of the enemy. Genius is the ultimate test of theory, and stands above it, incorporating elements of flexibility and imagination that allow those who possess it to transcend the dictates of "war on paper" in order to engage effectively in the real thing. "What genius does," Clausewitz insisted, "is the best rule," before which prescribed solutions must give way.[4]

Despite this Clausewitzian imprimatur, however, it is easy to see why a readiness to change one's mind is not much celebrated among professional soldiers, who are more inclined to admire perseverance in adversity, if only because it is a quality that is always in demand on the battlefield. They also know that the cohesion and coordinated mutual support that distinguish armies from mobs depend on a disciplined adherence to agreed plans and procedures, rooted in shared habits and values, on which the lives of others may depend. Genius as Clausewitz conceived it may incorporate adaptability, but it is not synonymous with it. Genius can also manifest itself in a willingness to carry on despite setbacks, rather than give up on a course of action that personal insight has shown to be more promising than prudence might suggest.[5]

Failures of adaptation in the military sphere arise in two forms. Bad armies are slow to adapt because they lack the intellectual penetration and institutional flexibility required to do so. But good armies may also be slow to adapt, precisely because they possess great confidence in the excellence of their established methods. Those methods may well be recognized as less than ideal in given circumstances, but are adhered to just the same, because those in charge believe present circumstances are not the only ones that matter. A lot of pressure is required to get a good army to change its mind, and the period in which that adaptive pressure builds up will often appear in retrospect to be one of missed opportunities and willful blindness to new facts. The adaptive moment, if and when it comes, will no less inevitably feel belated and hard-won by advocates of change, and rightly so, because to get their way they will have had to overcome the many sources of psychological and institutional inertia that are the natural counterpart of a willingness to inflict and endure the violence of war.

It is apparent, finally, that the limits of a belligerent's adaptive capacity offer a source of strategic leverage for the enemy, provided that enemy is able to recognize and exploit it. This is a trivial observation in the case of organized armies that plan against each other. Each seeks to exploit the other's weakness while mitigating its own, and knows that its opponent is doing the same, so that both prewar planning and war-time adaptation become self-consciously dialectical processes. In the conduct of irregular warfare, however, other, less symmetrical considerations intrude. In some ways the irregular warrior is a more naturally adaptable creature than his uniformed adversary. He is, at any rate, unencumbered by the preconceptions instilled by military training and discipline, and more familiar with local terrain and social conditions, whose effects loom large in this kind of fighting. Yet the irregular warrior may also carry a burden of unspoken assumptions, arising from cultural practices and inherited traditions, whose constraints can be as severe as those exercised by declaratory doctrine.

In addition, however great the tactical inventiveness of the irregular, it has almost invariably been balanced by a correspondingly narrow range of strategic resources. The Afghans whom the British encountered in their three attempts to subdue, or at any rate to intimidate, the country had only the most basic understanding of how the workings of international politics had rendered their position so precarious. Their metaphors did not run to games, but to the plight of "a goat between two lions,"[6] and they resisted the invader with all the elemental fury that image implies. Yet their fighting methods barely changed from one generation to the next. Their tactics also resembled, in their essentials, the tactics of other colonized nations—all goats together from the lion's point of view. Whatever adaptive pressures the British may have experienced in Afghanistan, they did not regard them as distinctive of that place and people, but as inherent in the externalities of the Great Game, and in the pursuit of empire generally.

The Soviet case is different. Afghanistan was still very largely a traditional society in 1979; yet "very largely" can conceal crucial differences, particularly about the way such societies are located in the international system. Afghanistan was still a goat facing a lion, but now other lions were prepared, not to risk their own safety, certainly, but at least to take an interest in the goat's fate. The fighting methods of the Afghans were also, still, very largely traditional; but not to the point where they were unable to employ new weapons and communications technologies—ranging from radios and man-portable missiles to the ordinary printed word. In the mid-nineteenth century literacy outside the cities of Europe and America was the preserve of narrow clerical and commercial elites. By 1979, in contrast, a third of Afghan men could probably read.[7] This sort of change does not necessarily make much of an impression on the strategic planners of powerful states. Yet it can make an enormous difference to the flexibility, coordination, and resilience of a national resistance movement, while securing it more firmly to its social base.

The adaptive pressures the Soviets confronted in Afghanistan were qualitatively different from those the British had faced, because the tactical resourcefulness of the *mujahid* was no longer deployed in a political vacuum, either within Afghanistan or internationally. His war now resonated within a global system whose reactions proved unmanageable to the Soviet leadership, and amplified their frustrations at the adaptive failures of the Red Army. At the same time, their ideological preconceptions caused them to cast aside the implied guarantee of limited liability that was part and parcel of the Great Game. The British had expected, and demanded, relatively little from the Afghans; and while one might feel that they overpaid for what they got, their pockets were deep, and they had reason to be satisfied in the end. The Soviets entered Afghanistan intending not just to alter the outlook of its government—that had been the British aim as well—but to remake its society in ways that were supposed to guarantee peace and quiet in perpetuity. Any British colonial administrator could have told them that is not how the game was meant to be played.

THREE ARMIES IN AFGHANISTAN: THE ANGLO-AFGHAN WARS

Over a period of eighty years the British sent three armies into Afghanistan. Although the long intervals that separate these campaigns make direct comparisons of their military methods difficult or even pointless, there is still something to be said for considering them as elements of a single, extended military engagement. They were in fact elements of a single, extended political engagement, whose aim was to secure the northwest frontier of British India from encroachment by the Russians, the Persians (as Russian proxies), and the Afghans themselves. Placing the three campaigns alongside each other makes it easier to judge their combined effect, and to gauge the adaptive pressures that weighed upon them, within the broader context of technological and organization change brought about by the Industrial Revolution.

The British sent their first army into Afghanistan in 1839 to depose its emir, Dost Muhammad.[8] Dost had offered the British his friendship in return for their assistance in recovering the former Afghan province of Peshawar, lost some years earlier to one of Britain's clients, Ranjit Singh, the Maharaja of Punjab. The British declined to risk their good relations with Singh, and Dost responded by receiving a Russian envoy in Kabul. This action alarmed Britain's representatives in India. They persuaded London that a policy of what would nowadays be called "regime change" was required, in support of an ostensibly more legitimate claimant to the Afghan throne, Shuja Shah Durrani, who had ruled there thirty years before.

The so-called Army of the Indus was a mixed force of British and British Indian troops numbering twenty-one thousand, under the command of Sir John Keane. It was a large force by Indian standards, and it advanced rapidly, seizing Kandahar without resistance (its governor having fled), and taking the purportedly impregnable fortress at Ghazni by the straightforward expedient of blowing up its main gate (the approach to the gate having been betrayed by an informer). Dost, deserted by his troops, took refuge in Bokhara. By August Shah Shuja had again been installed in Kabul. Keane declared the war over and went home, leaving behind a garrison of eight thousand to bolster Shuja's fledgling army and provide security for British diplomats who also remained behind.

The calamitous consequences of this decision have become a fable for our time, in which the fruits of an initially brilliant military success are squandered in an ill-conceived and poorly resourced effort to sustain an unpopular government. This is, as far as it goes, a perfectly accurate description of what happened next. The British force, divided between Kabul and a few major outposts (chiefly at Ghazni, Jalalabad, and Kandahar), was not large enough to maintain order in the countryside, where resentment at Shuja's restoration was nourished by his efforts to conscript soldiers and raise taxes to create a credible army of his own. Nor was it small enough to allow people to forget that the man on the throne owed his position to outsiders. Dost, seeking to rally opposition forces around himself, was captured and taken to India; but this setback was made good when his son, Akbar Khan, stepped forward to lead the opposition.

By the start of 1842 Shuja had concluded that he would be better off if the British, having become magnets of resentment, left the capital. Terms for their safe conduct were negotiated with Akbar, who immediately betrayed them. Of the forty-five hundred men (and about twelve thousand dependents and camp followers) who departed Kabul, all but a handful were slaughtered as they made their way toward Peshawar. Their fate came to symbolize the miserable unraveling of the whole venture, and would later be commemorated by Elizabeth Thompson, Lady Butler, a celebrated painter of military scenes, in a work called *The Remnants of an Army* (1879). It portrays the exhausted figure of William Brydon, an army surgeon who, in the popular imagination, was believed to be the sole survivor of the retreat.[9]

Brydon was not in fact the sole survivor. Nor did his miraculous escape mark the end of the fighting. About ninety prisoners survived the debacle of the British retreat from Kabul, including the army's commander-in-chief, William Elphinstone, a number of subordinate officers, and some civilians. Their rescue was one of the tasks assigned to a force known, sotto voce, as "The Army of Retribution," which relieved the surviving but besieged garrisons at Jalalabad and Kandahar. Then, together with these, it advanced on Kabul in a campaign of biblical ferocity, calculated, in the words of India's Governor-General, Edward Ellenborough, to deliver a "signal and decisive blow" by way of restoring Britain's authority. "We destroyed all the vineyards," Captain Augustus Abbott wrote in his journal, "and cut deep rings round trees of two centuries' growth," which he deemed a "lamentable" but "quite necessary" measure.[10] Major Henry Rawlinson, father of the future commander of British forces on the Somme, described gunshots coming from a village, whereupon about a hundred of its men were rounded up and "butchered."[11] Organized Afghan forces were brushed aside and relentlessly pursued. On September 16 the British flag once again flew over the citadel at Kabul. The city's Grand Bazaar, scene of the murder of British diplomats a year before, was burned to the ground.

Akbar, deserted by his army and assuming that the British intended to reassert control, sought to negotiate a personal amnesty (and a small pension) for himself—which may be why, having massacred thousands through treachery, he had spared the lives of the prisoners who had fallen into his hands. In the event, however, the British intended nothing of the kind. Honor had been restored, and although their preferred client, Shuja, had been murdered during the war, they simply released Akbar's father, Dost Muhammad, from prison and invited him to resume his place on the Afghan throne. Dost in turn promised that he would have no further truck with the Russians.

Dost, who died in 1863, was true to his word. That is sometimes overlooked, since his loyalty to the British was not unconditional: he tried one last time to seize Peshawar from the Sikhs (1848–49), then periodically conspired with them against the British, who had taken the opportunity of his failed attack to seize the province for themselves. It is also true that the Russian advance southward into Central Asia continued apace after the First Afghan War, gradually approaching the borders of Afghanistan itself. From London's perspective, however, this did not suggest that their efforts were a failure. On the contrary, it confirmed the value of Afghanistan as a buffer, the more so as Britain's position in India was also creeping steadily northward, drawing ever closer to the Russians through the incorporation of the Punjab, Peshawar, and the Sind. Perhaps the clearest signal of Dost's pragmatic loyalty to his deal with the British was that he made no effort to intervene in the Indian Mutiny of 1857, a step that British officials in India feared desperately but that did not happen.

The British sent their second army to Afghanistan in 1878, as an indirect consequence of the anxiety that Russia's steady southward advance inspired in Dost's successor, Sher Ali Khan. Sher Ali's regime depended from the start on arms and financial assistance from the British. As the Russians edged closer, however, he sought more concrete assurances of help in adversity. This the British would not provide, for reasons that, if nothing else, illustrate how different the Great Game looked when viewed from London than from Kabul.

The Treaty of Paris, which ended the Crimean War (1854–56), had included clauses neutralizing the Black Sea, a provision valuable to the British. In 1871 Russia took advantage of the uncertainty introduced into Great Power relations by the Franco-Prussian War to abrogate those clauses. By way of seeking recompense, the British negotiated a sphere-of-influence agreement with Russia the following year, in which the Russians agreed that Afghanistan was not part of their sphere. Britain declined to offer additional assurances to Sher Ali because it did not wish to disturb its deal with the Russians.

The Russians were less scrupulous. A few years later, at the Congress of Berlin (1878), they were obliged by the other European powers to give up territory they had won in a war with the Ottoman Empire, on the grounds that, if Russia were to retain all the gains it had made, the balance of power in Europe would be disturbed. Having acquiesced in this demand, the Russians sought to compensate themselves by pressing forward elsewhere, against the weak khanates of Central Asia. It was the success of this advance that brought them to the borders of Afghanistan. Once there they insisted that Sher Ali's government receive a Russian mission in Kabul, which arrived despite the emir's refusal to invite them. Britain, learning of the Russian move, demanded that their representatives be admitted as well. Sher Ali again refused. Although his view of the Great Game may not have encompassed the nuances of events in Paris and Berlin, he had no difficulty recognizing that, if Russian and British plenipotentiaries were ever to find themselves together in his capital, it might well result in the diminution, if not the partition, of his country.

It was Sher Ali's refusal the admit the British mission, which was turned back at the Kyber Pass, that sparked the Second Afghan War.[12] Instead of a diplomatic delegation, Britain sent an army of forty thousand men that entered the country toward the end of November. Afghan resistance crumbled. Sher Ali fled north, hoping to find assistance in Russia, but died at Mazar-e Sharif. His successor, Yakub Kahn, had no thought of further resistance. In May 1879 he concluded the Treaty of Gandamak, in which he conceded the British right to maintain diplomats in Kabul, and to control Afghanistan's international relations generally. In return Yakub received a renewed subsidy and general assurance (not unlike what Sher Ali had sought) of British aid if Afghanistan were attacked.

Like the First Afghan War, then, the second began as a textbook demonstration of imperial muscle-flexing. Also like the first, initial success proved fleeting once the army that achieved it had withdrawn. In September 1879 the British diplomatic mission in Kabul was massacred by mutinous elements of the Afghan Army, setting off a second round of fighting on terms little different from what had gone before. Once again the ruling emir was forced to flee his capital ahead of an advancing British column, commanded in this case by one of the outstanding British soldiers of the nineteenth century, Frederick Roberts. Roberts, lacking a government with which to parlay, installed Yakub's cousin, Abdur Rahman, as emir, a move that might be regarded as having recklessly set off a struggle for succession with Yakub's brother, Ayub, had such a struggle not been inevitable in any case.

Having chosen their man, however, and having obtained from him a promise to stick by the agreement that was supposed to have ended the war, the British had little choice but to defend him against Ayub's uprising. Ayub, for his part, did not flinch from confronting the British, defeating a small British force in a hard-fought action at Maiwand (July 27, 1880), and then laying siege to Kandahar. Roberts's brilliant relief of that city, following a march from Kabul in which a force of ten thousand men covered three hundred miles in twenty days, proved to be the culminating act of the war, securing Roberts's fame and sending Ayub into exile in Persia. Abdur Rahman remained on his throne, and the terms of the Treaty of Gandamak were reaffirmed—though afterward the British would choose to exercise their diplomatic rights from Jalalabad rather than Kabul, whose dangers for British diplomats had (at last) become obvious.

Britain gave up its right to control Afghan foreign policy as a consequence of the Third Afghan War, fought in the summer of 1919. By then the Great Game had drawn to a close, owing to the revolutionary dissolution of one of the contestants in the maelstrom of the Great War. It is perhaps not surprising that the destruction of the Russian Empire, and the near exhaustion of the British, should have tempted the Afghans to take what they could from the wreckage; though the British were in fact surprised when, in May 1919, a number of their border outposts and several small towns along the Northwest Frontier were attacked and overrun by Afghan forces.[13]

It is hard to know what the Afghan government expected to accomplish by this attack—unless it was precisely what it did accomplish, which was to demonstrate to the British that Afghanistan was no longer worth their trouble. It is true that, when the attack came, British forces in India were depleted, war-weary, and unprepared; yet they remained inherently superior to those of Afghanistan in training and equipment, and given that there was not the slightest possibility that His Majesty's government would stand by while the borders of British India were altered by force, the course of the subsequent fighting was a foregone conclusion. By early June Afghan forces had been driven back across the border, with moderate losses on both sides—a sign, perhaps, that something other than all-out war had been intended after all.[14] Shortly thereafter Britain agree that Afghanistan would henceforth enjoy the rights of a fully independent state, and withdrew the subsidy it had been providing since the 1860s. Afghanistan promised to respect the territorial integrity of British India, as demarcated by the so-called Durand Line of 1893—still the border (if that is the word) between Afghanistan and Pakistan, whose Pashtun populations continue to reach out across it toward each other.

It is fitting that the man who negotiated that most porous and ill-conceived international frontier, Henry Durand, Foreign Secretary of British India, should have been the son and namesake of the lieutenant who blew the gates off the fortress of Ghazni in 1839. The family connection is if nothing else a reminder of the strong elements of continuity that governed British policy in Afghanistan. The British never sought to integrate Afghanistan into their empire, because the territory was not valuable in itself, and because they recognized that its usefulness as a buffer between themselves and the Russians outweighed whatever advantages might come from ruling a people long famous for irascibility among their neighbors.

The British were prepared to pay for Afghanistan's cooperation, and did so throughout most of the period we have been describing. That substantial force was also periodically required reflected the weakness of the Afghan state, whose rulers were no less amenable to cooptation than the princes of British India, but who were often unable to make their writ run among the tribal chieftains and warlords of their country. All of Britain's major interventions in Afghanistan, including the last, defensive one, occurred during periods when the emir's grasp on power was conspicuously in doubt, a condition that tended to produce either military adventurism or ill-conceived diplomatic brinksmanship in Kabul.

The extended periods of relative calm in between were owed in substantial part to the British subsidies that accompanied the making of peace. These helped make up for the fact that the Afghan government always found it difficult to finance itself domestically. None of which should be construed as assigning blame for the Afghan wars to the Afghans. In the conduct of the Great Game, blame hardly comes into it anyway. Yet it is fair to say that a goat living between two lions is unlikely to prosper by setting the lions against each other. Abdur Rahman, whose metaphor this is, was the longest-ruling of Afghanistan's emirs while the game was underway, because he understood the goat's interests well.

The military means by which Britain sought to sustain its policy were straightforward in their conception. None of the operations it conducted in Afghanistan were aimed at anything we would recognize today as pacification or counterinsurgency. Their watchword was "shock and awe," not "hearts and minds." Britain's Afghan wars amounted to repeated demonstrations that Britain could put an army in the Afghan capital at will. Although the political leverage that could be acquired in this way was sometimes overestimated, the British did not interpret their difficulties as a reason to escalate their military effort, or alter their basic approach. The central expectation around which the conduct of all three of their armies revolved was that the native levies of the colonial world were no match for forces trained and equipped to conduct war in the European manner, and led by British officers. Nothing happened in any of the Afghan wars to change anyone's mind about this.

The massacre of the Kabul garrison in 1842 did not register as an exception to this rule. Its fate was attributed by the Duke of Wellington, still the senior officer of the British Army, to "the grossest treachery or the most inconceivable imbecility, and very likely a mixture of both, as they often go together."[15] Treachery was deemed natural to the Afghan character. The imbecility was laid at the feet of Elphinstone, the senior officer commanding. Elphinstone's lassitude has continued to impress scholars, above all his failure to rouse himself to impending danger in the wake of two episodes in which British diplomats under his protection were murdered. His failings have often been attributed to advanced age, despite the fact that, at sixty, he was a year younger than John Keane, whose energetic initial campaign in 1839 had gotten the British into Kabul in the first place.

Whatever the source of Elphinstone's incompetence, there is no question that it made matters worse than they needed to be. Yet it is also apparent that there were no military means at Britain's disposal that could have redeemed the political mistake of leaving a garrison in Kabul in the first place. This was a decision made by British authorities in India, and acquiesced in without demur by the army on the spot and by the government in Whitehall. All shared a basic complacency that the methods of colonial governance that had been perfected in India and elsewhere would work in Afghanistan. This is doubtless an illustration of how institutional learning can go astray, though it must be admitted that it is hard to imagine anyone coming forward with the argument that tried-and-true techniques of intimidate, divide, and rule would not work in Afghanistan because Afghans were too weak and divided already. In the final analysis the extemporized and speculative policy that led to the massacre of the Kabul garrison was another one of those fits of absence of mind for which British imperialism is justly famous. A better man than Elphinstone might have saved his garrison, but not the policy that had put it at risk.

All three of the British armies that went to Afghanistan were products of the last major war that preceded their deployments. In 1839 that last major war had ended a generation earlier at Waterloo, an event that inspired more self-confidence than self-criticism. Among historians the British Army in the decades between Waterloo and Balaclava is routinely characterized as "unreformed," because in contrast to its Continental allies, Britain had found no reason to alter its military institutions in order to defeat Napoleonic France. That being the case, it is not surprising that lesser adversaries made a limited impression. The weapons and fighting methods employed in Afghanistan were scarcely different from those employed in the Peninsular campaign, where both Keane and Elphinstone had served with distinction.

George Pollock and William Nott, who commanded the "Army of Retribution" in the war's final phase, had not fought Napoleon. They were contemporaries of Keane and Elphinstone, but both had spent their careers in India, which may explain why the forces they commanded proved more adept at the small unit tactics and individual pursuit that were required to engage their Afghan opponents. As J. A. Norris has observed, by the end, "[I]nfantrymen accustomed to the life and battles of the plains dashed up steep slopes to fight men who had lived among the mountains all their lives."[16] Such exploits were not expected to be required in a major European war, and, however admirable on the individual level, they made no critical difference to the outcome of this one, either: British infantrymen bent on retribution may have leapt over rocks and dashed up slopes to get at the Afghans, but the Afghans were running away already, as expected.

The massacre of the Kabul garrison that made retribution necessary made no immediate impact on British military institutions; but it did galvanize public interest in military affairs for the first time since Bonaparte had been sent packing.[17] In this respect it contributed to the emergence of a new and distinctly modern form of adaptive pressure—public opinion—with which the armed forces of advanced societies have had to contend ever since. In the decades separating the first and second Afghan wars that pressure would be periodically stoked by a series of anxieties and misadventures that felt (and mostly were) much closer to home: the invasion scare the followed the coronation of another Bonaparte as emperor of the French (1852), the Crimean War (1854–56), the India Mutiny (1857), and the startling performance of the Prussian Army in its wars against Denmark (1864), Austria (1866), and France (1870–71). The result, slow in coming but decisive in the end, was the Cardwell Reforms (1867–74), which were intended to provide Britain with an army up to Continental standards. They also contributed to the improved performance of the British Army in the Second Afghan War.

Like the "Great Reforms" undertaken in Russia at the same time (and the Prussian military reforms of half a century earlier), the thrust of the Cardwell Reforms was to increase the absolute size of the army by moving toward a system of short-service enlistment backed by a trained reserve; to improve the way soldiers were treated by abolishing the most brutal forms of corporal punishment; and to professionalize the officer corps by abolishing the sale of commissions and demanding a modicum of real expertise to achieve promotion. The impact of these changes was amplified by new technology, above all the replacement of muzzle-loaded muskets by breech-loaded rifles as the standard infantry weapon. Rifles increased the absolute firepower of the army, and thus facilitated the rapid movements en masse by Roberts's forces, which proved to be the Second Afghan War's most outstanding tactical feature. Such weapons were far better suited to warfare in broken country, because soldiers were able to carry more ammunition with them, and because a breech-loaded rifle can be fired from a concealed position or while lying down, which a muzzle-loaded weapon cannot. Any army that the British might have sent to Afghanistan in the 1870s would have benefited from these changes, for which their Afghan opponents had no answer.

To this must be added the dynamic leadership of Frederick Roberts, a graduate of Eton and Sandhurst who came up through the artillery, the most technical and least socially glamorous of the army's branches. Roberts was too senior to be considered a product of the Cardwell Reforms (which he personally disliked), and to that extent his role of the Second Afghan War was fortuitous. But even so, the Reforms had shifted the odds in favor of competent leadership, by ensuring that gentlemen generals like Elphinstone, who had languished on half-pay for twenty-five years before being dispatched to Kabul, would become a thing of the past.

Like the men who led the "Army of Retribution," Roberts had also spent his career overseas. Although the institution he served was busy reforming itself in light of the ever-more-demanding requirements of European warfare, he did what he could to bend it to the tasks of empire. His celebrated march from Kabul to Kandahar was only possible, for instance, because Roberts ordered that wheeled transport and heavy guns be left behind. But the institutional influence of such a bold move, superbly adaptive to its moment, was limited precisely because it would have been unthinkable against a more competent opponent. Like all previous British commanders in Afghanistan, Roberts sought to deliver rapid, heavy blows against an enemy that he (correctly) anticipated would be poorly prepared to deflect them. If he had been fighting the Germans or the Russians he would have sought the same thing; but he would have brought his big guns along. Roberts's exploits made him famous, but they could not change anyone's mind about what war in Europe would require.

It goes without saying that the army that fought Britain's Third Afghan War was utterly different from the first two, having been forged in the fires of the

Western Front. Its conduct against the Afghans was unexceptional, and confirms the impression that the exigencies of what was then called "small war" could place only limited adaptive pressure on the military institutions of a Great Power. The habits of mind acquired in four years of fighting the Germans could not easily be set aside, even in the remote hinterlands of Central Asia. The British scarcely needed to pull out all the stops to hold off the Afghans, yet they conducted a "strategic" bombing attack on Kabul just the same—by a single aircraft, admittedly, but even so, it's the thought that counts.

As already noted, the First World War ended the Great Game, most directly by destroying one of its contestants, and more subtly by altering the terms on which the politics of empire would be conducted in the future. The ambivalence with which the British greeted the emergence of the Bolshevik regime in Russia is part of the reason why they felt so little compunction about trading away their rights in Kabul, though it was easy enough to see that, absent continued British engagement, the government there was destined to fall under the sway of whoever finally won the day in Moscow.[18] In addition, the extraordinary and, for the first time, almost entirely public process by which peace was made, followed closely around the world, meant that thoughts of nation and revolution would finally make their way to the furthest corners of Europe's imperium.

Britain's Third Afghan War, like the first two, was touched off by a succession crisis in which the traditional tools of fratricide, vendetta, and tribal horse-trading were fully on display. Yet the eventual winner, Amanullah Khan, presented himself as a democratic reformer who had embarked on war with Britain to secure the rights and dignity of the Afghan nation. However difficult these claims might be to make good in practice, the fact that they were made at all was a sign that the ground on which the Great Game had been played was shifting under foot.[19] This was not a shift to which the British could adapt strategically without abandoning the basic tenants of their policy, which did not encompass any form of what we would today call "nation-building." On the contrary, the outstanding feature of British conduct throughout their long engagement in Afghanistan was their determination to avoid escalation, even if doing so meant accepting less than they bargained for in return for their sometimes costly military interventions. Not the least reason they emerge as the winner, or at any rate as a survivor, of the Great Game is that they knew when to quit.

THE SOVIET-AFGHAN WAR

When the British finally laid down their burden in Afghanistan they took comfort in the fact that, whatever might happen there next, it was not a country in which Bolshevism was likely to take root. This judgment, needless to say, has been thoroughly vindicated by events. For many years it was also widely accepted by the Soviet leadership, which expected, and received, deference and good relations from Afghanistan's rulers, but did not imagine that conditions there would give rise to a worker's paradise anytime soon. With the passage of time those conditions began to change, however. Amanullah and his successors were all top-down modernizers cut from the same cloth as Ataturk, Nasser, and the Pahlavi Shahs of Iran. By the 1970s their efforts had set Afghanistan on its way toward acquiring the trappings of a modern state, a development to which the Soviets reacted by cautioning against any undue leaning toward the West as a model for development, or as a source of aid.

It appears to have been one especially emphatic warning, issued by Leonid Brezhnev to Afghanistan's president, Daud Mohamed, during the latter's visit to Moscow in April 1977, that set events on a path that would bring the Red Army to Kabul. Daud took umbrage at his brusque handling by Brezhnev, and reached out to the Americans, who invited him to visit the United States and offered increased financial assistance to his country. It was to bring an end to such activity that Afghanistan's communists overthrew Daud's government in a bloody coup in April 1978. Afterward they immediately turned against each other—an updated re-enactment of the lethal palace politics that had periodically pulled the British into the country in the past. In Moscow the resulting spectacle was viewed with dismay, as being bad for the communist brand in the developing world, and because it offered a potential opening for increased American interference in Central Asia.

Conditions in Afghanistan also posed a puzzling test for the so-called Brezhnev Doctrine, which bound the Soviets to intervene to prevent the overthrow of communist governments around the world, but did not specify what to do if the communist leader of a state was deposed and killed by a communist rival. In the West the Soviet invasion would be viewed, almost inevitably, as an expansionist move. Although the complexity (and opacity) of prewar political maneuvering in Kabul and Moscow leave room for doubt, the best evidence suggests that the war was begun to break up a murderous quarrel among rival groups of Afghan communists, whose inability to compose their differences was leading to the collapse of political and social order in a country the Soviets regarded as part of their natural sphere of influence.[20]

In appraising the adaptive capacity of the Red Army in Afghanistan, it is well to begin by considering what its forces were trying to accomplish, and how they were organized to do it. The immediate tasks were, first, to replace the regime in Kabul with one more amenable to Moscow's advice, and less prone to the kind of extremism that had inspired the blossoming of anticommunist resistance movements throughout the country; and second, to assist the Afghan Army in controlling the major towns and highways, after which it was assumed that Afghan forces would be able to complete the pacification of the countryside on their own. The Red Army had accomplished similar missions in the past, or imagined that it had. The examples of Hungary in 1956 and of Czechoslovakia in 1968 were recalled as precedents, though neither of those operations had actually involved much fighting. To the extent that objections to intervention were raised among Soviet military and political leaders, they turned not on the vestigial nature of the military plan, but on the diversion of military resources from the army's main task, which was to remain ready to fight the Americans. This objection was set aside, however, on the grounds that neither the time nor the effort required would be great enough to matter in the larger scheme of things.

These aspirations to swift decision and limited liability were reflected in the composition of the 40th Army, dispatched to Kabul in December 1979, and always known officially as "The Limited Contingent of Soviet Forces in Afghanistan." The 40th was an extemporized force consisting of motorized rifle and air assault formations, supplemented by Special Forces, engineering, intelligence, and support units, and well equipped with tanks and other fighting vehicles. It also included an integral air component of fighter-bombers, helicopters, and transports, a unique feature suggestive of the 40th's free-standing and, in Soviet terms, "expeditionary" nature.

By way of acknowledging that the expedition was to Afghanistan, extra effort was made to include units with soldiers from the USSR's Central Asia territories, despite their generally low standard of training and readiness. They were sent nevertheless because their personnel, drawn from groups that were minorities in Afghanistan, were presumed to be more familiar with local customs, languages, terrain, and so on. This ham-fisted gesture in the direction of cultural adaptation actually tended to heighten animosity on the ground, by inadvertently cutting across the remarkably fine-grained ethnic rivalries of the

Afghan population. Such matters need not have been quite as mysterious as they proved to be. The Moscow Institute of Oriental Studies was and is a major center of learning about Afghanistan; but apart from conscripting its staff to serve as interpreters, no effort was made to take advantage of the institute's expertise when it came to crafting Soviet strategy.

The 40th Army was accompanied by another, composed of hundreds of civilian policy advisers, educators, engineers, managers, and so on, whose purpose was to transform Afghanistan into a model of socialist development. Their presence testifies to the fact that the Soviets did not expect their intervention in Afghanistan to be an exercise in brute force. On the contrary, by their own lights they wished the Afghans well, and were surprised when neither the Afghans nor the larger world gave them credit for that. By even the most charitable reckoning, however, it is difficult to reconcile the impulse to social transformation embodied by this second, nameless army, with the ostensibly straightforward military goals assigned to the Limited Contingent. Ideologically the Soviets were naturally attuned to the idea that war is a social phenomenon. They took for granted that the only way to achieve security in Afghanistan, and to protect themselves from the consequences of future disorder there, was to alter the social relations of its people. That they should have found themselves embroiled so quickly in war with and among those people cannot have been entirely due to misfortune, mission creep, or other similarly contingent phenomena. The Red Army went into Afghanistan as a police force, and as an agent of revolution. There is no adaptive process by which the competing requirements of those two missions could have been reconciled.

Better planning might have brought this looming contradiction to the attention of those in charge, true enough; which begs the question where the impulse to better planning would have come from, and what "better" would have meant. The debate within the Soviet leadership on the eve of the Afghan War was whether to intervene military or clandestinely (the preferred KGB option). The form that military intervention would take once it was chosen was not in doubt, and the speed with which the 40th Army had to be thrown together once military action was chosen ruled out detailed reflection as to its composition. The Soviets had never conducted a campaign of counterinsurgency in their history, and did not anticipate one in this instance because they believed that a show of force would suffice to discourage resistance from a population they regarded as amenable to top-down instruction but incapable of bottom-up self-organization. Although the Red Army had been born amid the internecine violence of the civil war that followed the Bolshevik seizure of power, its identity and sense of purpose had been forged by its epic struggle with the Nazis, and by decades of planning for war with the United States and its NATO allies. These experiences provided the doctrinal baseline from which its efforts to adapt to the Afghan quagmire derived. All accounts of Soviet military performance in the war emphasize what a difficult and incomplete process this proved to be.[21]

The extent of Soviet doctrinal conservatism should not be exaggerated. The Soviet military press collected data and analyzed the distinctive conditions prevailing in Afghanistan from the very outset of the war. Soviet Special Forces, airborne, reconnaissance, and airmobile troops gradually became adept practitioners of the tactics of the *mujahedeen* themselves—ambush, infiltration, booby-traps, and close-quarter fighting on foot—and were much feared on that account. The motorized rifle forces that constituted the bulk of the 40th Army's strength also learned to fight in smaller and more dispersed formations; to take precautions against ambush and harassing raids, and also to conduct them; and to perform cordon and search operations, which were previously unknown in the Red Army. On the whole, however, their efforts at tactical adaptation focused on getting the most out of their weapons, especially their fighting vehicles, whose fire-power seemed to offer the best chance of success against a more agile but less well equipped enemy. Among the most distinctive tactical innovations pioneered by the 40th Army was the small armored formation called the *bronegruppa*, in which a handful of armored personnel carriers and tanks were deployed as independent fire support for dismounted infantry.

The fact that such a measure would attract any attention at all testifies more to the limits than to the extent of the 40th Army's adaptability. Even in more conventional settings the Red Army was never known for encouraging initiative among junior and field-grade officers, and it is fair to say that the new forms of adaptive pressure applied by the *mujahedeen* was insufficient to bring about any dramatic change in this regard. It goes too far to imagine that the Red Army was made up of unthinking automatons, a Cold War mythology that mars a number of early and otherwise useful accounts of the war.[22] Yet even if the 40th Army's leadership had been more determined to reshape itself to the task at hand, the institutional obstacles in its way would have been formidable. The Limited Contingent, after all, was no more than a small projection of a much greater institution that was determined not to let itself to be bent out of shape by an operation that, even as it appeared to be going terribly wrong, could not be allowed to detract from the larger requirements of national defense.

Thus, for instance, the 40th Army was always dependent on logistical and training systems that remained tailored to the requirements of war in Europe. This placed severe limits on the extent to which it could have further dispersed its forces or adopted more aggressive small-unit tactics, even if it had wished to do so. More generally, the Soviet war in Afghanistan was fought against a background of budgetary stringency, arising from the growing realization within the Soviet ruling class that the country was not keeping up economically with the West. This understanding would prove foundational to Mikhail Gorbachev's climactic tenure as General Secretary, in which the war would be wound down and the USSR with it; but its weight was already felt in the last years of Brezhnev's rule, and it was bitterly resented by the senior officer corps. They were prepared to employ the agony of the Afghan War to symbolize the pusillanimity of their civilian counterparts, but neither they nor anyone else argued that the war could be won only by committing substantially more resources to it. On the contrary, any increases in the Soviet defense budget in the 1980s would have been spent pursuing the "military-technical revolution," a concept the Soviet military leadership promoted as being crucial to holding off the Americans.[23]

Absent more resources, the only adaptive course that made sense for the 40th Army was to improve its ability to deliver military blows against an elusive adversary. It had not been sent to Afghanistan to conquer and occupy the country, and was never configured to accomplish this: at its height the Limited Contingent numbered 109,000 men, limited indeed if the aim is to control fifteen million people in a country the size of France. It entered Afghanistan under rules of engagement that allowed it to fire only in self-defense, and in order to rescue Soviet citizens taken prisoner by the enemy. Mounting casualties drove adaptation at the tactical level, but did not point toward any theory of strategic victory other than the one with which the 40th had started, which was to reestablish native political authority at the center and allow it to do its work. Failing that, it kept fighting until otherwise instructed. Often this meant fighting the same battle over and over—the 40th Army conducted nine operations in the Pandsher Valley alone. If asked, its leadership would surely have said that victory would come when the enemy realized that the forces of the Soviet Union could not be defeated in the field. In fact this proved to be true, as it had for the Americans in Vietnam; but, as in Vietnam, it made no difference.

The first Politburo meeting to consider withdrawing from Afghanistan occurred in October 1985.[24] By then the 40th Army had already done enormous damage to the fabric of Afghan society, a fact that seems to have counted against withdrawal, to the extent that it heightened the reputational risks associated with the war, above all in the developing world. In the West the Soviet intervention was condemned from the outset as an act of illegal aggression; and then, as it dragged on, as a campaign of extraordinary and indiscriminate violence. That a good deal of indiscriminate violence was involved is undeniable, though its extraordinary character, relative to other protracted campaigns of counterinsurgency, may be disputed.[25] The Soviets, in any case, had prepared a cover story for the illegality—they said they had been invited to intervene by the Afghans themselves; but they had none for the rest of the war. Counterinsurgency was a form of warfare characteristic of imperialists, in which communists did not engage because they were, by definition, on the side of the people. The misery of Afghanistan for the Soviet leadership was not just that the war had gone on too long, but that they had ended up on the wrong side.

The task of extracting the 40th Army from Afghanistan fell to Mikhail Gorbachev, who became General Secretary of the Communist Party in March 1985. Gorbachev had had no role in the decision to go to war, and believed from the start that it was a reckless idea. His career until then included only limited involvement with the military leadership, however, and while he shared the belief of most Politburo members that the country had been let down by its most prestigious institution, he had ample reason to tread lightly, and none to hold up the army to public ridicule for having failed. Gorbachev and his colleague, Soviet Foreign Minister Andrei Gromyko, also believed that the war had gone on so long because the Americans had wished it to. They thought American support for the *mujahedeen* was not calculated to win the war but to ensure its continuation. Neither believed the United States would force the issue by direct intervention. On the contrary, it wished the war to go on forever. That was reason enough to end it.[26]

Establishing an appropriate political context to cover the military withdrawal would take years, however,[27] during which the adaptive challenges facing the 40th Army shifted from those required to inflict maximum damage on the enemy and, by extension, his "social base," to the reduction of casualties and the management of risk. In practice this meant greater reliance on artillery and aircraft. Their longer-range fires were used to fill the vacuum, both tactical and psychological, created by the Soviet withdrawal, which began on the ground in the summer of 1986, when six Soviet regiments (fifteen thousand men) were sent home. No one doubted that more would follow.

It is against this background that the celebrated decision by the United States to arm the Afghan insurgents with man-portable Stinger anti-aircraft missiles must be judged. The Stinger was an especially sophisticated version of a form of weapon that first appeared in Afghanistan in 1984, and was familiar in its essentials to Soviet military planners. The US military objected to distribution of Stingers to the Afghans out of what proved to be a well-founded fear that the weapons would make their way into Soviet and Iranian hands. The Stingers supplied in Afghanistan had a range of about three miles, and could strike targets at altitudes between six hundred and just over twelve thousand feet. Within that lethal envelope they are more effective than other missiles of similar type, though like any weapon they require skill in application. Their introduction in the last months of 1986 produced a spike in Soviet aircraft losses, and eventually accounted for the destruction of dozens of Soviet aircraft.[28]

Stingers could be evaded, of course. Helicopters could fly lower, to avoid being silhouetted against the sky. Fixed-wing aircraft could fly higher, or at night, and take additional precautions on takeoff and landing. But evasion also required skill, and the necessity of engaging in it made the use of all air assets a more nerve-racking and exhausting business, whereas in the past the air had been a realm of freedom in which the Soviets could operate with confidence. Within eighteen months of the Stingers' introduction Soviet aircraft losses had dropped back to their previous level; but the sense of "commanding" the air had been lost, further eroding whatever feelings the Soviets still possessed that they were in control of events.[29] The Stingers did not contribute to the Soviet decision to leave the war. Nor did they present the 40th Army with an insurmountable adaptive challenge in military terms. But they did deepen the demoralization of Soviet forces already facing the prospect of withdrawal, with nothing more than a Potemkin Village "government of national reconciliation" to show for their trouble. Most galling of all, the Stingers showed that, even as the Soviets struggled to escape the quagmire into which they had stumbled, the Americans were still playing the Great Game. You could almost hear them laughing.

HISTORY LESSONS

The last soldiers of the Limited Contingent left Afghanistan in February 1989. By then upward of 600,000 Soviet men and women had served in-country, of whom 15,000 died, a high but not a remarkable figure given the length of the war and the severity of the fighting. Another 50,000 were wounded, 10,000 seriously. To this, however, must be added the astonishing toll taken by illness and disease, to which over 400,000 fell victim at one time or another.[30] The last figure, as many students of the war have observed, is beyond the experience of other modern armies, and raises doubts about Soviet capacities that extend far beyond the military sphere. The subsequent collapse of the Soviet Union itself, to which the defeat in Afghanistan must have contributed something, if only by way of helping to undermine public faith in the government, can make the 40th Army's difficulties look so overdetermined that the question whether it might have done better seems beside the point.

Yet that it could have done better is obvious, for reasons that the earlier British experience helps to make clear—though only if that experience is viewed in the light in which it appeared to contemporaries. The historiography of the Great Game has long been dominated by tales of individual suffering and heroism, which have tended to obscure the unromantic manner in which the game was conducted by the governments concerned. To this has now been added a growing literature, inspired by the disheartening Russian and American experiences of more recent times, that portrays Afghanistan as something like the Bermuda Triangle of international politics, a place where great empires go to die. Britain's experience has been tacitly assimilated to this metanarrative, despite the fact that its empire did not die there. On the contrary: although the travails of Britain's armies in Afghanistan sometimes rivaled the worst that the history of war has to offer, the distinctive feature of their efforts was their strategic success. The fact that their tactical adversities resemble those of the armies that came later is even more telling if this difference is kept in mind.

Success is, of course, a relative concept, and the more easily achieved if one's goals are modest to begin with. This is part of the story in Britain's case, but it is no less true for the Soviets, whose aspirations in Afghanistan were no different: to establish a government in Kabul friendly to their interests and open to their persuasion, particularly regarding relations with the outside world. The Soviet understanding of that interest was clouded by the ideological baggage of their own revolutionary heritage, which tempted them to suppose that a truly friendly government could only be based upon a society whose values were harmonious with their own. But even that did not condemn them to failure. The Soviets did not invade Afghanistan expecting to create such a society by force. They simply believed that, once sufficient force had been applied to reestablish political authority, the opportunity for social transformation would present itself. This proved to be wishful thinking, but it need not have been a harbinger of strategic disaster.

Once the Soviets recognized that the reestablishment of political authority in Afghanistan was a more formidable task than they had supposed, their ideological preferences cut against the impulse to accept the limits of their power, and leave. But the greatest obstacle to that decision lay in their failure to locate the Afghan War correctly in relation to the United States, whose furious reaction seems to have led Soviet leaders to conclude that the stakes in Afghanistan were actually higher than they had thought. Gorbachev's offhand insight that the Americans did not care about winning in Afghanistan, and would have been glad to see the war go on forever, reversed the prevailing Soviet understanding of where American interests lay, and in so doing pointed the way out.

Gorbachev was not the first Soviet leader to recognize the quagmire that Afghanistan had become. Even Brezhnev had been alarmed by the unintended entanglement of Soviet forces with the *mujahedeen*; but he and his senior advisers felt unable to pull back from the brink because he believed doing so would be seen as a concession to the Americans, and would strengthen America's prestige.[31] That might have been true for a moment, but the moment would have passed, and the primary relationship on which the Soviet Union's place in the world depended would have resumed its familiar course, in which the outcome in Afghanistan would have played a steadily diminishing part. Of the four general secretaries who fought the Afghan War, Gorbachev was the only one to judge the reputational risk of its continuance correctly in relation to broader Soviet interests. That is why he was able to end it.

It is now commonplace to suppose that failure in Afghanistan arises from an inability to master and adapt to its "human terrain." Yet even if one concedes that that terrain is exceptionally complex, the fact remains that its mastery only matters, and is probably only possible, for those who wish to rule and to live there; which is to say, for the Afghans themselves. For everyone else, the essential task has always been to judge the true extent and limits of their interests, and to view them steadily in their correct strategic proportions, even in the face of grave and confounding tactical pressures. The British did this well for the most part, the Soviets only at the end. By then it was too late.

NOTES

1. Arthur Conolly, *Journey to the North of India, Overland from England, through Russian Persia, and Affghaunistaun*, 2 volumes (London: Richard Bentley, 1834).

2. David Halberstam, *The Making of a Quagmire: An Uncompromising Account of Our Precarious Commitment in South Vietnam by the Pulitzer Prize-Winning Reporter for the New York Times* (New York: Random House, 1965). Later editions have the less daunting subtitle *America and Vietnam during the Kennedy Era*.

3. It took only a month for the Soviet invasion to acquire the "quagmire" brand. See James A. Phillips, "Afghanistan: The Soviet Quagmire," Heritage Foundation Report (Washington, DC, October 1979), http://www.policyarchive.org/handle/10207/bitstreams/8908.pdf.

4. Carl von Clausewitz, *On War*, ed. and trans. by Michael Howard and Peter Paret (Princeton: Princeton University Press, 1976), bk. 1, ch. 3.

5. See, for instance, Clausewitz's *The Campaign of 1812 in Russia*, in which he rejects the commonplace idea that Napoleon was foolish to persevere in an effort that was doomed from the start; in Carl von Clausewitz, *Historical and Political Writings*, ed. and trans. by Peter Paret and Daniel Moran (Princeton: Princeton University Press, 1992), 201–4.

6. Abdur Rahman Khan, Amir of Afghanistan (1880–91), in David Dilks, *Curzon in India* (London: Rupert Hart-Davis, 1969), 40.

7. A rough but reasonable estimate. About 18 percent of the total population were literate at the time of the last Afghan national census, in 1978 (UNESCO Institute for Statistics, http://www.indexmundi.com/facts/afghanistan/literacy-rate#SE.ADT.LITR.ZS; compare World Bank, World Development Indicators, http://data.worldbank.org/country/afghanistan). The vast majority of that 18 percent would have been men. As of 2011 the CIA estimates that 28 percent of adult Afghans can read: 43 percent of the men, and something over 12 percent of the women (World Fact Book, https://www.cia.gov/library/publications/the-world-factbook/geos/af.html).

8. The best general history of the war is still J. A. Norris, *The First Afghan War, 1838–1842* (Cambridge: Cambridge University Press, 1967). Jeffery J. Roberts, *The Origins of Conflict in Afghanistan* (Westport, CT: Praeger, 2003), surveys Britain's long engagement in Afghanistan by way of setting the stage for the Soviet invasion.

9. Butler's painting is in the Tate Gallery. A second painting commemorating the death march from Kabul was done twenty years later by William Barns Wollen, entitled *Last Stand of the 44th Regiment at Gundamuck, 1842* (1898). It is owned by the Essex Regiment Association.

10. Augustus Abbott, journal entry of August 24, 1842, in Charles Rathbone Low, ed., *The Afghan War, 1838–1842: From the Journal and Correspondence of the Late Major-General Augustus Abbott* (London: Richard Bentley and Son, 1879), 330.

11. In Sir J. W. Kaye, *History of the War in Afghanistan*, 3 vols. (London: Richard Bentley, 1857–58), 3: 329.

12. Britain's Second Afghan War has been poorly served by modern historians; but see the relevant sections of Edgar O'Ballance, *Afghan Wars, 1839–1992: What Britain*

Gave Up and the Soviet Union Lost (London: Brassey's, 1993); Brian Robson, *The Road to Kabul: The Second Afghan War, 1878–1881* (Stroud: Spellmount, 2003); and Rodney Atwood, *The March to Kandahar: Roberts in Afghanistan* (Barnsley: Pen and Sword, 2008).

13. In addition to O'Ballance, *Afghan Wars, 1839–1992*, see Brian Robson, *Crisis on the Frontier: The Third Afghan War and the Campaign in Waziristan, 1919–1920*, 2nd ed. (Stroud: Spellmount, 2007); and Army Headquarters India, General Staff Branch, *The Third Afghan War 1919: Official Account* (1926; reprinted London: Naval and Military Press, 2004).

14. The British committed a total of eight divisions, plus eight independent brigades and a good deal of aircraft, to the expulsion of the Afghans. Altogether they suffered 235 KIA and 615 wounded. Afghan casualties are unknown and presumably higher, but not dramatically so. For the British figures, see George Molesworth, *Afghanistan 1919—An Account of Operations in the Third Afghan War* (New York: Asia Publishing House, 1962), vii.

15. See the account of a conversation between the Duke of Wellington and Charles Greville, March 14, 1832, in Greville, *Memoires: A Journal of the Reign of Queen Victoria from 1837 to 1852*, 3 vols. (London: Longmans, Green, and Co., 1885), 2: 89.

16. Norris, *The First Afghan War, 1838–1842*, 414.

17. Hew Strachan, *Wellington's Legacy: The Reform of the British Army, 1830–54* (Manchester: Manchester University Press, 1984), 20.

18. On the evolution of British policy toward Russia following the 1917 revolution, see Richard H. Ullman, *Anglo-Soviet Relations, 1917–1921*, 3 vols. (Princeton: Princeton University Press, 1962–72); for Afghanistan, Persia, and Central Asia, 3: 317–94.

19. This change was apparent to contemporaries. See Herbert Sidebotham, "The Third Afghan War," *New Statesman*, August 16, 1919, http://www.newstatesman.com/200607170060.

20. The case for the war as an act of calculated aggression, as opposed to reluctant policing, turns in large part on the question of whether the KGB had a role in stimulating the overthrow of Daud, a move for which, as Rodric Braithwaite has noted, reliable evidence has not emerged. See *Afgantsy: The Russians in Afghanistan, 1979–89* (Oxford: Oxford University Press, 2011), 37. Braithwaite's book is now the best general history of the war as viewed from the Russian side. It has a good brief account of how it got started, as does William Maley, *The Afghanistan Wars*, 2nd ed. (New York: Palgrave Macmillan, 2009).

21. See, for instance, Scott R. McMichael, *Stumbling Bear: Soviet Military Performance in Afghanistan* (London: Barssey's, 1991); Lester W. Grau, *The Bear Went over the Mountain: Soviet Combat Tactics in Afghanistan* (Darby, PA: Diane, 1996); The Russian General Staff Authors' Collective, *The Soviet-Afghan War: How a Superpower Fought and Lost*, ed. and trans. by Lester W. Grau and Michael A. Gress (Lawrence: University

of Kansas Press, 2006); Gregory Feifer, *The Great Gamble: The Soviet War in Afghanistan* (New York: Harper Collins, 2009).

22. See, for instance, McMichael, *Stumbling Bear*, 121–25 and passim. One of the many merits of Braithwaite, *Afgantsy*, is that such notions are set aside.

23. In 1984 Marshal N. V. Ogarkov, chief of the Soviet general staff, argued in the military journal *Krasnaya Zyezda* [*Red Star*] that new sensors, guidance systems, and communications technology were transforming conventional warfare, and that it was imperative for the Soviets to keep up with the United States in mastering these things, rather than relying upon nuclear weapons to counter a potentially decisive conventional superiority. See Williamson Murray and MacGregor Knox, "Thinking about Revolutions in Warfare," in Murray and Knox, eds., *The Dynamics of Military Revolution, 1300–2050* (Cambridge: Cambridge University Press, 2001), 3n.

24. Anatoly Chernyaev, Notes from the Politburo meeting of October 17, 1985, http://www.gwu.edu/%7Ensarchiv/NSAEBB/NSAEBB57/r17.pdf.

25. See Braithwaite, *Afgantsy*, 331–33.

26. See the minutes of the Politburo meeting on November 13, 1986, http://www.gwu.edu/%7Ensarchiv/NSAEBB/NSAEBB57/r18.pdf.

27. See Artemy M. Kalinovsky, *A Long Goodbye: The Soviet Withdrawal from Afghanistan* (Cambridge: Harvard University Press, 2011).

28. The 40th Army headquarters in Kabul reported in April 1987 that 62 man-portable missile launches had been detected in 1984, 141 in 1985, and 847 in 1986, for a total loss of 26 fixed-wing aircraft destroyed. In the first quarter of 1987, when the Stingers were at their most effective, 86 launches accounted for an additional 18 aircraft. Report of the 40th Army Headquarters, dated April 20, 1987, in Alexander Lyakhovsky, *Tragediya i Doblest' Afgana* [*The Tragedy and Valor of the Afghan Veteran*] (Moscow: Iskon, 1995), http://www.wilsoncenter.org/sites/default/files/AfghanistanV2_1974–1989.pdf (page 258 of 463). American sources at the time estimated that the Soviets and their Afghan allies had together lost about 100 aircraft from all causes prior to the introduction of the Stingers, and between 150 and 200 in 1987. In 1988 total losses fell to around 50. See Diego Cordovez and Selig S. Harrison, *Out of Afghanistan: The Inside Story of the Soviet Withdrawal* (Oxford: Oxford University Press, 1995), 198.

29. See the comments by the Soviet civilian Alla Smolina, in Braithwaite, *Afgantsy*, 203–4.

30. Grau, *The Bear Went over the Mountain*, xiv; compare Braithwaite, *Afgantsy*, 329–31.

31. Lyakhovsky, *Tragediya i Doblest' Afgana*, http://www.gwu.edu/~nsarchiv/NSAEBB/NSAEBB57/r14.doc.

3 Into the Great Wadi: The United States and the War in Afghanistan

James A. Russell

> *Don't be afraid to make stuff up as you go on. One of the best traditions of our military throughout its history has been its ability to adapt to changing battlefields. Enforce the nonnegotiables of our trade (the Army Values, human rights, Laws of War, etc.) but give your Soldiers the latitude to be creative in accomplishing their missions. Soldiers find out before most commanders what works and what doesn't.*[1]

The United States had no idea what it was getting itself into when it invaded Afghanistan in 2001. Afghanistan had been celebrated as the graveyard of the Soviet Army during the 1980s, but it held no particular allure to US military planners prior to the September 11 attacks. While the Clinton administration had struck Afghanistan in 1998 with cruise missiles in a futile attack directed at Osama bin Laden, it was still regarded as a strategic backwater—at best. Foot soldiers arriving on the heels of the fleeing Taliban in late 2001 and 2002, however, would eventually rediscover the truths of Afghanistan that had been painfully learned by other foreign invaders: it was a place of extremely forbidding terrain with equally forbidding and complex social and political forces. These complexities would combine to make it one of the most difficult environments ever faced by US military units. The United States indeed did leap into the great Wadis and mountain valleys of Afghanistan as had others before it and quickly found itself overwhelmed by the complexities of this distant land.

It is difficult to generalize about the American military experience in Afghanistan, except perhaps to note the timeless nature of the experiences of soldiers fighting in wars as referenced in the quotation above by a US Army company commander operating in Kandahar in 2005–6. From the perspectives of those on the ground, this timeless truth perhaps remained the only constant of this experience over ten years of fighting—that unit commanders were constantly faced with the need to adapt and change to suit varied circumstances on the ground.

In examining the war as a national-level case of adaptation and innovation, this chapter argues that the war lurched through distinctive phases in which a complex series of factors affected the level of national commitment to achieve strategic objectives. That level of commitment dramatically affected tactical operations by the land forces.[2] The war happened in three distinctive phases. Phase one consisted of the invasion, which featured the imaginative if limited application of military forces. During phase two, from 2002 to 2009, the war sputtered along in a period of strategic neglect—a period when the Taliban re-established themselves after their rout in phase one. In phase three, after the election of President Obama, the United States finally attempted to align the strategic, operational, and tactical approach to the war and dramatically increased its level of effort. When the epitaph of the US military experience in Afghanistan is written, the disconnects between the strategic, operational, and tactical levels of the war must certainly represent a defining feature of the experience.

In addition to the disconnects between strategy, operations, and resources throughout these phases, the critical systemic factor driving the innovation and adaptation of US units was the ebb and flow of enemy activity—always higher in the summer months than in the winter. The central role of the enemy in the adaptive process is a timeless feature of all wars—and Afghanistan proved to be no exception. If the Taliban and its allies had been defeated strategically in 2001, no adaptation and innovation by the United States would have been necessary. The insurgency, however, has become steadily more dangerous from mid-2005 to the present, making it increasingly more difficult for units in the field. While more comprehensively dealt with by Antonio Giustozzi in Chapter 10 of this book, the Taliban and other groups resisting the United States and its coalition partners never lost their will to continue fighting, nor the means to do so—despite assertions to the contrary early in the conflict by senior officials in the Bush administration.[3] Like the United States and its partners, the Taliban's war also moved through different phases as a result of a complex series of factors in the strategic interaction with the International Security Assistance Force (ISAF) and the United States. The strategic interaction with the adversary played a decisive and powerful shaping role in the US national-level case of military adaptation.

This chapter examines the factors that shaped US military operations in Afghanistan. It will first address a series of adaptive pressures exerted on US military operations from the outset of the war. These were strategic adaptive pressures that profoundly shaped the conduct of operations. The chapter will also identify "adaptive moments," which were more operational in nature and time-bound than the general strategic-level adaptive pressures. Lastly, the chapter will examine organic-level adaptive efforts by units in the field as they struggled to cope with the difficult environment. It will illustrate the organic-level adaptive processes by examining operations by several units of the 4th Brigade/25th Infantry Division operating in Khost, Paktiya, and Paktika provinces in the winter of 2010.

That the US military adapted in Afghanistan is to state the obvious. It entered the war with a conventionally structured land force overwhelmingly trained and equipped to conduct fire and maneuver operations. By 2011, the land force had systematically reoriented itself to the requirement of fighting irregular war and conducting counterinsurgency operations. While the formal structure of the conventional forces did not substantially change during the course of the war, it is clear that both the US Army and Marine Corps systematically prepared themselves for irregular warfare with new training and tactics, techniques, procedures, and training, and, after 2006, drew upon new joint doctrine to guide counterinsurgency operations in the field.[4]

The argument in this chapter suggests, however, that the process of adaptation and innovation did not necessarily follow from a logical and rational process that saw the rear echelon issue reports and directives that flowed down the command hierarchy to the executing military elements that, in turn, structured their outputs to meet newly stated priorities. As will be discussed later in the chapter, the military departments indeed were subjected to a veritable flood of national and organizational directives to change their strategic focus during the course of the war.

The process of adaptation in the field proved much more dynamic and complex. While the bureaucratic reorientation of the Defense Department indeed did happen, these bureaucratic processes collided with the realities of war, which, in Afghanistan, posed (and continues to pose) formidable challenges to tactical commanders. It proceeded in parallel with field-level adaptation and learning as battalion commanders and their subordinates determined what did and did not work in the field, and were not shy about restructuring their organizations and operations to improve performance in battle. In 2009, the rear-echelon efforts became synchronized with the field-level processes as the United States increased its forces in Afghanistan—forces that had learned their trade on the difficult battlegrounds of Iraq.

Evidence from the author's ongoing research and other studies conducted in Afghanistan suggests significant variation in the tactical approach across the theater during the course of the war—particularly from 2002 to 2009.[5] This was partly due to the lack of a unified, strategic approach to the war. Drawing upon case study research covering the period, researchers at the Center for Naval Analyses found the following: "There were campaign plans and strategy documents, but they were often contradictory and not stringently enforced or clearly communicated. For every unit that focused on the population, there were others that did not. Operations at the tactical and operational levels were not nested within a single strategic framework. Units often worked at cross purposes, and approaches changed from one commander to the next."[6] This lack of alignment profoundly affected operations at the tactical level, since commanders had little idea of the relationship of their operations to broader operational and strategic objectives. In June 2006, for example, an interagency government team examining the role played by provincial reconstruction teams found that the "lack of explicit guidance led to confusion about the civilian and military roles in the US-led Provincial Reconstruction Team (PRT). Without a shared understanding of respective roles and responsibilities, individual experience, skills, and leadership style, personality played a disproportionate role in determining the direction of PRT activities."[7]

These varied tactical approaches and confusing command arrangements undoubtedly affected organizational learning and adaptation in the field. All units move through learning cycles in their deployments that drive the adaptation and innovation process. These learning cycles are shaped by deployment cycles, organizational leadership, state of the insurgency, the local environment, and a host of other variables. It is also the case that variation at the tactical level may be a more general function of irregular warfare conducted in complex environments like Afghanistan that can stretch over many years and that requires a wide variety of different organizational competencies. In Afghanistan, variation in unit-level organizational outputs reflected the different social, political, and military dimensions of Afghanistan's complex environment. Each valley constituted its own area of operations and required its own peculiar form of military operations. Stated differently, uniformity in military operations in an environment like Afghanistan is virtually impossible.[8] Requirements for these variations placed immense importance on individual unit learning, adaptation, and innovation so units can deliver the right mix of organizational outputs. As was the case in Iraq, the model of adaptation and innovation in Afghanistan exhibited by US military forces functioned as a dynamic, dialectical interaction of top-down and bottom-up processes that changed in emphasis over the course of the war.

STRATEGIC ADAPTIVE PRESSURES
Evolving Mission Creep

From 2001 to 2011, the United States significantly increased and expanded its level of effort—in itself an important adaptation. Importantly, that effort included more than just the commitment of additional troops. The United States reached across the range of its governmental capacities to involve a diverse array of civilian and military organizations in the war effort. This increased scale of effort dramatically affected tactical operations in that they made possible an ever-widening and diverse range of activities by military and civilian personnel. It is worth emphasizing that in October 2001, the United States went to war to achieve narrowly defined counterterrorism objectives with a vaguely defined humanitarian component. By 2011, the war had evolved into arguably the most ambitious social, political, and military engineering project ever attempted by the United States.[9] Over the decade, the American military footprint increased from approximately 1,000 Special Forces troops in 2001 to nearly 100,000 mostly conventional troops and several thousand civilians from various agencies by mid-2011. This increased level of effort and broadened strategic focus in itself represented a monumental adaptation that dramatically affected the conduct of units in the field.

What began as a campaign by Special Forces working under an overarching doctrinal concept of a foreign internal defense mission to topple the Taliban gradually became replaced by a systematic attempt to build an entirely new national political structure down to the district and subdistrict level. Entire new national government ministries were designed and built from the ground up. Last but not least, the United States and its coalition partners organized a train and equip program to build an indigenous security force on a scale not seen in recent US history since the Vietnam War.[10]

Reflecting these expanded missions, by 2011 brigade level headquarters around the country included expertise and task specialization from myriad bureaucracies and organizations throughout the US government, focusing on such areas as cryptological and signals interception and various additional intelligence-related support, agricultural development, law enforcement and rule of law, border patrol and customs, special teams to address the threat from improvised explosive devices,[11] police training, economic development, reconstruction, and civil-military relations. Such task-specialized expertise drawn from both contractors as well as military and civilian organizations represented the most comprehensive collection of organizational expertise assembled by the United States in a war since its ill-fated experience in Vietnam.

Despite attempts and much progress in involving various US government agencies in this massive effort, many of the tasks associated with these diverse activities fell to (or were made possible by) American military organizations. National Guard units from the Midwest worked on crop development and irrigation; military personnel tried to design and build a new Afghan defense ministry from the ground up; units trained their Afghan counterparts while simultaneously conducting their own operations; military teams headed by captains built information operations campaigns on their own; platoon leaders and company commanders conducted civics lessons with their Afghan counterparts to build local governance capacities; and battalion commanders provided general security for a host of development-related activities, to name but a few of these activities.[12]

The United States tumbled into this buildup of organizational capacities gradually over time. It cannot be said that this approach to fighting the war happened through clearly articulated strategic objectives at the war's outset that required steadily evolving and increasing levels of effort. Instead, the approach evolved iteratively on an ad hoc basis, dictated in part by strategic circumstance framed by the war in Iraq as well as deteriorating circumstances on the ground in Afghanistan. Deteriorating circumstances on the ground resulted from a combination of the lack of coalition troops, the weakness and corruption of the central government in Kabul, and the resilience of the Taliban and its affiliated insurgent groups. These factors produced a process of iterative, ad hoc adaptation in the field that lacked strategic focus and that systematically compromised the broader application of US military power—particularly from 2002 to 2009.

National Level Politics and Policy Preferences

Public opinion in the United States had little impact on the conduct of the war, although national-level politics did influence decision-making by political leaders at various points. With the exception of the fall of 2001, when the American public and its elected representatives overwhelmingly favored the decision to invade Afghanistan in response to the 9/11 attacks, there is no evidence that public opinion (either for or against) shaped decision-making by political and military leaders during the next decade of the war.[13]

A characteristic of US military interventions since the advent of the volunteer force (and the demise of the "citizen soldier") is the lack of engagement by a largely uninterested general public. A feature of US post–Cold War military interventions is that they have not been accompanied by tax increases or other measures like the reintroduction of conscription that might provoke a negative public reaction. In keeping with past practices, the Bush administration financed both the Iraq and Afghanistan wars "off budget" with borrowed money and sought no increases in federal revenues to finance the war.

The case of innovation and adaptation in Afghanistan cannot be considered in strategic, military, and political isolation. The strategic choices of elected leaders represented the most important strategic adaptive factor that affected military operations in Afghanistan. At the risk of stating the blindingly obvious, the shadow cast by the Iraq War over Afghanistan was and remains deep to this day. The war in Iraq became justified as part of the so-called global war on terror, in which the invasion of Afghanistan represented only the opening phase. The wars thus became linked and justified at least rhetorically and conceptually during the early period of both wars, and the Bush administration's strategic priority of removing Saddam Hussein decisively shaped military operations in Afghanistan,[14] by deeply affecting the level of commitment to the Afghanistan War.[15] As early as 2002, as it became clear that the United States intended to invade Iraq, operations and troop levels in Afghanistan were limited as the Bush administration prepared for the invasion. The focus on Iraq came through conscious decisions at the senior levels that relegated the war in Afghanistan to secondary importance. The war in Iraq consumed ever-increasing quantities of personnel, money, and political attention of senior US decision-makers that could have been used on Afghanistan. It was not until June 2010 that the United States had more troops in Afghanistan than in Iraq (94,000 troops in Afghanistan and 92,000 in Iraq). By the end of 2010, the United States had drawn down to approximately 50,000 troops in Iraq and increased troops in Afghanistan to their highest levels, of nearly 100,000 in early 2011. The inverse relationship between troop levels in these respective wars is illustrated in Table 1.

TABLE 1. US Troop Levels in Iraq and Afghanistan, FY 2002–12[16]

Fiscal year/ Country	Afghanistan	Iraq	Total	Percentage change		
				Annual	Since FY2003	Since FY2008
FY2002	5,200	0	5,200	NA	NA	NA
FY2003	10,400	67,700	78,100	1402%	NA	NA
FY2004	15,200	130,600	145,800	87%	87%	NA
FY2005	19,100	143,800	162,900	12%	109%	NA
FY2006	20,400	141,100	161,500	-1%	107%	NA
FY2007	23,700	148,300	172,000	7%	120%	NA
FY2008	30,100	157,800	187,900	9%	141%	NA
FY2009	50,700	135,600	186,300	-1%	139%	-1%
FY2010	63,500	88,300	151,800	-19%	94%	-19%
FY2011	63,500	42,800	106,200	-30%	36%	-43%
FY2012	63,500	4,100	67,500	-36%	-14%	-64%

Indeed there are eerie parallels between the evolving US approaches in Iraq and Afghanistan. In both cases, the United States initially believed that force could be applied quickly and relatively easily, representing a profound misunderstanding of the environment in which each war was fought. Reflecting this belief, in both cases, Bush administration leaders quickly and mistakenly declared that "major combat operations" had ceased relatively early after the toppling of each respective government. The wars in some sense became mirror images of each other for the United States as, in both cases, the security environments gradually deteriorated after the initial invasion for a variety of reasons.

By 2006, it was generally recognized that Afghanistan was slipping backward as a result of incompetence and corruption in the Karzai government and a strengthened insurgency. From 2006 onward, the security environment dramatically deteriorated as the levels of violence increased steadily each year through 2009, reaching a high of one thousand violent monthly incidents in the summer of 2009.[17] In September 2008, Admiral Michael Mullen, Chairman of the Joint Chiefs of Staff, seemed to sum up the gloomy mood in testimony in which he bluntly declared, "I'm not convinced that we're winning it in Afghanistan."[18]

National-level politics clearly played a role in shaping military operations, particularly after 2009, when military leaders requested a significant increase in troop levels to turn the tide of the war. Civil-military relations and domestic political considerations undeniably shaped the decision by the Obama administration to increase troops significantly in 2009. The Obama administration clearly did not want to be seen as an administration that refused to provide its commanders with the forces needed to achieve national objectives. Perhaps most important, it did not want to be seen as an administration that had lost the war in Afghanistan. These political factors shaped decision-making by the Obama administration during the 2009–11 period.

While campaigning against the war in Iraq in 2008, President Obama immediately embarked on a strategic review of US strategy and policy in Afghanistan after acceding to office. Those reviews eventually resulted in an increase of fifty thousand troops. As noted by Obama in December 2009, US objectives in Afghanistan were the following: (1) to deny Al Qaeda a safe haven; (2) to reverse Taliban momentum and prevent it from overthrowing the government; and (3) to strengthen the capacity of Afghanistan's security forces to take over from ISAF. In conjunction with the troop increases and the articulation of these objectives, Obama also announced his intention to start drawing down the US presence in the summer of 2011 and set 2014 as a target date for full transition to the Afghan security forces.

In conjunction with the increase in military forces, the Obama administration also tried to boost civilian and interagency participation in the war significantly. By September 2011, the Obama administration had deployed more than a thousand civilians (a threefold increase since 2009) into Afghanistan, working primarily on law enforcement, governance, and economic development.[19]

Strategic Contradictions and the Revolution in Military Affairs

Both the Iraq and Afghanistan wars were fought under the rubric of strategic confusion and aggressiveness that stemmed from the 9/11 attacks and the resulting conceptual confusion created by identifying these operations as part of the so-called war on terror. The conceptual confusion obscured and arguably disconnected the link between strategic objectives, military operations, and available resources. While the United States significantly increased its defense budget in the aftermath of the 9/11 attacks, these additional resources were not committed on a systemic basis to support combat operations in both wars. Ironically, at a time when defense spending in the United States increased, both wars were under-resourced for several years following the invasions. The military interventions in both Iraq and Afghanistan suffered from a clearly articulated link between the ends and means that left military commanders with no clear idea of how to address competing priorities with scarce resources.

The lack of connectivity between ends and means can be explained partly by a contradiction at the heart of the Bush Doctrine. That doctrine, enunciated in the 2001 *National Security Strategy* report, called for the United States to intervene militarily around the world more aggressively than it had in the past in order to forestall threats to the homeland like the one that had materialized in the 9/11 attacks. The new and more muscular approach to these interventions was packaged under the concepts of pre-emption and preventive war.[20] The calls for a more aggressive posture on military interventions came from a group of neoconservative advisers: Vice President Dick Cheney, Secretary of Defense Donald Rumsfeld, Deputy Secretary of Defense Paul Wolfowitz, Undersecretary of Defense for Policy Doug Feith, and others. During the first term of the Bush administration, these advisers exerted significant influence on the application of force in Iraq and Afghanistan.

The contradiction within the doctrine was that these advisers eschewed the prospect of "nation-building" that stemmed from a general ideological skepticism of government-administered social engineering projects.[21] So while these advisers championed the idea of military interventions, they did not want the US military to become responsible for the aftermath. As was the case in Iraq, the result in Afghanistan was that commanders were never provided with the troops necessary to perform the myriad missions that political leaders had given them—which gradually expanded over the course of the war.

This contradiction was never resolved, leaving military commanders with inadequate resources and no guidance on how to address competing strategic priorities. During the fall of 2001 and winter of 2002, American political leaders enunciated a gradually widening series of strategic objectives but with no clear idea of the role that the military was supposed to play in achieving them. President Bush initially defined the war objectives quite narrowly, telling the American people on October 7, 2001, that the United States sought to "disrupt the use of Afghanistan as a terrorist base of operations, and to attack the military capability of the Taliban regime."[22] In congressional testimony in December 2001, Assistant Secretary of State Christina Rocca told senators that in addition to the counterterrorism objectives, the Unites States sought to "assist the people of Afghanistan restore freedom, prosperity and good governance to their country."[23] Rocca identified additional "principles" for a new government: it should be inclusive of all Afghanistan's ethnic and religious groups; it should protect the human rights of all its citizens, including women; and it should not export illegal drugs. In December 2001, the United States and its NATO allies agreed as part of the Bonn process to establish a democracy in Afghanistan. The

Bonn process established a series of deadlines for local and national elections (among other things) that had immense implications for military units, which were responsible for providing security for the elections.

In May 2002, President Bush stated: "Peace will be achieved by helping Afghanistan develop its own stable government. Peace will be achieved by helping Afghanistan train and develop its own national army, and peace will be achieved through an education system for boys and girls which works."[24] The role of the US armed forces in achieving these objectives remained unclear, and the Bush administration never provided them with the resources to accomplish these objectives.

Curiously, the Bush administration released no other comprehensive statement about US strategic objectives. In announcing a modest increase of thirty-five hundred troops in 2008, President Bush stated: "The mission of these forces will be to work with the Afghan forces to provide security for the Afghan people, protect Afghanistan's infrastructure and democratic institutions, and help ensure access to services like education and health care." He further elaborated that the United States would double the size of the Afghan Army over a five-year period and "work to increase the involvement of the Afghan tribes. ... [We] will once again encourage Afghan security forces and Afghan tribes to take a leading role in the building of a democratic Afghanistan."[25] Bush also indicated that more civilian advisers would be deployed to assist in economic development.

Both the Iraq and Afghanistan wars reflected the Bush administration's ideas on the role of force as an instrument of national policy and the ways it should be applied in pursuit of national objectives. Embracing the intellectual rubric of the "revolution in military affairs," high-level officials in the Bush administration believed that a new generation of standoff munitions and sensors could be applied with a relatively small footprint on the ground under the concept of "effects-based operations." Rather than traditional fire and maneuver by conventional forces to defeat the enemy on the battlefield, these operations sought to target enemy centers of gravity selectively to achieve broader strategic effects. This model of applying force held forth the alluring (but false) promise of low costs in terms of resources and casualties. Following the 9/11 attacks, regime change was an openly stated objective that became married to Revolution in Military Affairs (RMA) ideas as a template to use of force in states harboring terrorists or developing dangerous weapons that could threaten the American homeland. Officials like Defense Secretary Donald Rumsfeld and Andy Marshall, director of net assessments, championed this new approach against a conservative bureaucracy that they believed opposed this new way of war and that still embraced a more attritional and campaign style of war that necessitated large numbers of troops on the ground.[26]

The confidence in the alleged new way of war became married with another policy preference that had been voiced by George Bush while campaigning for the presidency. Candidate Bush had criticized the Clinton administration for committing US forces in too many open-ended missions that had little to do with military operations. His criticisms reflected a sentiment in some quarters that the US military had become engaged in too many "military operations other than war" during the 1990s. Candidate Bush called for scaling back those missions, and his Defense Secretary expressed open disdain for the nation-building activities that had become the focus of the US military during the 1990s. While eventually swept aside by the tide of events inside both countries, this policy and intellectual predisposition dramatically affected the ways in which force was applied at the early stages of each war, which, in turn, helped lay the seeds for the return of the Taliban.

As military commanders searched for a command organization in 2002, they quickly confronted an informal "personnel cap" on the numbers of military personnel that could be deployed to the country. In May 2002, the United States formally established a combined joint task force headquarters (CJTF 180) to synchronize operations across all the services and with ISAF. Military leaders charged with establishing the headquarters all reported a general understanding at the Central Command, or CENTCOM, and the Joint Staff of a force cap of seven thousand military personnel, which in turn limited the size of the CJTF staff. No explicit guidance on this initial force cap was ever published. Commanders were also well aware of the lack of interest in "nation-building activities." Brigadier General David Kratzer, selected to administer the humanitarian relief operation in 2002 with the 377 Theater Support Command, recalled that the CENTCOM Commander General Tommy Franks, "told me directly, with his finger in my face, that I would not get involved in nation building."[27]

The mismatch between the resources available to the CJTF and its main lines of effort were obvious. Those priorities were the following: combat operations; support the Afghan National Army (ANA); support the ISAF, civil-military operations, and information operations. The CJTF clearly recognized that they needed to conduct full-spectrum operations in Afghanistan. But with a headquarters staff of several hundred, it simply lacked the forces to pursue all the command priorities aggressively. This fundamental mismatch characterized this phase of the war, in which national-level policy and strategic preferences limited the ability of commanders to perform these missions effectively. Confidence in the RMA had guided the application of force by Special Forces in the invasion, but the lack of interest in taking responsibility for the aftermath only ensured that the mess that followed was dropped into the laps of the conventionally structured land forces.

Institutional and Bureaucratic Factors

During the decade of 2001 to 2011, US civilian authorities subjected their military institutions to an unprecedented level of direction to change their way of fighting.[28] Following the 9/11 attacks, the Bush administration released a veritable flood of strategy documents setting new priorities for its armed forces. These priorities flowed from a national strategy that spelled out a requirement for military interventions around the world under a rubric of preventative war. These documents emphasized the need to reorient the nation's armed forces away from conventional conflict toward fighting irregular warfare and combating terrorism—believed to be the most likely environments encountered in these interventions. The strategy documents became operationalized in various Defense Department reports, directives, and guidance directing its military departments to reorient their priorities to the demands of fighting the so-called war on terror.[29] The two wars in Iraq and Afghanistan became the deadly proving ground for these new ideas in which the land forces (particularly the army) slowly but surely wrenched themselves away from their institutional preferences for fire and maneuver operations aimed at destroying the enemy to full-spectrum operations. The war in Afghanistan certainly represented one powerful factor that drove this reorientation and institutional adaptation.

This change in institutional focus of the US military happened through deliberate bureaucratic efforts in the rear echelon, in combination with learning and innovation in the field. A host of internal planning documents and directives were produced by the Defense Department as the wars in Afghanistan and Iraq progressed. In September 2004, the Joint Forces Command published the *Stability Operations Joint Operating Concept* to guide commanders in the field. This document continues to be released in newer versions. In November 2005, the Department of Defense published DoD Directive 3000.05, "Military Support for Stability, Security, Transition, and Reconstruction (SSTR) Operations," which established stability operations as a core mission for the Defense

Department. That guidance was subsequently updated in September 2009 to further reinforce and align intraorganizational support for the mission.[30] In December 2006, the US Army and Marine Corps developed new counterinsurgency doctrine, FM 3–24, to better prepare their forces for the demands of war in Afghanistan and Iraq. In September 2007, the Joint Forces Command and the Special Operations Command published an initial *Irregular Warfare Joint Operating Concept* that was updated most recently in May 2010.[31]

The new doctrine and concepts for operations were both developed in parallel with wartime learning that had to some extent already seen the US land forces transition to irregular warfare by 2007.[32] There can be no doubt that the combination of learning in the wartime environment and deliberate organizational efforts at the national level combined to reorient the military departments toward developing the capabilities originally called for by the Bush administration in 2001 to combat terrorism and fight irregular war in Afghanistan.[33]

ADAPTIVE MOMENTS
Toppling the Taliban

The removal of the Taliban by the United States must be regarded as an adaptive moment during the war in which force was successfully applied to achieve relatively limited objectives. The strategic and policy preferences of the Bush administration clearly were operationalized successfully in the invasion plan. As previously noted, the invasion plan reflected Defense Secretary Donald Rumsfeld's preference for the "new" way of war, which emphasized small numbers of ground troops and Special Forces projected over great distances working with host nation resistance groups. The role of these Special Forces, or operational detachments, with expertise in the foreign internal defense missions, was to liaise and train local militia groups, advise them, and coordinate close air support with precision-guided munitions.

While the decisions to apply force in this way indeed reflected Bush administration policy and strategic preferences, a good deal of thought had gone into the decision to minimize the numbers of troops used in the invasion. War planners consciously drew upon historical and cultural experts (including former Soviet generals) in drawing up the campaign plan. Those experts mostly advised policy-makers that the appearance of large numbers of foreign invaders invariably led to armed resistance—a constant in Afghan history. The Bush administration made the decision to use CIA paramilitary teams and special operations forces after the first interagency meetings at Camp David following the 9/11 attacks.[34]

During the initial campaign planning, CENTCOM staff drew upon principles of effects-based operations to identify critical centers of gravity for the United States and the Taliban. Planners believed US domestic and international political support was the key center of gravity for the coalition and sought to ensure that military action took place with the support of the international community.[35] The Bush administration assiduously built a comprehensive international coalition to support the campaign. By 2003, more than thirty countries were contributing to the operation. Planners realized that Taliban military forces consisted of a collection of many disparate groups with a loosely defined command structure.[36] They sought to build a campaign plan that aimed not to seize physical terrain but to attack what they saw as the Taliban's main weakness—its lack of cohesion as a military force.

The CENTCOM campaign plan also sought to conduct a simultaneous humanitarian relief operation as the Taliban was being removed. The plan included few specifics about what role the US military would play when the Taliban was removed. There was a general expectation that there would be an international humanitarian operation to help rebuild the country. The plan made no reference to the US military's role in this phase of operations and made no references to a US-commanded effort to rebuild the country's political and military institutions.[37]

The connection between strategic-level preferences and direction during the war planning and execution was undeniably strong and direct during the invasion. The Bush administration consciously—and successfully—decided to make the Afghan invasion a test-bed for Secretary Rumsfeld's vision of a new US way of war that minimized the US footprint and that, at the operational level, drew upon established Special Forces unconventional warfare doctrine, which proved to be a perfect fit for the mission. The 5th Special Forces Group, based at Fort Campbell, Kentucky, had been training for operations in this part of the world and was well versed in Afghanistan's language and culture. It provided much of the Special Operations Forces (SOF) manpower for the invasion and continues to support current operations in Afghanistan.

Extraordinary latitude was given the Special Forces teams in executing their missions, which was consistent with SOF unconventional warfare doctrine. Captain Dean Newman, Operational Detachment Alpha (ODA)-534 team leader, later lauded the wide latitude given to him to carry out this mission, commenting that his "entire mandate consisted of a handful of PowerPoint slides that told him to conduct unconventional warfare, render Afghanistan no longer a safe haven for terrorists, defeat al-Qaeda, and coup the Taliban. ... We were given an extraordinarily wonderful amount of authority to make decisions."[38] The work of Special Forces operational detachments working in combination with the Central Intelligence Agency's JAWBREAKER teams are a story of creativity, imagination, and flexibility as these teams marshaled various local Afghan militias to drive the Taliban from power in December of 2001.[39]

In the end, however, the adaptive moment during the invasion was just that—an adaptive moment. The limitations of the approach to using force to topple the Taliban remained unsuited to cope with the aftermath—which got dropped into the laps of undermanned conventional units. The lack of troops also proved critical in December 2001, when the United States was forced to rely on indigenous Afghan forces in the battle of Tora Bora, which allowed Osama bin Laden to escape over the border into Pakistan. While the Taliban had been driven from power during the invasion, its leadership structure remained intact. It continued to receive safe haven and support across the border in Pakistan. And while the Bonn Agreement produced a plan to transition Afghanistan to democracy, it remained unclear whether or how this would work at the national level. In Phase II, the disconnects between the strategic, operational, and tactical levels of the war began to open—sowing the seeds for the re-emergence of the Taliban in 2006–8.

Fissures between these three levels began to open in 2002. What seemed like a good fit during the invasion—using Special Forces in combination with local militias—demonstrated serious shortcomings after the Taliban was removed and a new government was installed in Kabul. The limitation facing the United States emerged early in 2002 in Operation Anaconda, in which it became clear that the Taliban had not been destroyed and that Afghanistan's physical terrain posed significant hurdles to sustained military operations over great distances. During Operation Anaconda the Taliban and its Al Qaeda affiliates demonstrated that they had learned some significant lessons from the fall of 2001, and US forces confronted a well-organized, motivated opponent during the operation.[40]

In March 2002, the United States airlifted approximately seventeen hundred soldiers drawn from the 10th Mountain Division and teamed them with several hundred Afghan militia soldiers in the two-week operation to dislodge an estimated one thousand Taliban and Al Qaeda fighters from Paktiya Province.

The operation was commanded by Combined Joint Task Force Mountain at Bagram airfield. Over two weeks of sustained combat, the force dislodged the determined resistance and, at the time, it was seen as a victory that had destroyed a center of Taliban resistance in the country. The operation, however, demonstrated the limits of what could be accomplished with such a small military footprint. US troops withdrew following the operation and returned to their bases at Bagram and Kandahar airfield, and could not remain in the area. It was a pattern that would be repeated during Phase II around the country until troop levels increased significantly in 2009 and 2010.

In interviews following the operation, the local population stated plainly that the Taliban infrastructure had remained in place and that many fighters had simply fled across the border into Pakistan to fight another day, or melted back into the local population.[41] During the summer of 2002, a series of military operations launched under the rubric of Operation Mountain Lion in the provinces of Khost, Paktika, and Paktiya further reinforced these tendencies: Taliban forces refused to engage US and coalition forces directly, resorting to tried and true guerrilla tactics, waiting for US forces to complete their operations before returning to the contested areas.

Development of a Counterinsurgency Campaign in 2003

The land forces saw their mission in late 2002 and 2003 transitioning to stabilization operations. During this early phase of the war, US military leaders made attempts to align their command elements and develop a nationwide counterinsurgency campaign. The introduction of Provincial Reconstruction Teams in early 2003 to coordinate economic development activities was seen as an indication of this transition and certainly represented an important step to align economic development activities with military operations. In May 2003, Defense Secretary Donald Rumsfeld told reporters, "We have concluded we're at a point where we clearly have moved from major combat activity to a period of stability and stabilization and reconstruction activities. The bulk of this country today is permissive, it's secure."[42] Rumsfeld's remarks masked a slowly deteriorating security situation. As would later become the case in Iraq, it became painfully obvious to military commanders that the Taliban had not been defeated politically, and levels of violence started increasing in the fall of 2003 in which the Taliban directly targeted aid workers, Afghan police, and army units.

Assuming the position of the new theater military command in Afghanistan, Combined Forces Afghanistan, or CFC-A, in the fall of 2003, Lieutenant

General David Barno found himself competing with the demands for Iraq in the establishment of his headquarters staff. He sought to reorient US military operations away from the search and destroy missions that had characterized missions in the spring and summer of 2003. Barno set about establishing an integrated counterinsurgency campaign plan in the fall of 2003 that embraced the requirement for protecting the Afghan population and that called for full-spectrum military operations. Barno's plan sought to disperse his limited maneuver units in critical areas to engage the locals and focus on protecting the Afghan population.[43]

By the end of 2003, the informal troop cap had been dispensed with, and the US troop presence reached approximately 10,500. While Barno's ideas on employing force reflected classic ideas of counterinsurgency, the reality was that entire provinces were in some cases policed by a single battalion. The main maneuver element came from a single brigade combat team—the 1st Brigade, 10th Mountain Division. However much Barno sought to protect the population—a central principle of counterinsurgency—he lacked the troops to perform this mission and never requested additional forces.

During this period of the war, the adaptation and innovation meter had clearly shifted in Afghanistan to the operational level, in which there was a conscious effort to achieve a unity of effort. Barno co-located with the embassy in order to align headquarters staff with the political activities of the embassy, and he tried hard to integrate the activities of the numerous military organizations operating in the country. But despite his attempts to fill in the blanks at the operational level, strategic circumstances determined that he would never have the resources to execute his plan properly at the tactical level.

Shift to Population-Centric Warfare in 2009

In May 2009, seeking a new direction in the war, the Obama administration replaced General David McKiernan with General Stanley McChrystal, a commander with significant experience in special operations. According to various reports, senior officials believed the move was seen as necessary to reorient US operations more directly toward counterinsurgency operations.[44] McChrystal set about conducting his own assessment of the situation on the ground. His subsequent August 2009 report requested forty-four thousand additional troops, warned of the possibility of failure, and recommended reorienting US operations toward protecting the population—a central tenet of counterinsurgency theory.[45]

Importantly, these reviews at the strategic and operational levels were followed by guidance promulgated throughout the civilian and military hierarchies both in Washington, DC, and in the theater in Afghanistan. The reviews and the decisions taken from them resulted in the alignment of the strategic and operational levels of the war in Afghanistan—the first time these levels had been aligned since the invasion of 2001. Taken in conjunction with the gradual drawdown of US forces in Iraq, these steps had a dramatic impact in the field. The additional troops came with significant prior experience in conducting counterinsurgency operations in both wars.

The adaptive moment in 2009 must be seen as a complex one. Many things changed about the way the land forces fought the war: a new emphasis on protecting the population; a conscious effort to minimize civilian casualties through more restrictive rules of engagement; alignment of the civilian and military lines of effort; a concerted effort to build up the Afghan National Security Forces (ANSF), army, and police; and an emphasis on nonkinetic activities such as economic development. Yet the war in this phase also became much more aggressive and kinetic in nature as the land forces moved into contested areas like Kandahar and Helmand.

Not only did operational commanders finally have the troops needed to spread out and actually protect the population, but the United States also flowed many of its Intelligence, Surveillance, and Reconnaissance (ISR) platforms, which had been deployed in Iraq, into the war. Specialized army units like Task Force Odin (a package of ISR platforms) deployed into Afghanistan, providing ground units with a new level of surveillance and intelligence support.[46] The use of ISR and intelligence collection capabilities fed into a targeting process that had been refined and used to great effect in Iraq against insurgent groups.

Military operations under both McChrystal and his successor, General David Petraeus, integrated Special Forces more comprehensively into the counterinsurgency campaign. This included initiatives like the Village Stabilization Program (VSP) in the spring of 2010, which sought to build security from the ground up in addition to the top-down efforts at the national level.[47] The program inserted twelve-man special operations teams into villages around Afghanistan to develop local police and militias to provide security for their villages. This initially focused on nine villages in Arghandab District located just to the north of Kandahar in southern Afghanistan.[48] These programs initially received only grudging support of the Afghanistan government because of fears that that program would create forces not directly connected to the central government in Kabul. By fall of 2011, a program that had started with the nine villages in 2010 had expanded to include one thousand US Special Forces from the Combined Joint Special Operations Task Force (CJSOTF-A) in more than 103 locations covering 23,300 square kilometers.[49] The Afghan local police initiative proceeded in tandem with the VSP, in which Special Forces trained local villagers in basic policing techniques. By the fall of 2011, the units were operating in forty-eight districts with a force of approximately 8,100. As of this writing, the program is set to expand to 30,000 local police operating in one hundred districts.[50]

Special Forces became more integrated into tactical operations after 2009 in other ways as well. In Iraq, both the conventional and special operations forces had refined a targeting process built on intelligence collection coupled with link-nodal analysis that provided a detailed picture of insurgent networks. In Iraq, the targeting centered on "high value targets," or HVTs, within the insurgent networks.[51] In Afghanistan, the Joint Special Operations Command and conventional forces took the basic methodology developed in Iraq and aggressively applied it in Afghanistan, particularly after 2009 when McChrystal took over control of the war effort. The primary intelligence collection methods were intercepted phone calls, surveillance by unmanned drones, and pinpointing the locations of cell phones.[52] After 2009, the targeting broadened from HVTs associated with the Taliban leadership structure to anyone thought to be contributing to the Taliban war effort.[53]

Mounting night raids against suspected insurgents represented only the end result of collection, analysis, and packaging requests sent up the chain of command for approval. The targets were compiled in the Joint Prioritized Effects List, which consisted largely of individuals identified by their cell phone numbers.[54] Before McChrystal arrived in May 2009, the United States was mounting an estimated 20 raids per month. Within five months, the number had increased to 90 per month. By the spring of 2010, those numbers had increased to 250 per month, an increase 12.5-fold in the course of one year. Gauging the impact of these raids on the will of the enemy to continue the fight is difficult, and some suggested that the raids were in fact counterproductive because of the widespread popular disapproval of them among Afghans. Some figures suggest that the raids may have been responsible for more civilian deaths in Afghanistan than IEDs.[55] Regardless, after Petraeus took over, the number of night raids continued its rapid ascent, reaching 600 per month in the summer of 2010. According to some sources, those raids further increased to more than 40 a night by April 2011.[56]

From 2009 onward, both the rear echelon and forward headquarters attempted to achieve unity of effort in the war. Upon assuming command in the summer of 2009, McChrystal released a series of directives aimed at unifying conduct of the war at the operational level. On August 9, he and US ambassador Karl Eikenberry released the Integrated Civil Military Campaign Plan, which addressed the roles and missions of military and civilian organizations and clearly nested those activities in strategic objectives. On August 25, 2009, General McChrystal released ISAF Commander's Counterinsurgency Guidance, which identified population protection as the main military objective. McChrystal told his commanders: "We will not win by simply killing insurgents. We will help the Afghan people win by securing them, by protecting them from intimidation, violence, and abuse, and by operating in a way that respects their culture and religion. That means that we must change the way we think, act, and operate."[57] McChrystal followed this up on August 29, ordering all units in the field to partner with their Afghan counterparts to build host nation capability. On November 10, he released Counterinsurgency Training Guidance, directing that commanders and their subordinates "master counterinsurgency theory and practical implementation."[58]

After replacing General McChrystal in June 2010, General Petraeus reiterated the commitment to counterinsurgency operations through his release of Counterinsurgency Guidance Tactical Directive (Rev 2). The guidance sought more clearly to articulate roles and missions at the operational level and reaffirmed the commitment under McChrystal to reduce civilian casualties. Figures released by ISAF covering the period from July 2008 to September 2010 showed a reduction from 140 casualties per three months at the beginning of the reporting period to 110 over the final reporting quarter. In February 2011, General Petraeus and the American ambassador to Afghanistan published the Integrated U.S. Government Civilian-Military Campaign Plan in a clear attempt to spell out roles and missions in the carrying out of the myriad activities underway in the country. The plan specifically sought to nest tactical operations by military units and supporting activities by civilian agencies with the operational and strategic levels of the war.

In the field, the increased troop levels allowed the United States to deploy significant forces to execute the same sort of "clear, hold, build" approach to counterinsurgency that it had used in Iraq. Forces were deployed into the most dangerous areas of the country in Helmand Province and Kandahar. In February 2010, ISAF launched Operation Moshtarak in Helmand Province, using Afghan, British, and US forces to wrest control over various areas of the province that had been controlled by the Taliban. The operation was seen as a test of McChrystal's new approach of clearing areas to be followed by Afghan security forces and government officials. The template played to mixed success in Marjah—a focal point of the operation.[59]

In parallel with the offensive in Helmand, ISAF launched Operation Hamkari in November 2010 to secure Kandahar—the historic center of the Taliban in Afghanistan.[60] Like the operation in Helmand, ISAF designed an integrated civil-military campaign plan to clear the area of the Taliban with a combination of coalition and Afghan troops and simultaneously launch economic development projects. Most important, the campaign template called for Afghan government officials to flow into the cleared areas in order to prevent the Taliban from reasserting their political authority. As was the case in Operation Moshtarak, military operations aimed at clearing the Taliban were successfully conducted, but the follow-up stage featuring Afghan-government police and administration was mixed.[61]

There can be no question that the "adaptive moment" in the summer of 2009 decisively reoriented US military operations in the field—even if the final impact of those operations remains unclear. As noted, that reorientation included a move toward population-centric operations on a variety of levels. While the land force mounted a variety of large-scale operations in Helmand and Kandahar, the Special Forces greatly expanded their operations to improve local defense forces in the south and east. In parallel with these initiatives, the US ramped up its kill and capture operations aimed at the insurgent networks throughout the country. Last, but not least, the increase in civilian advisers represented an important step to emphasize the nonlethal spectrum of operations. Whether this adaptive moment and its reorientation can successfully reverse the tide of the war remains uncertain.[62]

BOTTOM-UP ADAPTATION TO COUNTERINSURGENCY

4/25 Operations in Khost, Paktika, and Paktiya Provinces, 2009–10

In Khost, Paktika, and Paktiya provinces, McChrystal's approach to reorienting military operations toward counterinsurgency had been anticipated and was being refined and adapted by the 4th Brigade 25th Infantry Division in

2009–10.[63] As noted at the outset of this chapter, wartime adaptation and innovation are clearly shaped by bottom-up and top-down processes. The 4/25 provided an excellent example of this phenomenon, in which Counterinsurgency (COIN) doctrine, operational guidance provided by ISAF, and a brigade-level campaign plan were liberally and differently interpreted by the brigade elements and applied in creative ways on the battlefield. Each maneuver element in the brigade structured its operations in ways to support this top-down guidance, but showed significant differences in structuring their organizational outputs in pursuit of the command's objectives.

To integrate civil and military operations in ways that reflected McChrystal's subsequent guidance, the 4/25 commanding officer, Colonel Michael Howard, established an interagency "Board of Directors" for all operations in the three provinces that sought to coordinate and synchronize all the US government (USG) organizations working in the provinces. Howard built this command structure before McChrystal's guidance had been promulgated. This organizational structure featured formal participation by the State Department, the Agency for International Development (USAID), the Department of Agriculture, Provincial Reconstruction Teams, Agribusiness Development Teams, Law Enforcement Professionals, a Counter-IED Task Force, Human Terrain Teams (HTTs), and other organizations. Operations in each battalion province were coordinated through a "Team Khost" framework, with weekly meetings between all organizational participants to jointly plan and review activities in each of the provinces. The management structure at the brigade was replicated in each battalion S9 office, in which the PRT, USAID, State, and civil affairs formed an integrated civil-military team, working side by side on a daily basis.

The brigade made significant strides in integrating capabilities from other government agencies and special operations forces to maximize its offensive capabilities against the insurgents. Disruption of enemy operations was a shared function across the interagency. While this is not necessarily new to 4/25, the targeting fusion cell provided a particularly good example of how wartime circumstances can lead to complex, flattened organizational hierarchies that (mostly) share information and bring strategic/national level collection capabilities to bear on the tactical fight. The brigade successively layered on a series of specialized organizational competencies in its S2 staff that helped in the free flow of information across classification and organizational domains.

Most of the 4/25's battalions exhibited strong evidence of being learning organizations, whose approach on the battlefield evolved significantly over the course of their deployments, although sometimes events interfered with that process. For instance, several of the battalions expressed frustration at not being able to conduct the kind of COIN operations for which they had trained until the fall of 2009—nearly six months into their deployment cycle. The disappearance of a Private Bowe Bergdhal in the summer of 2009 and the intensive operational tempo that followed this incident turned various units away from their main lines of COIN effort. Second, the Afghan national elections in August 2009 also imposed requirements on battalion commanders that some felt diverted them from their COIN operations.[64] Still, illustrative of the significant bottom-up adaptation is the way in which, in December 2009, the 1st Battalion, 501st Regiment, or 1-501, restored security in the town of Yahya Kheyl in Western Paktika using a soft cordon operation based on an example from the Vietnam War that had been studied by the battalion during its predeployment training. The unit initially air assaulted Afghan and American elements in to the beleaguered district center in the town. Following the insertion, the battalion and its Afghan Army counterparts surrounded the town. The ANA built several redoubts, or forts, on the ingress and egress routes. They allowed people to leave, but not return through the cordon. They then mounted information operations using loudspeakers featuring Afghan interlocutors to urge the insurgents to lay down their arms or leave the city. They emphasized that they did not want to have a pitched battle that would destroy the city. The village elders subsequently pledged to help manage the security environment—negotiating primarily with the ANA.[65]

In eastern Paktika, the 3rd Battalion, 509 Infantry applied a systems-based methodology called Tactical Combat Advisory Planning Framework (TCAPF) to help the unit identify local sources of instability. The methodology required platoons to provide responses to several basic questions about the sources of instability in a particular village. The quality of the data gathered under the program has steadily improved in 1-509's Area of Responsibility (AOR), and the unit used the data to guide its approach in each village. In some towns, security was the paramount concern of the local population, while in others irrigation and water disputes seemed more important. The unit used this survey research to guide its operations throughout its deployment.

Brigade leadership set the tone for a focus on information operations, which has been aggressively executed in units such as the 1-509, which successfully involved company commanders working with Afghan cultural advisers and a growing cadre of Afghan media personnel to tailor broadcasts in eastern Paktika—a capacity that did not exist before the 4/25 arrived in theater.

CONCLUSION

As noted at the outset of this chapter, it is hard to generalize about the US national case of adaptation and innovation in Afghanistan. Military operations in the field clearly were shaped by myriad top-down and bottom-up processes operating in parallel. Strategic level adaptive pressures exerted a profound influence on the overall conduct of the war. Operational adaptive moments also clearly affected operations—particularly after 2009. Throughout the war, units adapted organically from the bottom up as they sought to conduct operations within the constraints affecting them from the top-down adaptive pressures. Strategic-level predispositions and choices of policy-makers in both the Bush and Obama administrations exercised a profound impact on military commanders. The lack of troops, which resulted both from the decision to invade Iraq and the lack of interest in nation-building activities, clearly deprived ground commanders of the means to conduct operations in pursuit of the myriad objectives that had been assigned to them. The lack of troops represented an overwhelming and systemically distorting and constraining adaptive pressure on ground commanders.

From 2002 through 2009, operations at the tactical level were not specifically nested in clearly articulated operational and strategic objectives that were well understood throughout the chain of command. During this phase of the war, adaptation and innovation functioned on a more ad hoc basis as tactical commanders structured their operations on their own in the absence of strong strategic and operational guidance.

In 2009, with the appointment of General McChrystal, the strategic, operational, and tactical levels came back into alignment. The commitment of more troops by the Obama administration and the strategic commitment to Afghanistan happened in parallel with a clear realignment of the operational level of the war by McChrystal and his successor, Petraeus, centered on the overarching objective of securing the population in parallel with building host nation institutions and capabilities. The military campaigns in Helmand and Kandahar were structured as integrated civil-military campaigns.

As noted by Captain Bill Davis at the outset of this chapter, a constant throughout these phases was the process of tactical adaptation by military commanders in response to local circumstances. In the case of 4/25 operations in Regional Command (RC) East in 2009 and 2010, the linkages between low-level adaptation, the reassertion of operational control by McChrystal, and the attempt to build more integrated civilian and military capacities were clearly apparent. Consistent with the flow of authority in the US military, the brigade commander gave his battalion commanders significant latitude to conduct operations appropriate to their circumstance within an innovative civil-military command structure to frame company- and battalion-level operations. These operations provide a good example of the dialectical nature of adaptation and innovation in war in which organic-level processes at the lowest levels in military units feed up into and are supported by broader objectives set at higher levels in the organizational hierarchy.

Clearly in the Afghanistan War the application of force by military organizations was most effective when the strategic, operational, and tactical levels of the war were properly aligned. An enduring and timeless lesson of the war in Afghanistan is that adaptation and innovation feature both top-down and bottom-up processes and can be most effective when these processes are integrated and mutually supportive.

NOTES

1. Bill Davis, Commander C/173 BSB, 173rd ABCT, operating in Bagram and Kandahar, February 2005–February 2006, as quoted in "Insights from OEF: Commanding in Afghanistan," *Army Magazine*, February 2007, 62.

2. All the US military departments furnished manpower to fight the war in Afghanistan. But the bulk of combat operations in the field fell to the Army and Marine Corps, with Special Forces in a supporting role. This paper refers to these organizations as the "land forces."

3. Secretary of Defense Donald Rumsfeld told Larry King in December 2001 that the Taliban "was gone." In September 2004, President Bush asserted, "The Taliban is no longer in existence."

4. *FM 3-24, Counterinsurgency* (Department of the Army, December 2006).

5. See also Jerry Meyerle, Megan Katt, and Jim Gavrilis, *Counterinsurgency on the Ground in Afghanistan: How Different Units Adapted to Local Conditions* (Alexandria, VA: Center for Naval Analyses, 2010); Michael Moore and James Fussell, *Kunar and Nuristan: Rethinking U.S. Counterinsurgency Operations*, Afghanistan Report I, Institute for the Study of War, Washington, DC, July 2009: available at http://www.understandingwar.org/files/Afghanistan_Report_1.pdf.

6. Meyerle, Katt, and Gavrilis, *Counterinsurgency on the Ground in Afghanistan*, 17.

7. *Provincial Reconstruction Teams in Afghanistan: An Interagency Assessment* (Agency for International Development, PN ADG 252, Washington, DC, June 2006), 10: available at http://pdf.usaid.gov/pdf_docs/PNADG252.pdf.

8. Author's observations in studying 4/25 operations in RC East in the winter of 2010, in which there was immense variation in battalion-level operations and activities. Also emphasized by Meyerle, Katt, and Gavrilis, *Counterinsurgency on the Ground in Afghanistan*, 17–18.

9. One could argue that the United States attempted a similarly ambitious effort in Iraq. Unlike Afghanistan, however, Iraq had been governed by a strong central government for fifty years prior to the invasion of 2003. It also had a large standing army and established government ministries. Education levels and the complex tribal, ethnic, and sectarian division in Afghanistan were additional factors that made the US social and institutional engineering more ambitious than in Iraq.

10. For background, see *Year in Review: November 2009–November 2010*, NATO Training Mission—Afghanistan.

11. The counter-IED effort mounted by the United States in Iraq and Afghanistan represents a singularly important adaptation all on its own that lies outside the scope of this chapter. A massive research and development effort was mounted by various arms of the US government in response to the IED threat in Afghanistan and Iraq. According to one count, the counter-IED effort involved as many as twenty-three separate government offices and seventy-three companies. In 2006, the Defense Department formed an entire new organization called the Joint Improvised Explosive Device Defense Organization (JIEDDO) to coordinate the effort. By the spring of 2010 the organization was reported to have thirty-six hundred employees and to have spent an estimated $17 billion on its programs. See Adam Higginbotham, "US Learns to Fight Deadliest Weapons," *Wired Magazine*, July 28, 2010. According to the *FY 2010 JIEDDO Annual Report*, Congress provided more than $13 billion for the Joint IED Defeat Fund from FY 2007 to 2010.

12. Chronicled in secondary-source reporting and directly observed by the author in his research conducted on site with the 4th Brigade, 25th Infantry Division operating in Khost, Paktika, and Paktiya provinces in the winter of 2010.

13. Glen Greenwald, "Public Opinion and the War in Afghanistan," Salon.com, September 16, 2010. As noted by Greenwald, officials seemed unmoved by public opinion generally opposing the wars in Afghanistan and Iraq: available at http://www.salon.com/2010/09/16/afghanistan_39/.

14. The author participated in several meetings in the Defense Department in the winter/spring of 2000–2001 in which Bush administration appointees investigated ways to increase the pressure on Saddam Hussein, including the option of declaring the whole of Iraq a no-fly zone.

15. This was acknowledged by Secretary of Defense Robert Gates in reporting by

Hans Nichols, "Gates Tells Lawmakers That Iraq War Is Hurting Afghanistan Mission," Bloomberg, October 1, 2007.

16. Chart as used in Amy Belasco, *Troop Levels in the Afghan and Iraq Wars, FY2001–FY2012: Costs and Other Potential Issues,* Congressional Research Service Report R40682, July 2, 2009, 9: available at http://www.fas.org/sgp/crs/natsec/R40682.pdf. The chart does not include the data from 2010 and 2011, when President Obama authorized an increase of 51,000 troops during two reviews in 2009 that brought US troop levels to just under 100,000 in early 2011.

17. Trends in insurgent violence as presented by Major General Michael Flynn, director of intelligence, ISAF, "State of the Insurgency: Trends, Intentions and Objectives," December 22, 2009.

18. Nancy A. Youssef, "Top Military Officer Warns That US Isn't Winning in Afghanistan," McClatchey Newspapers, September 8, 2008.

19. Jason Uckman, "Cost of 'Civilian Surge' in Afghanistan $1.7 Billion," *Washington Post,* September 8, 2011.

20. James J. Wirtz and James A. Russell, "U.S. Policy and Preventive War and Preemption," *Nonproliferation Review* (Spring 2003): 113–23; Walter B. Slocombe, "Force, Preemption and Legitimacy," *Survival* 45, no. 1 (2003): 117–30; Robert S. Litwak, "The New Calculus of Preemption," *Survival* 44, no. 2 (2002): 53–80; Charles W. Kegley and Gregory A. Raymond, "Preventive War and Normative Global Order," *International Studies Perspective*s 4, no. 4 (November 2003): 385–494; Peter Dombrowski and Roger A. Payne, "The Emerging Consensus for Preventive War," *Survival* 48, no. 2 (July 2006): 115–36; John Lewis Gaddis, "Grand Strategy in the Second Term," *Foreign Affairs* 84, no. 1 (January/February 2005): 2–15; Mark Trachtenberg, "Preventive War and US Foreign Policy," *Security Studies* 16, no. 1 (July 2007): 1–31.

21. Background on the neoconservatives is available in Francis Fukuyama, "The Neoconservative Moment," *National Interest* 76 (Summer 2004); See also "After Neoconservatism," *New York Times Magazine,* February 19, 2006; James Mann, *Rise of the Vulcans: The History of Bush's War Cabinet* (London and New York: Viking-Penguin, 2004).

22. President George W. Bush, "Statement to the Nation," October 7, 2001.

23. Senate Foreign Relations Committee, "The Political Future of Afghanistan," S. Hrg. 107-236, 107th Congress, 1st Session, December 6, 2001, Prepared Statement of Christina Rocca, Assistant Secretary of State for South Asian Affairs, 7.

24. President George W. Bush, speech at the Virginia Military Institute, April 17, 2002: available at http://transcripts.cnn.com/TRANSCRIPTS/0204/17/se.02.html.

25. President George W. Bush, Distinguished Lecture Program, National Defense University, September 9, 2008: available at http://georgewbush-whitehouse.archives.gov/news/releases/2008/09/print/20080909.html.

26. For background on the RMA, see Jeffrey McKitrick et al., "The Revolution in

Military Affairs," *Future Airpower and Strategy Issues*, September 1995: available at http://www.airpower.au.af.mil/airchronicles/battle/bftoc.html. For a more recent treatment, see Barry D. Watts, *The Maturing Revolution in Military Affairs*, Center for Strategic and Budgetary Assessments, Washington, DC, 2011.

27. Donald P. Wright et al., *A Different Kind of War: The United States Army in Operation Enduring Freedom (OEF) October 2001–September 2005* (Fort Leavenworth, KS: Combat Studies Institute Press, 2009), 202.

28. Chronicled in David Ucko, *The New Counterinsurgency Era* (Washington, DC: Georgetown University Press, 2009).

29. No US administration in recent history has ever publicly explained its strategic priorities in such detail. See *The National Security Strategy of the United States of America*, The White House, Washington, DC, February 2006 and 2001; *The National Strategy to Combat Weapons of Mass Destruction*, The White House, Washington, DC, December 2002; *The National Strategy for Homeland Security*, The White House, Washington, DC, July 2002; *The National Strategy for Combating Terrorism*, The White House, Washington, DC, February 2003; *National Military Strategy*, Joint Chiefs of Staff, Washington, DC, 2004; *National Defense Strategy*, Department of Defense, Washington, DC, March 2005; *The National Strategy for Maritime Security*, The White House, Washington, DC, September 2005. This list is by no means exhaustive but provides a flavor of the unprecedented attention paid by the Bush administration to such issues.

30. The September 2009 version of Department of Defense Directive 3000.5 can be accessed online at http://www.dtic.mil/whs/directives/corres/pdf/300005p.pdf.

31. The latest version of the JFCOM/SOCOM irregular warfare Joint Operating Concept can be accessed online at http://www.au.af.mil/au/awc/awcgate/irregular/iw_joc2_0.pdf.

32. James A. Russell, *Innovation, Transformation and War: U.S. Counterinsurgency Operations in Anbar and Ninewa Provinces, Iraq, 2005–2007* (Palo Alto: Stanford University Press, 2011).

33. This is particularly emphasized in Ucko, *The New Counterinsurgency Era*.

34. Bob Woodward, *Plan of Attack* (New York: Simon and Schuster, 2004).

35. Wright, *A Different Kind of War*, 42.

36. The Taliban's troops consisted of Afghan and non-Afghan Pashtuns as well as Al Qaeda–affiliated personnel. The foreign element of the Taliban (25 percent) were generally recognized to be the best trained and most reliable.

37. Wright, *A Different Kind of War*, 47.

38. Ibid., 83.

39. Gary Schroen, *First In: An Insider's Account of How the CIA Spearheaded the War on Terror in Afghanistan* (New York: Ballantine, 2005); Charles H. Briscoe et al., *Weapon of Choice: US Army Special Operations Forces in Afghanistan* (Fort Leavenworth, KS: Combat Studies Institute Press, 2003).

40. Steven Biddle, *Afghanistan and the Future of Warfare: Implications for Army and Defense Policy* (Carlisle, PA: US Army War College Strategic Studies Institute, 2002).

41. Wright, *A Different Kind of War*, 200.

42. "Rumsfeld Declares Major Combat Over in Afghanistan," Fox News.com, May 1, 2003: available at http://www.foxnews.com/story/0,2933,85688,00.html.

43. For details, see Lieutenant General David Barno, "Fighting the 'Other War': Counterinsurgency Strategy in Afghanistan, *Military Review* (September/October 2007): 32–44.

44. Ann Scott Tyson, "General David McKiernan Ousted as Top U.S. Commander in Afghanistan," *Washington Post*, May 12, 2009.

45. See David Galula, *Counterinsurgency Warfare: Theory and Practice* (London: Pall Mall, 1964); Mao tse-Tung, *On Guerilla Warfare*, 2nd ed. (Champaign: University of Illinois Press, 2000); Robert Taber, *War of the Flea: The Classic Study of Guerilla Warfare* (Potomac, MD: Potomac Books, 2003); David Kilcullen, *Counterinsurgency* (Oxford: Oxford University Press, 2010).

46. For a summary of the task force's capabilities, see Defense Industry Daily, "Task Force Odin: In the Valleys of the Blind," June 15, 2009.

47. For details and background on various stages of the program, see Anand Gopal and Yochi J. Dreazen, "Afghanistan Enlists Tribal Militia Forces," *Wall Street Journal*, August 12, 2009; Rajiv Chandrasekaran, "U.S. Training Afghan Villagers to Fight the Taliban," *Washington Post*, April 27, 2010; Sean Naylor, "Program Has Afghans as First Line of Defense," *Army Times*, July 20, 2010: available at http://www.armytimes.com/news/2010/07/army_specialforces_072010w; Yaroslav Trofimov, "U.S. Enlists New Afghan Forces," *Wall Street Journal*, July 1, 2010; Colonel Ty Connett and Colonel Bob Cassidy, "Village Stability Operations: More than Village Defense," *Special Warfare* (July/September 2011): available at http://www.soc.mil/SWCS/SWmag/archive/SW2403/SW2403VillageStabilityOperations_MoreThanVillageDefense.html.

48. For details on implementing these programs and the difficulties of balancing the local focus with the need to involve district-level government, see Andrew R. Feitt, "The Importance of Vertical Engagement in Village Stability Operations," *Small Wars Journal*, November 1, 2011: available at http://smallwarsjournal.com/jrnl/art/the-importance-of-vertical-engagement-in-village-stability-operations; see also Rory Hanlin, "One Team's Approach to Village Stability Operations," *Small Wars Journal*, September 4, 2011. For a comparative look at VSO in several different cases, see Matt Dearing, "Formalizing the Informal: Historical Lessons on Local Defense in Counterinsurgency," *Small Wars Journal*, December 1, 2011: available at http://smallwarsjournal.com/jrnl/art/formalizing-the-informal-historical-lessons-on-local-defense-in-counterinsurgency.

49. *Report on Progress and Stability toward Security and Stability in Afghanistan*, Department of Defense, October 2011, 67–68.

50. Ibid., 68–69.

51. Christopher J. Lamb and Evan Munsing, "Secret Weapon: High Value Target Teams as an Organizational Innovation," *Strategic Perspectives* 4, National Defense University Press, Washington, DC, March 2011.

52. Use of these techniques in Iraq is detailed in Gareth Porter, "How McChrystal and Petraeus Built an Indiscriminate 'Killing Machine,'" Truthout.org, September 26, 2011; Mark Urban, *Task Force Black* (London: Little Brown, 2010); Lt. Col. Anthony Shafer, *Operation Dark Heart* (New York: Thomas Dunne Books, 2010). See also Russell, *Innovation, Transformation and War*, 155–59.

53. Porter, "How McChrystal and Petraeus Built an Indiscriminate 'Killing Machine.'"

54. Ibid.

55. Gareth Porter, "ISAF Data Show Night Raids Killed over 1,500 Civilians," Inter Press Service, November 2, 2011.

56. *The Cost of Kill/Capture: Impact of Night Raid Surge on Afghan Civilians*, Open Society Foundations, Kabul, Afghanistan, September 19, 2011: available at http://www.soros.org/initiatives/washington/articles_publications/publications/the-cost-of-kill-capture-impact-of-the-night-raid-surge-on-afghan-civilians-20110919/Night-Raids-Report-FINAL-092011.pdf.

57. ISAF Commander's Counterinsurgency Guidance, Headquarters, International Security Assistance Force, Kabul, Afghanistan, August 2009.

58. ISAF Counterinsurgency Training Guidance, Headquarters, International Security Assistance Force, Kabul, Afghanistan, November 2009.

59. Dexter Filkins, "Prize on the Battlefields of Marja May Be Momentum," *New York Times*, February 20, 2010; see also Nancy Montgomery, "One Year after Offensive, Signs of Progress in Marja," *Stars and Stripes*, February 14, 2011; Brett Van Ess, "The Fight for Marjah: Recent Counterinsurgency Operations in Southern Afghanistan," *Small Wars Journal*, September 30, 2010: available at http://smallwarsjournal.com/jrnl/art/the-fight-for-marjah; Dion Nissenbaum, "Knocked out of Power in Afghan Town, Taliban Turn to Intimidation," McClatchy Newspapers, Washington Bureau, March 14, 2010; Rajiv Chandrasekaran, "Commanders Fear Time Is Running Out in Marja; Operation Intended as Model Falters amid Resistance from Taliban," *Washington Post*, June 10, 2010.

60. Details of military operations in Operation Hamkari are covered in Carl Forsberg, "Counterinsurgency in Kandahar: Evaluating the 2010 Hamkari Campaign," Afghanistan Report 7, Institute for the Study of War, Washington, DC, December 2010.

61. In one indication of the lack of local enthusiasm for the Afghan national government, locals in Kandahar showed little interest in joining the Afghan National Army—despite chronically high unemployment in the province. See Ray Rivera, "Afghan Army Attracts Few Where Fear Reigns," *New York Times*, September 6, 2011.

62. For a mixed review of the series of actions undertaken in 2009 to reorient US operations, see Rudra Chaudhuri and Theo Farrell, "Campaign Disconnect: Operational

Progress and Strategic Obstacles in Afghanistan, 2009–2011," *International Affairs* 87, no. 2 (2011): 271–96.

63. This section draws exclusively upon the author's research on 4/25 in Khost, Paktiya, and Paktika provinces conducted in January 2010. The brigade conducted operations in RC East from January 2009 through January 2010.

64. Interviews by author in P2K in January 2010.

65. Ibid.

4 ISAF and NATO: Campaign Innovation and Organizational Adaptation

Sten Rynning

In strategy as in human evolution, to paraphrase Charles Darwin, it is the one who is most responsive to change who survives. The North Atlantic Treaty Organization (NATO) has so far survived the Afghan campaign, where it commands the International Security Assistance Force (ISAF), but the general perception is that the alliance is bruised by the experience and could suffer further political fragmentation as the campaign wears on. NATO, a conglomerate of forces, may fundamentally be ill suited to fight the highly flexible and dynamic insurgency in Afghanistan, and Machiavelli's adage—that "nothing is so unhealthy or unstable as the reputation for power that is not based upon one's own forces"—might once again ring true.[1] However, while there is no question that a decade of war in Afghanistan has been bruising, NATO has in fact managed to change in response to the campaign.[2] If NATO survives, it may thus be because it remains fairly fit.

We should start by noting that NATO change has been uneven. NATO has managed big change (innovation) on the ground but moderate change (adaptation) in Brussels. It has innovated because insurgent warfare in Afghanistan challenged NATO at its political and military core. The outcome in terms of campaign design and organization is a new way of war, and NATO has consciously sought this change. In Brussels, campaign innovation runs into a larger NATO puzzle of political commitments and operational diversity. NATO adaptation is still an achievement, though, and we shall see how it has affected many dimensions of NATO.

Where criticism is warranted is in relation to NATO's limited ability to align its many adaptation processes and, more important, engage in the kind of political thinking that defines and sustains a strategic engagement. One might argue that the most committed ISAF allies have been drivers of ISAF innovation and that NATO adaptation results from compromise politics in the Brussels headquarters. However, there is more to it than this traditional challenge of managing alliance diversity. Afghanistan sowed seeds of confusion in the NATO body politic because ISAF is on the one hand a case of collective defense against an enemy (Al Qaeda and affiliates) but on the other a case of collective security whereby the international community writ large mobilizes against spoilers of a globalized and liberal world. Western countries have since the 1940s been prone to confuse the two—collective defense and collective security—to the effect that guiding concepts such as threat, enemy, aggressor, and peace become ambiguously entangled.[3] NATO has begun the process of sorting out these concepts and thus its underlying approach to the war, but it remains a work in progress.

NATO TACTICAL LEVEL: ISAF CAMPAIGN INNOVATION

If we define innovation as a new way of war, as we do in this book, then ISAF has come to represent innovation for NATO. The alliance has not fought wars as it does in Afghanistan, and it is the nature of the campaign and the grueling prospect of defeat that have driven NATO to innovate its campaign thinking and organization.[4]

NATO is a collective framework and not a state, and this book's distinction between strategic and operational adaptation therefore needs to be contextualized. If we look only at the Afghan campaign, NATO has innovated both at the strategic level (strategy, forces, and equipment) and the operational level (preparation and conduct of operations). Innovation has taken place specifically in the command organization running from the Kabul ISAF headquarters (strategic command) to regions (operational command) and down to provinces and districts (tactical command). However, if we broaden the view to include the full toolbox and strategic perspective of the alliance, we then encounter adaptation. This is so because the ISAF campaign from an alliance perspective is all tactical: ISAF as tactical level is Afghanistan; the operational level is the joint force command in Brunssum, the Netherlands; and the strategic level is the Brussels Headquarters (HQ). At these operational and strategic levels we do not encounter innovation but adaptation, which is the topic of the later sections. In this section we focus on the campaign/tactical level and the case for innovation.

Although ISAF innovation to an extent has happened gradually, it is possible to identify a moment of major change that runs from the spring of 2009 to January 2010. This moment in the campaign—though it constitutes most of a year, it is a short span of time in the campaign's history, which in early 2012 passed the decade mark—saw ISAF agree to a reform of the command structure that strengthened the commander's (COMISAF) strategic role, reinforced the operational command underneath him, and provided for a more comprehensive and integrated campaign effort. It amounted to a new way of war: a counterinsurgency (COIN) campaign that hitherto had been foreign to NATO. This required the near-fusion of two operations—ISAF proper and the parallel Operation Enduring Freedom (OEF) run by a US-led coalition—and a wider confrontation with the track record of coercing by air power, which NATO did in Bosnia and Kosovo. In contrast, the ISAF commander and his forces were on the ground, fighting real battles, and connecting military and civilian efforts—and all this within NATO. ISAF retained its name and thus its formal focus on "security assistance," but the ground effort of fusing battle and governance efforts was new.

The first new component was the NATO Training Mission-Afghanistan (NTM-A), which was agreed to at NATO's April 4, 2009, summit, stood up over the summer, and was ready for command in mid-November.[5] It was what NATO had done in Iraq as well—with the NTM-I—except that the mission in Afghanistan was an integral part of the security operation (ISAF), where in Iraq it had operated in parallel with the international security coalition (Operation Iraqi Freedom, or OIF). Moreover, NTM-A was a bigger affair and more widely endorsed within the alliance. In practice the OEF training mission Combined Security Transition Command-Afghanistan (CSTC-A) was folded into ISAF. Training was still largely dominated by the United States but now within the alliance portfolio of responsibilities. It was testimony to more coherent and strategic thinking within the alliance, though there were limits. CSTC-A had done it all, building up a whole new national defense organization including the Ministry of Defense—with all its policy, planning, and management facets, recruitment, training of both recruits and officers, and operational and logistical infrastructure. NATO was ready for most of this but not all. It notably did not want to handle the ministerial development tasks, not because they were deemed unimportant but because they overlapped with the kind of civilian development work that NATO allies could not agree to as a matter of principle. Paradigmatic comprehensive thinking thus inhibited pragmatic comprehensive action. CSTC-A remained therefore as a separate command but now with a much reduced role. To ease matters, NATO agreed to double-hat the new NTM-A commanding general as CSTC-A commander as well. An American—Lieutenant General William Caldwell IV—became commanding general of both commands in November 2009.

NATO's move allowed for a smoother integration of the regular training mission (NTM-A/CSTC-A) on the one hand and operational training (Operational Mentor and Liaison Teams/Embedded Training Teams, or OMLT/ETT) on the other.[6] Training and operations were run by different commands but now within a more coherent command organization—with greater unity of command. Previously, COMISAF had either no authority at all over training, which was the case up until 2008, or authority by virtue of a double-hat, which is what COMISAF General McKiernan gained in the fall of 2008. Now in 2009 the organizations fused, which was long overdue, and ambitions therefore increased.

Special Operations Forces (SOF) were always part of the Afghan campaign, as contributing allies deployed them to help secure their main forces and sometimes to take part in the counterterrorist—or kill-or-capture—missions that were carried out under US/OEF command. SOF did not fit in easily with the ISAF mission of security assistance, however, and SOF was therefore not an explicit component of the ISAF command structure, just as ISAF-OEF cooperation on counterterrorist matters was strictly limited and in any case covert. Things changed in 2009. A special operations "element" was added to the ISAF chain of command as a third pillar next to the training mission and the main security operations.

This brings us to the top of the chain of command, where a division of labor was introduced between COMISAF and a new commander of ISAF's newly created Joint Command (IJC). In 2009 the four-star COMISAF was General McChrystal and the three-star COMIJC General Rodrigues; in 2011 they were General Allen and Lieutenant General Scaparrotti, respectively. The logic of this division was inspired by the US experience in Iraq, where generals Petraeus and Odierno with success focused their efforts on political-strategic and operational issues, respectively, and the idea behind the ISAF reforms were to free up COMISAF and integrate him more systematically in the Kabul politics that would be decisive to the outcome of the COIN campaign.

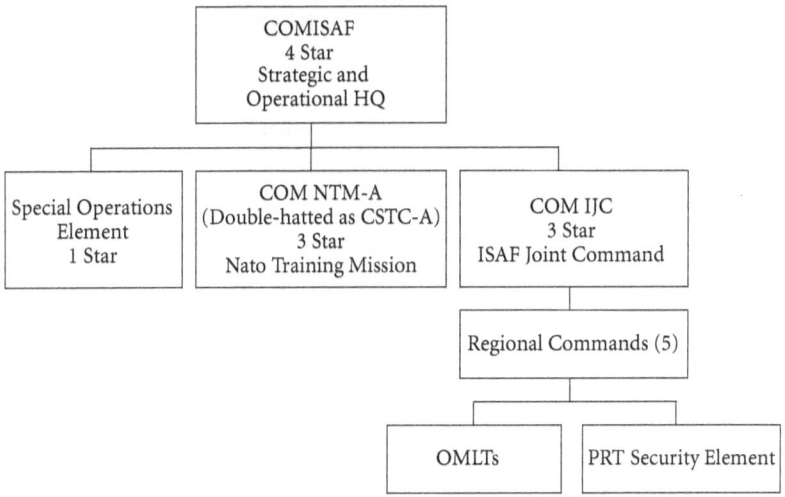

FIG. 1. New ISAF Command, 2009[7]

NATO gave its approval to NTM-A at the summit in April 2009; the IJC gained its approval at a defense ministerial meeting in June 2009;[8] and both became fully operational on November 12, 2009.

The collective framework went beyond military command and control, though, and COIN again provides the raison d'etre for the extension. At issue is the interface between the military and civilian sides of the campaign, an interface that in NATO has given birth to a so-called Comprehensive Approach that elsewhere—in the European Union (EU) and United Nations (UN)—is defined as a human security approach. The vehicle for organizing this interface in Afghanistan has been the Provincial Reconstruction Teams (PRT), which, as the name indicates, operate in Afghanistan's provinces. In Kabul the vehicle has been coordination between the security provider (NATO/ISAF), civilian agencies led by the UN Assistance Mission to Afghanistan (UNAMA), and the Afghan government.

NATO had provided for a couple of mechanisms of coordination in 2003 as ISAF got under way: one was a PRT Executive Steering Committee (ESC); another was the alliance's Senior Civilian Representative (SCR).[9] Neither had much authority, though, as the PRT ESC evolved into a talking shop, while the SCR became the North Atlantic Council's reporting agent in Kabul. In 2009, as NATO reinforced the ISAF command structure, it was clear that the entire

Afghan campaign would surge militarily but then hand off—transition—to an Afghan lead, which in turn required enhanced mechanisms of civil-military cooperation (CIMIC).

NATO's response came in the context of a large Afghanistan conference held in London, January 2010—just months after the new command structure had been stood up—where NATO announced the reinforcement of the SCR office: its core staff of civilian advisers was upgraded from six to twenty-four, and the SCR himself was placed at the four-star level, on par with COMISAF. His operational responsibility would henceforth be to help the Afghan government define a framework for terminating the PRTs—and thus establish Afghan governance—and an underlying strategy for realizing the transition. This became the *Inteqal* strategy that now guides the effort to transition to full Afghan leadership by 2014. What was new was the definition of the SCR as an active actor of political change in Kabul. Previously a reporting agent, he was now a real actor with a strong political mandate from ISAF's masters (the NATO allies). His level of influence was the PRTs, even if he did not control it directly (the nations did) because to nourish governance and development, the Afghan government needed them.

This advance into the domains of political business was new for NATO. It had not previously sought to create a political peace while providing for security. To be sure, the SCR was not involved in the reconciliation and reintegration program that may become the backbone of a new political settlement in the country, and which NATO endorses, but the reinforced SCR signaled that NATO was more than a military actor. NATO had in fact tried to strengthen the SCR in late 2007 and early 2008. Back then NATO, the UN, and others had sought to fuse their senior civilian offices to thus coordinate their efforts—and gain leverage in Kabul (the reported candidate for the post was British Paddy Ashdown). President Karzai smelled a rat and vetoed the idea, however, and so NATO had to settle for minor reinforcements.[10] By early 2010, the game had changed and NATO could go further. British diplomat Mark Sedwill took over from Italian Fernando Gentilini as SCR, and COMISAF General McChrystal and he formed a formidable political-strategic team in Kabul.

With this set of changes running from the inclusion of an SOF command element in the chain of command to enhanced civilian leadership and COMISAF-SCR partnering, NATO put to rest a number of disputes that had bedeviled its Afghan mission. It concerned notably a split between allies in favor of a traditional peace support operation and thus a comparatively modest and strictly military role for NATO, and allies in favor of an activist campaign to build up local allies and track down and incapacitate local enemies. This split was reflected in the ISAF-OEF divide and the banning of the term COIN from NATO. NATO had no COIN doctrine and did not do COIN campaigns, quite simply. The new ISAF command organization of 2009 represented a full reversal. NATO now geared ISAF for COIN. COMISAF General McChrystal outlined his campaign plan in the late summer of 2009, which amounted to a classical COIN campaign, and NATO defense ministers unequivocally endorsed it on October 23, 2009, at their meeting in Bratislava, Slovakia.[11] They also endorsed a strategic concept for so-called Phase 4 planning, which became the transition strategy and which underscored the necessity of thinking outside the military box.[12]

There is reason to emphasize that innovation is not equal to success. The Afghan campaign wears on as ISAF forces prepare for transition to full Afghan security lead. IJC has been a command improvement but has had to struggle to leave an imprint on an established campaign organization characterized by disunity. The IJC's first annual assessment, of October 2010, concluded that the overall IJC campaign plan was being followed by the regional commands, but also noted bottom-up complaints regarding the provision of resources and notably the IJC's insufficient involvement in making security, governance, and development ends meet.[13] Moreover, the COIN heydays of generals McChrystal and Petraeus are over, and COMISAF General Allen has become more of a security transition manager. The same goes for Simon Gass, who took over as SCR from Mark Sedwill in early 2011. The *Inteqal* transition strategy designed through 2010 was the height of a renewed COIN campaign that will prove short. The mid-2012 end of the surge initiated by President Obama in late 2009 means an end to the fighting season. ISAF forces will henceforth focus fully on preparing the Afghan security lead. This might still be COIN, but it is COIN in transition and not the kind of fully fledged campaign that was possible in, say, 2004 or 2005 when patience and money were in greater supply.

CAUSES OF INNOVATION

Among the shaping factors driving what is after all a case of innovation are international politics in the shape of international reactions to Barack Obama's presidency. It was a game changer because alliance diplomacy had exhausted its potential for renewal for as long as President George W. Bush remained in office, which he did until January 2009. The controversial war on terror that got underway under President Bush's first term (2001–4) and included the Iraq War of March 2003 had divided the allies, and the antagonism that these years of friction generated proved deep-seated. Still, the desire to maintain the alliance as a functioning, operational partnership likewise proved deep-seated, and we detect a widespread desire in 2007–8 to mend fences and improve cooperation on the ground. The US presidential elections provided the framework for anticipated change. Within a short span of time following President Obama's assumption of office, the allies were ready for the conceptual shift that COIN represented and that informed the ISAF innovation of 2009.

We shall look into the politics of this shift in a moment, but first we must place the politics of the US presidency in context. A momentum for change had been building for several years, which explains the relative ease with which President Obama could convince his NATO and ISAF allies that COIN was the proper framework of thought. This momentum was driven by other shaping factors that we may label prospective defeat, geopolitical upheaval, and advocacy for commitment. The first factor is largely operational and related to the Afghan battlefield; the second relates to a combination of international and alliance politics; the last to the type of consistent advice that reached policymakers from the NATO institution.

INNOVATION'S SHAPING FACTORS

Prospective defeat became an issue as ISAF fanned out into the southern and eastern provinces of Afghanistan and encountered stiff armed resistance. The United States had, since 2001, operated under OEF banners in the eastern provinces that along with Pakistan's Northwest Frontier Province and Federally Administered Tribal Areas make up the regional base of Al Qaeda and affiliates, and US forces continued to dominate what became ISAF's Regional Command East (RC-E). The real shock to ISAF was the southern enlargement of the campaign, therefore, and it involved the entry of Canadian, British, Danish, Dutch, Australian, and Romanian forces into the provinces of Kandahar, Helmand, Uruzgan, and Zabul. In these provinces Western security forces suffered a total of 21 casualties in the early years of 2001–4. From 2005 to 2007, which comprises the preparation of ISAF expansion and its implementation, they suffered a total of 224 casualties, which amounts to a tenfold increase.[14] The situation spiraled out of control, in other words, as the "security assistance" mission evolved into a type of war. There were no official government structures to assist, and since the environment was too dangerous for civilian agencies to operate in, ISAF forces found themselves saddled with development and governance responsibilities though they were meant to be in the lead of security operations only. To relieve forces of these duties, ISAF brought forth Provincial Reconstruction Teams (PRTs), which were intended as interagency shops that, protected by ISAF forces, would allow diplomats and aid workers to get governance and development going.[15]

The problem was that there were effective governance structures in place, though they did not connect to the central government but rather to illicit trade in drugs, notably, and sometimes the Taliban. To get the PRT effort off the ground, ISAF needed either to convert existing strongmen or sideline them. It was not a simple affair, as the British experience in Helmand illustrates. The local strongman and governor, Sher Mohammad Akhonzada, was judged an obstacle to progress by the British given his infiltration into the Helmand narcotics industry and power structure. One US officer who operated in northern Helmand prior to the ISAF takeover noted that Helmand's "drug militias were the finest fighting organization we ever faced, and I've faced a few."[16] US forces opted to bypass these militias in the early days of the conflict, but Britain's ISAF force decided to tackle them head on.[17] Britain thus insisted to President Karzai in Kabul that he sack Sher Mohammad Akhonzada as governor. Akhonzada was indeed sacked in December 2005, but the dash for renewal soon turned sour. Helmand militias turned on ISAF forces, and northern Helmand turned into a fighting zone. District centers such as Musa Qala, Sangin, Kajaki, and Nawzad that were meant to become hubs of development instead became cauldrons of asymmetrical and deadly warfare.

The predictable result was dissent within the ISAF campaign, and dissent produced drift. It was obvious that ISAF forces needed to protect the PRTs—the campaign's acclaimed "spearhead"—but striking the right balance between PRT and development on the one hand and combat operations on the other was both difficult and controversial. ISAF forces were de facto constantly driven to chase local militias and Taliban insurgents to a greater extent than they wished for. In 2006 the British COMISAF, General Richards, wanted to establish Afghan Development Zones (ADZs) in a classical COIN or oil spot strategy, but that meant a concentration of troops in key population areas. To deal with outlying areas such as Musa Qala, he encouraged negotiations with local elders and authorities who were judged to be unrelated to the growing insurgency. But these deals were feeble. Militias and insurgents found it easy to overrun villages once ISAF forces left them. COMISAF Richard's successor,

US General McNeill, took over in February 2007, determined to teach the insurgents a lesson, a toughened approach that earned General McNeill the nickname "bomber McNeill." ISAF thus got drawn—perhaps lured—into a kinetic campaign, and the ADZ concept and approach faltered on the ground. The problem was both conceptual and material. Conceptually, the ISAF allies talked up COIN but pursued a different and distinctively kinetic campaign on the ground. Materially, the allies simply did not resource the campaign adequately both in military and civilian-diplomatic terms. By 2008, therefore, they were headed for defeat.

Geopolitical upheaval would naturally follow from defeat in Afghanistan. Perhaps NATO could have survived such a defeat. After all, the United States has broad geopolitical interests in its European presence via NATO, and European allies remain committed to the partnership. However, NATO's future would have been uncertain, and the perception within the headquarters and among observers of it was that the alliance was at risk of losing its will and ability to confront dangers in common. This author had the opportunity in September 2008 to ask a range of key NATO ambassadors whether they felt that NATO's survival as an alliance was at stake, and the unanimous and unequivocal answer was yes. Had the NATO mission failed in Afghanistan, it, the United States, and select allies operating in a coalition would have continued a hunt for terrorists in eastern Afghanistan. But the cost to the "grand alliance" could well have been great. A worst-case scenario, feared by some in NATO, was that the transatlantic relationship would have been reconfigured into a bilateral US-EU relationship that would have seen NATO's military structures folded into the EU and the termination of the North Atlantic Council as a collective body of transatlantic consultation and deliberation.

As a geopolitical shaping factor we are dealing with alliance dynamics that feed off international security developments. These dynamics refer to a combination of burden-sharing and reputation. On the one hand was the view that some European allies contributed too little to the campaign and had sunk into a mellow state of foreign policy thinking that emasculated their expeditionary warfare capacity. On the other hand was the view that the United States contributed too little to the alliance in terms of both political consultations and forces (most US forces were OEF and not ISAF) and had gained an addiction for Euro-bashing. There was plenty of bad news from the field to feed either of these stereotypes, and Iraq was a particularly sore issue. In 2005–7 US forces in Iraq were struggling with a vicious insurgency, and US policy-makers were looking for European assistance in Afghanistan, the other main theater of conflict. Some Europeans, however, felt that the United States had bungled Afghanistan in the first place and left Europeans to handle it in order to go to war in Iraq.

Defeat thus cast an ominous shadow over the alliance in the course of 2007–8: the Afghan campaign could lead to defeat at the hands of insurgents, and Afghan failure would set off a wider geopolitical chain reaction that could spell the end of the alliance. Through these difficult years there was a source of consistent advice for commitment and change, and this was the NATO institution as such—our third shaping factor. By NATO "institution," I refer not only to the organization—composed of the Secretary General, the international staff, and the whole military chain of command—but also to the network of officials that branch out into national defense and foreign ministries that connect to NATO and incorporate alliance business into the national security and defense agenda. From this NATO institution emerged a pervasive sense that something was not right and policy was in need of correction, though disagreement abounded in respect to the right policy. The network was obliged to address the issue in preparation for the many NATO ministerials—with at least two foreign minister meetings and three defense minister meetings per year. On top of this there were the NATO summits involving heads of state and government, and these summits have run with considerable frequency (in 2002, 2004, 2006, 2008, 2009, 2010, and 2012).

The 2002 Prague summit came prior to NATO's ISAF command but did focus on the need for transformation, including expeditionary warfare, which was essential to a campaign such as ISAF. The 2004 Istanbul and 2006 Riga summits were directly engulfed by the Afghan issue. In 2004 NATO was preparing for ISAF expansion but needed countries to invest in PRTs and everything that came with them—from supporting military forces to teams of diplomats and development experts. It was an exercise in frustrated diplomacy that saw NATO commit too few forces to meet the minimum force requirements drawn up by its own military authorities, a pattern that was to repeat itself in the coming years. In 2006 NATO had a blueprint for ISAF expansion but was—again—hampered by too few forces and notably a lack of critical enablers such as helicopters, as well as a multitude of national caveats. Someone needed to drive the alliance forward, and Secretary General de Hoop Scheffer—who took office in January 2004—did try. Scheffer dominated the summits' press coverage with his explanations that Afghanistan was winnable but needed greater efforts: he became a public advocate of ISAF commitment. Putting pressure on allied governments may have been needed but was also controversial, and it is widely acknowledged that Scheffer spent a considerable part of his political capital in these early years of his term (he stayed on until mid-2009).

ISAF was permanently on the agenda of North Atlantic Council (NAC) meetings, which is to say every week in the headquarters. The meetings are run by the permanent representatives who have delegations at the HQ to prepare their contributions but who also work closely with their national foreign and defense ministries. With weekly briefings and discussions, the "institution" was thus in operation, though it can be difficult to gauge from the outside how substantial some of these discussions were. The virtue of having ISAF on every meeting agenda is that the operation draws attention and thus perhaps commitments; the danger is that the issue becomes a routine matter. This danger appears to have set in during 2007, when ISAF had been built up and ran into the national insurgency. The NAC ceased to think strategically about the campaign at this point, one centrally placed officer of the strategic military headquarters argued, instead showing an "enormous appetite" for tactical briefings.[18] The Military Committee (MC) and its staff should be doing these briefings, but the NAC wanted operational insights and therefore tended to pull in SHAPE (Supreme Headquarters Allied Powers Europe), which had the effect of muddling the command structure: SHAPE was supposed to focus sharply on the campaign, while the MC was supposed to prepare the NAC's strategic reflections.

The sum total of these shaping factors—prospect of operational defeat, geopolitical upheaval, and institutional preoccupation—came together in mid-2007 and created a momentum for change. The outcome was first the Bucharest summit of April 2008 and the alliance commitment to a comprehensive political-military plan for the campaign, and then secondly the full turn to COIN thinking that came in the course of 2009, as we have seen. The Bucharest summit of 2008 was important because it saw the adoption of a Comprehensive Strategic Political-Military Plan (CSPMP) for ISAF. This classified plan focused on all lines of the campaign—security, governance, and development—and defined seventeen "desirable outcomes." It then elaborated on the role of NATO (in lead of security) and the need for broad international cooperation (because other organizations are in the lead of governance and development).[19] It was significant not so much because it was a brilliant plan—which it probably was not: it would be continuously updated and adjusted to reality on the ground—but because it represented NATO's recognition that it needed to engage the campaign holistically and strategically. At Bucharest, ISAF (which includes many non-NATO members) also issued the "Strategic Vision" that underlay the CSPMP effort. Moreover, NATO had invited the full ISAF family to its summit—from individual allies to partner organizations, including the UN, the World Bank, the EU, and the World Food Program.

The CSPMP was an achievement. NATO had begun ISAF as a Kabul operation in 2003 (when NATO's command started), then decided in 2003–4 to expand ISAF but within the mission statement of "security assistance," and then in 2008, following the better part of three years of critical fighting on the ground, had come to the recognition that it needed to reconceptualize the campaign in comprehensive terms and be more strategic. It was the first important step toward the kind of COIN campaign with which we have become accustomed, but which NATO for some time after 2008 continued to subsume under its Comprehensive Approach.

CHALLENGES REMAIN

The advance was considerable but also a prelude to new challenges. The decision-making phase was over; now the time had come to act strategically. The CSPMP momentum spoke in favor of strategic action, but three conditions, we now know, worked against it. One was the lingering ideological divide between the United States and some European allies regarding the role of democratic ideals in alliance policy. The Bush administration saw these ideals as a leitmotif; others saw them as pragmatic ideals; and the difference erupted at the Bucharest summit where especially Germany felt that the alliance leader went further than it had promised to in its efforts to issue invitations for NATO membership to Ukraine and Georgia. This confrontation dominated the media's coverage—and ensured that Russia's role as a critical supply route to Afghanistan remained underdeveloped. The continuing differences persuaded most alliance leaders that major initiatives had to await the US presidential elections in the fall of 2008, as mentioned, and Bucharest therefore commenced work on a short Declaration on Alliance Security that would become the first step in the revision of NATO's Strategic Concept. The official initiation of the reform process would take place at the sixtieth anniversary of the Strasbourg-Kehl summit in April 2009 and with a new US president.

President Obama thus was brought into NATO's spotlight, but the first year of his presidency—2009—proved to be ambiguous. The alliance celebrated its unity, for sure, not least at the April summit, but guidance for the Afghan campaign proved slow in coming, and Washington ended up largely sidetracking NATO. President Obama at first emphasized a wider and predominantly counterterrorist campaign, which was the Af-Pak strategy of focusing more on terrorist networks operating in the region rather than the Afghan insurgency, but then got drawn back to countering this insurgency. It happened in two steps: first as the new ISAF commander General McChrystal over the summer worked out a classical COIN campaign plan for Afghanistan (see above)—a plan that leaked and challenged President Obama—and then as the President spent almost the entire fall of 2009 going through his Afghan plan anew, effectively ending up endorsing McChrystal's approach and plan.

NATO was not central to this process. Naturally, COMISAF General McChrystal was in NATO's chain of command and thus consulted with the NAC on this evolving approach. However, as the campaign commander, he had considerable leeway in defining the campaign plan. Contrast this with the situation of General Wesley Clark who, as Supreme Allied Commander Europe, was in charge of NATO's air campaign in the 1999 Kosovo War. Clark had to seek NAC approval for target sets and even individual targets in the bombing campaign.[20] Moreover, on the big strategic issues Washington was the center of gravity. President Obama's original Af-Pak plan was American made, though NATO endorsed it in April; his fall campaign review was likewise national and led to noticeable frustrations among allies who thought they had gone far in redesigning the campaign (defined here as a case of innovation), only to find themselves marginalized. NATO did endorse the new plan and provide an additional ten thousand troops for ISAF—a contribution Obama in fact calculated with and had coordinated closely with Secretary General Rasmussen[21]—but during the review NATO got briefed rather than consulted. There was of course a degree of comfort in this for the allies because as long as the US president took ownership of the war—a war of necessity, as Obama phrased it—the allies were in a sense off the strategic hook. US ownership would also ease the burden-sharing debate, which was contentious as always. However, their commitment to strategic engagement stood, and had the United States run strategic policy through the NAC, NATO might have gained some real strategic leadership capacity.

Bush's ideology and Obama's policy process are not the only factors that explain NATO's arrested momentum toward strategic impact in 2008–9. The insurgency itself, our third factor, was intractable and in fact soon challenged the CSPMP. With the CSPMP NATO had a roadmap—the aforementioned oil spot strategy—but it presupposed that building up the new Afghanistan as outlined in diplomatic agreements was a valid and feasible goal. Was the Bonn agreement of 2001—the source of Afghanistan's constitution and thus political institutions as well as the framework for the national development strategy that ISAF and others paid lip service to—still valid? Should ISAF and/or President Karzai negotiate with the Taliban to reach a new settlement? Was such reconciliation desirable? At this level of high politics the allies were divided. Moreover, Karzai had doubts as to whether the allies should be involved in reconciliation. Such a process formally began in mid-to-late 2010, but it was national (that is, Karzai) led, and NATO has no policy on constitutional adjustment or bargaining. To the contrary, NATO maintains a neutral role by stating that it accepts reconciliation as long as it takes place in respect of the 2004 constitution.

We thus come to the conclusion that there were certain drivers behind NATO's acceptance of campaign innovation in the shape of a revamped campaign concept and command structure—an innovation that began in late 2008 and was completed in early 2010: the drivers were operational and geopolitical, with both of them promising defeat and upheaval in the case of continuity, as well as institutional in the sense that a network of NATO actors and organizations became fundamentally preoccupied by the deteriorating campaign. Another conclusion is that there were limits to renewal, notably in the direction of fulfilling the promise to be fully strategic in the approach to the campaign. Transatlantic ideological differences, Obama's preference for national policy reform, and a bloody and exhausting insurgency saw to it. Britain's former ambassador to Afghanistan, Sherard Cowper-Coles, argues that the campaign lacked a political strategy because it involved no negotiations with the adversary.[22] Cowper-Coles thus tends to agree with President Karzai, who laments that the forced decision in 2006 to fire Governor Sher Mohammad Akhonzada of Helmand—which is where this section began—was a political mistake that fed the southern insurgency.[23] The NATO and ISAF campaign certainly lacked strategic direction but, importantly, NATO had improved in its approach to the war. The result was ISAF innovation. What we should make of this curious case of strategic paucity and campaign innovation is the topic of the next section.

NATO STRATEGIC AND OPERATIONAL ADAPTATION

Campaign innovation has had a ripple effect inside the alliance. We can trace this effect from two of the core factors of innovation, the central command reorganization in Kabul with its division of a four- and three-star HQ and then the reinforcement of the senior civilian representative. From here we see how NATO's general command structure has become more expeditionary and how NATO's wider organization and toolbox have become more population-centric and more geared to CIMIC. However, NATO is a multifaceted alliance. Multiple political interests and operational needs shape the agenda, and no single campaign can singularly define NATO transformation. NATO adaptation has thus followed ISAF innovation. Still, adaptation is important, and to gauge it we should turn to the new Strategic Concept of November 2010.

The novelty of the 2010 Strategic Concept lies in its three "essential core tasks" that are put on par: collective defense, crisis management, and cooperative security.[24] This is new because the two previous Strategic Concepts (1991 and 1999) introduced a hierarchy of tasks beginning with collective defense. Hierarchy has given way to plurality, therefore. It reflects a fundamental lesson of Afghanistan that even though the core mission may be related to collective defense (and thus NATO's Article 5), it will not succeed unless the NATO is as good at managing a crisis environment and partnering with stakeholders as it is at fighting enemies. It shows how far the alliance has moved since the 1990s, even in relation to the 1999 Kosovo intervention when the reigning conception was that if you could do wars, then you could also do military operations other than wars (MOOTW).

If we look first at the task of "collective defense," we see how Afghanistan has shaped the reform of the integrated command structure—the backbone of the collective defense capability. The operational chain of command—Atlantic Command Operations (ACO)—was slimmed down and reconfigured in June 2011, and the notable changes involved the operational level of command.[25] Where NATO previously had three Joint Force Headquarters (JFHQ) (in Brunssum, Naples, and Lisbon), it now reduced the number to two (Brunssum and Naples) and, critically, redesigned them. NATO's JFHQ will now, for the first time in alliance history, be able to deploy "up to a major joint operation into theatre."[26] With this, NATO will not again find itself in a situation where its JFHQ (in this case Brunssum) becomes a widely perceived bureaucratic obstacle to the running of the campaign by the campaign headquarters (ISAF in Kabul), based on strategic directions from the supreme headquarters (SHAPE). The Brunssum JFHQ might actually have done a good job in relation to ISAF, as officials in NATO pointed out to the author, but Afghanistan has simply been an eye-opener.[27] Compared with earlier operations in the Balkans, Afghanistan is a much larger affair and far removed from NATO territory. In the Balkans,

NATO needed only small campaign HQs, and they were located close to the parental JFHQ (in Naples). In Afghanistan, ISAF HQ has grown very large and can rival JFHQ Brunssum in many tasks. Moreover, given modern technology, it is as easy for ISAF HQ to communicate directly with SHAPE as with JFHQ Brunssum.

NATO is now drawing the lesson. It discards the old design for JFHQ expeditionary warfare—the so-called Combined Joint Task Force HQ, which was supposed to be located within the JFHQ but as a detachable and deployable element—because it proved difficult to realize and because Afghanistan has demonstrated a need for a more thorough approach. The JFHQ will instead be fully deployable. In the future, the ISAF HQ will be an integral part of the Brunssum JFHQ, and ISAF HQ will thus be able to communicate directly with SHAPE without short-circuiting the lines of command. This rationalization of the command structure might have been coming in any case, given the faulty nature of the CJTF HQ concept, but ISAF provided the direction and the impetus for the 2011 reform.

The command structure also consists of a forward-looking pillar—Allied Command Transformation (ACT)—which has not been touched by the recent reforms. However, it is worth noting that while the strategic ACT HQ in Norfolk continues to look for a role, not least now that the US sister organization, the Joint Force Command, has closed, other ACT command components—the Joint Warfare Center (JWC/Norway), the Joint Force Training Center (JFTC/Poland), and the Joint Analysis and Lessons Learned Center (JALLC/Lisbon)—have experienced a type of revival given Afghanistan. JWC and JFTC have been actively engaged in training and preparing the headquarter components that deploy to Afghanistan.[28] JALLC is in close touch with ISAF HQ and structures lessons learned from the campaign, lessons that are then fed notably to JWC and JFTC. This is a type of adaptation insofar as these three components would be lingering if they were to take their cues from ACT HQ in Norfolk: Afghanistan has to a great extent become their rationale. The challenge for the ACT pillar is to break this JWC-JFTC-JALLC Afghanistan loop and integrate their planning, thinking, and training into a wider alliance framework, which has yet to happen but which the alliance is trying to remedy.[29]

Collective defense involves not only the integrated command structure but also the set of national forces that must carry out military tasks. These forces fall outside the collective frame that is dealt with here, and they are the topic of the other chapters in this book. But it is important to note that NATO continues to plan for the expeditionary forces that allies must deliver. Given the Afghan experience, the issue of expeditionary force planning has become uncontested. No one evokes in-place forces as opposed to deployable forces any longer: it is simply no longer relevant or legitimate. There has been a debate on NATO's continuing focus on territorial Article 5 missions, a debate driven by allies in proximity to Russia, but this debate has been solved by emphasizing the Article 5 capability of expeditionary forces such as the NATO Response Force. Moreover, the role of SOF has been further recognized. As we saw earlier, SOF is now overtly located within the ISAF command structure; the same has happened at the strategic SHAPE HQ. NATO has set up a SOF HQ serving as an advisory body to the supreme commander and a coordinating center of excellence bringing together allied and partner SOF.[30] SOF in Afghanistan have been critical in performing a range of protective, intelligence-gathering, and kill-or-capture missions in Afghanistan, and officials in NATO HQ see the enhanced SOF HQ as a direct outcome of ISAF.[31]

With these adaptations in the military domain we turn to the second of the Strategic Concept tasks, crisis management. The distinction between managing a crisis and fighting enemy forces is important in Afghanistan: NATO must be able to do both, but they are separate disciplines. Crisis management in Afghanistan has run primarily through the PRTs, but the enhanced office of NATO's senior civilian representative in Kabul has become increasingly critical as the PRTs begin to fold into Afghan leadership. In a departure from previous alliance consensus, established at the Riga summit in 2006, NATO has realized that it needed to take charge in these matters. Where NATO allies such as France sought to prevent NATO incursions into civilian crisis management, which it saw as a reserved domain for the EU, they now acknowledge that if NATO is the only actor on the ground, it will need a certain capacity to kick-start development and governance. In the Strategic Concept of 2010 we therefore see an alliance agreement to build "an appropriate but modest civilian crisis management capability" and to train civilian specialists from member states made available for rapid deployment.[32] These initiatives represent the institutionalization of the PRT experience, which was new when it began in 2004–5 and which became an issue of controversy as NATO began debating its wider and principled implications. As mentioned, in Riga in 2006 NATO could only agree to the need for comprehensiveness, not a NATO role in crisis management; by 2010 that had changed. NATO adapted, and Afghanistan had been the engine of change.

NATO's ambition is to lead in crisis management if necessary but then to hand over responsibility to other organizations such as the UN or the EU when possible. Hence, NATO's capacity will be "appropriate but modest." The ambition thus requires good working relations with precisely the UN and the EU, and in this direction we witness a degree of adaptation but mostly deadlock. Afghanistan may in fact have undermined UN-NATO relations to a degree. There have been improvements, notably with the 2008 secretariat declarations that gave cover to interstaff relations, and NATO now has a civilian representative at UN headquarters in New York. However, this is just one person. In Afghanistan, NATO commands more than 100,000 troops and the UN mission (UNAMA) has largely run aground. The unit running UNAMA, the UN's Department of Peacekeeping Operations, is busily running 80,000 peacekeeping troops in Africa, and it tends to see Afghanistan as a NATO issue.[33] Relations to the EU are fully blocked given the Cyprus issue and its reverberations. The EU's operational effort in Afghanistan (it does foot a big part of the development and governance bill) has been limited to police academy training and with very modest results, and given the Cyprus deadlock, NATO and EU personnel on the ground are not formally allowed to exchange information and intelligence.

The crisis management dimension of NATO has thus seen significant adaptation when it comes to the organizational development of the alliance, with this movement quite clearly reflecting Afghan lessons, but quite limited adaptation at the level of partner relations. This brings us to the third dimension of the Strategic Concept—namely, cooperative security. With the concept, NATO declared its ambition to be more networked and more flexible, issues that again have direct connections to the Afghan experience, and it has resulted in a number of new partnership initiatives.[34] Two of these can be directly traced back to Afghanistan.

One is the preparedness to dialogue with any nations that "share our interest in peaceful international relations."[35] In Afghanistan NATO has been hampered by its constricted interpretation of its role, which has been strictly limited to the UN mandate of "security assistance" to the Afghan government. NATO has not been able to develop a dialogue with neighbors such as Pakistan, China, and Iran, though these neighbors are critically important to NATO's mission. Such dialogue instead takes place in various ad hoc bilateral and multilateral settings—such as the December 2011 Bonn conference meant to sustain Afghan transition politically—but NATO has now come to the recognition that, at least, it should give itself the opportunity to dialogue broadly. NATO has therefore agreed to a "wider engagement" and a policy of "enhancing consultations in flexible formats"—which is also referred to as the "28+N" format.[36] The reach and modalities of this reform remain open to question: some regionally focused allies see little purpose for a NATO-China dialogue; others fear that extensive consultations will dilute the heralded coupling of consultations (Article 4) and collective defense (Article 5). While NATO partnership policy thus has adapted as a result of the Afghan experience and become more geared to flexible political consultations, the impact hereof remains uncertain.

Another initiative that can be traced back to ISAF concerns the structural role of partners in operational planning and decision-making. ISAF has brought change to NATO where non-NATO ISAF partners now are full participants in ISAF planning and decision-shaping in NATO (NAC decision-making remains for NATO members). This ISAF model will now be made a general model for all NATO operations. NATO has adopted a Political-Military Framework to this effect but will now have to institutionalize it and notably define the boundaries between partner involvement in operations and routine activities. The boundary is easy to draw in big operations such as ISAF or KFOR (Kosovo) but less so in small operations or where NATO contributes only marginally to larger operations.

In relation to partnerships, there is an overall desire to be more flexible and more political (that is, less driven by military operational needs). However, there is also a significant degree of uncertainty as regards the real potential of these reforms given internal disagreement and the need for time to flesh out new policy. How the partnership policy will evolve in relationship to the "comprehensive approach"—which is at the core of the Afghan operation—is therefore "anybody's guess," as one NATO official put it.[37] The critical issue is that more widespread and flexible consultations could open or widen the North Atlantic Council: some allies will then lose influence and the comforting belief that NATO remains a tightly knit insiders' club for territorial defense.[38]

The sum total in relation to NATO adaptation as a whole is a degree of political-organizational flexibility that will surprise the alliance's most ardent detractors. The centerpieces are command structure reform in the shape of expeditionary operational commands (JFHQ), an ability of transformation command to support expeditionary warfare (the JWS-JFTC-JALLC trio), and an uncontested expeditionary blueprint for force planning; an emerging crisis management capability; and a partnership policy that may underpin the alliance's involvement in the high politics of operational management. The alliance needed all of this in Afghanistan years ago, and in the absence of political-military tools it resorted to innovation within ISAF. These innovations have now trickled into the organization as distinct ISAF imprints on future organizational capacity.

CONCLUSION

ISAF has been NATO's toughest mission to date, and one can doubt that NATO, having been at the brink of defeat and alliance disintegration in 2006–8, will be tempted to undertake another major land operation far from its territory. ISAF innovation and NATO adaptation are testimonies to strategic agility, but one should not mistake agility for ambition. Judging by NATO's Strategic Concept, NATO will henceforth focus more on "new" threats related to cyber, proliferation, and radicalism than old school military threats, and NATO will engage these new threats far more politically than in the past. In this ambition there is strategic foresight related to global trends but also pragmatism inspired by military exhaustion created by Afghanistan and also budgetary exhaustion following the financial crisis that began in 2008.

ISAF innovation happened along a number of military and civilian dimensions and resulted in a campaign design that was wholly new to NATO. The alliance had never before had such an imposing campaign commander (COMISAF) liberated from day-to-day operational issues—delegated to IJC HQ—and operative in the political-strategic milieu of Kabul. Nor had the alliance run crisis management units such as the PRTs—the "spearhead" of the campaign—and fated its involvement on developments of which it was not in charge (development and governance). As other organizations failed to take charge, NATO needed to invest leadership with the senior civilian representative, which it did, and which was new. NATO now has a COIN doctrine and, more important, a political openness to COIN thinking, something that was taboo three to four years ago, which we see in the CSPMP (from 2008), the subsequent willingness to support a COIN campaign plan, and also the willingness to adapt the organization in ways that hitherto had not been possible.

It speaks to NATO's disadvantage that it for long was resistant to change. ISAF was and continues to be engaged in a highly dynamic campaign, and though it is not possible to argue that NATO should have foreseen its nature, it is reasonable to argue that NATO should have been more dynamic in its response. It took the prospect of operational defeat and geopolitical upheaval supported by a pervasive institutional preoccupation with the conflict to generate change. If NATO changed, it was to rescue itself and prevent defeat in Afghanistan.

The prevalent role of the United States in pushing for change through 2009 and in providing solutions (picked from the Iraq experience) is double-edged, as noted. The United States helped solve campaign and alliance problems, and it invested significant resources—via the Obama surge—to carry through this new momentum. However, in its preference for national strategic planning the Obama administration also contributed to the strategic deficit in NATO, which in turn has undermined its efforts to ask more of the NATO allies. NATO allies are not likely to deliver more on the occasion of burden-sharing conferences and speeches by US officials; they are far likelier to deliver if they feel strategically involved, and, one might add, if the campaign has prospects of success. In recent years there has been a deficit on both accounts.

What does this tell us about NATO adaptation? Like ISAF innovation, it is the outcome of operational, geopolitical, and institutional pressure for change. However, where ISAF innovation has been completed, NATO's adaptation is a work in progress. The operational and geopolitical pressures for change have subsided, and NATO's multifaceted interests and engagements can thus intervene in the reform process. Operation Unified Protector in Libya has likewise intervened, capturing headlines to such an extent that Afghanistan appeared to be done and dusted for NATO. Such a view is mistaken, because Afghanistan continues to be a challenge for the kind of politically relevant and agile alliance that NATO would like to be. If NATO does not step up to the challenge of adding political to military engagements—and we see the challenge in Afghanistan, Libya, and elsewhere NATO has operated—NATO adaptation will derail and the ambition of Strategic Concept will remain unfulfilled. This is the continuing lesson of ISAF.

NOTES

1. Niccolò Machiavelli, *The Prince* (Oxford: Oxford University Press, 2008), 50.
2. Sten Rynning, *NATO in Afghanistan: The Liberal Disconnect* (Stanford: Stanford University Press, 2012); Alexander Mattelaer, "How Afghanistan Has Strengthened NATO," *Survival* 53, no. 6 (December 2011–January 2012): 127–40.
3. See Arnold Wolfers, "Collective Defense versus Collective Security," in *Discord and Collaboration: Essays on International Politics* (Baltimore, MD: Johns Hopkins University Press, 1991), 181–204.
4. In this case, "NATO" refers to the collective framework through which individual allies and partners—the ISAF nations—channel their force contributions and nationally designed campaigns.

5. NATO, *Summit Declaration on Afghanistan*, Press Release (2009) 045, April 4, 2009.

6. The OMLT—Operational Mentoring Liaison Teams—were the equivalent of US Embedded Training Teams (ETT). Their mission was to train operationally: to embed with Afghan army battalions (*Kandaks*) moving into the field, mentor them, and help them call in air support or medical evacuation as needed.

7. Source: ISAF website (available at http://www.isaf.nato.int/en/isaf-command-structure.html), combined with author's background interview, ISAF HQ, Kabul, September 29, 2009. Reprinted from Rynning, *NATO in Afghanistan*.

8. NATO Secretary General Jaap de Hoop Scheffer, *Press Conference*, June 12, 2009: available at http://www.nato.int/cps/en/SID-DB30D745-DEB726B1/natolive/opinions_55630.htm, accessed December 1, 2011.

9. The Afghan government was formally the chair of the ESC, with international actors serving as cochairs.

10. The SCR was redefined as the SCR of the NAC and not merely the secretary general of the alliance, and gained a greater say in PRT affairs. For more on this, see Rynning, *NATO in Afghanistan*, ch. 5; see also Sherard Cowper-Coles, *Cables from Kabul* (London: Harper Press, 2011), 136–37.

11. US secretary of defense Robert Gates held back, it should be noted, because President Obama was still reviewing his administration's Afghanistan plan and thus was not ready to commit one way or the other at this point. As the United States for long had been prepared in principle to wage COIN in Afghanistan, the significance of the Bratislava meeting related to NATO. See "NATO Ministers Endorse Wider Afghan Effort," *New York Times*, October 23, 2009. For COMISAF's campaign plan, see General McChrystal, "Commander's Initial Assessment," August 30, 2009: available at http://media.washingtonpost.com /wp-srv/politics/documents/Assessment_Redacted_092109.pdf?sid=ST2009092003140, accessed December 1, 2011.

12. This turn of events also put in motion work on a general NATO COIN doctrine, which was finalized in January 2011: *Allied Joint Doctrine for Counterinsurgency (COIN)*, AJP 3.4.4.

13. Author's interview with senior civilian adviser to IJC, Gloucester, England, September 2011.

14. Calculated with help of http://www.conflictmonitors.org/countries/afghanistan/facts-and-figures/casualties/security-forces, accessed December 1, 2011.

15. The PRTs originated in Afghanistan, though by American fiat inside OEF. The first three PRTs were opened in the spring of 2003—in Gardez, Bamyan, and Kunduz—and they were American. The fourth was British and opened in Mazar-e-Sharif. ISAF expansion began to be planned for in late 2003 by NATO, and the political guidance of the North Atlantic Council was that expansion must happen via PRTs: they were the center of gravity, in other words, while military forces were auxiliary. Such was the plan. NATO North Atlantic Council, "Final Communiqué," December 4, 2003, para. 4:

available at http://www.nato.int/cps/en/natolive/official_texts_20271.htm, accessed December 1, 2011.

16. Tom Coghlan, "The Taliban in Helmand: An Oral History," in Antonio Giustozzi, ed., *Decoding the New Taliban* (London: Hurst, 2009), 119–53.

17. The United States posted a mere 120 SOF forces to Helmand prior to 2005–6 to chase terrorists, not drug lords. For an account of these early years and the US and British dilemmas and problems, see Ahmed Rashid, *Descent into Chaos* (London: Penguin, 2008), 321–32. Britain was in addition to Helmand lead also the international lead on counternarcotics for all of Afghanistan following a division of labor established within the international community in early 2002.

18. Background interview by author, November 2009.

19. See Rynning, *NATO in Afghanistan*.

20. Gen. Wesley K. Clark, *Waging Modern War: Bosnia, Kosovo and the Future of Conflict* (New York: Public Affairs, 2001), 220–42. Incredibly, 64 percent of the fixed targets in the air campaign were approved by the NAC. US General Accounting Office (GAO), *Kosovo Air Operations: Need to Maintain Alliance Cohesion Resulted in Doctrinal Departures*, GAO-01–784 (Washington, DC: GAO, July 2001), 8.

21. Rynning, *NATO in Afghanistan*, ch. 6.

22. Cowper-Coles, *Cables from Kabul*, 115–20. Cowper-Coles's criticism of the lack of strategy is widely shared. However, his point of view that more should be done to negotiate with the Taliban remains controversial.

23. Ibid., 155.

24. NATO, *Active Engagement, Modern Defence*: available at http://www.nato.int/lisbon2010/strategic-concept-2010-eng.pdf, 2, accessed December 1, 2011.

25. NATO, *Backgrounder on NATO Command Structure Review*, June 9, 2011: available at http://www.nato.int/nato_static/assets/pdf/pdf_2011_06/20110609-Backgrounder_Command_Structure.pdf, accessed December 1, 2011.

26. Ibid.

27. Interviewed by author in NATO HQ, November 16, 2010.

28. The ISAF HQ is organized around core components from NATO's force structure (NFS), which is to say that corps headquarters deploy to Kabul to take command on a rotational basis. On top of this core are a litany of command add-ons.

29. NATO issued a Political Guidance in March 2011 defining a new Level of Ambition, which was then translated into Military Committee Guidance by June 2011 (MC 400/3). These documents, classified, define the parameters of capability improvement and transformation, and one would naturally expect that they integrate lessons from operations and strategic foresight—thus bridging ACO and ACT—touching notably on issues of counter-IED, medical support, logistics, and intelligence gathering capabilities. The 2011 planning will be followed in 2012 by a Capabilities Requirement Review and a Defense and Deterrence Posture Review.

30. For more on this new SOF HQ, see http://www.nshq.nato.int/NSHQ/, accessed December 1, 2011.

31. Interviewed by author, November 16, 2010.

32. NATO agreed back in 2008 to establish rosters—a database—of national civilian experts available for crisis management operations, which other international organizations had done as well. What was new in 2010 was the agreement actually to train some of these experts, who typically are unprepared for work in high-risk environments. See Niels Henrik Hedegaard, "NATO's Institutional Environment: The New Strategic Concept Endorses the Comprehensive Approach," in Jens Ringsmose and Sten Rynning, eds., *NATO's New Strategic Concept: A Comprehensive Assessment* (Copenhagen: DIIS, 2011), 75–82.

33. At least judging by NATO experience. Interview with NATO official, April 7, 2011.

34. See paragraphs 28–35 of the Strategic Concept. NATO adopted a package of partnership policy reforms at the Ministers of Foreign Affairs meeting in Berlin, April 14–15, 2011, as a follow-up to the Strategic Concept ambition. For this package, see http://www.nato.int/cps/en/SID-F043DE52–27A2A4D5/natolive/events_72278.htm, accessed December 1, 2011.

35. Ringsmose and Rynning, *NATO's New Strategic Concept*, para. 30.

36. Sections VII and VIII of *Active Engagement in Cooperative Security: A More Efficient and Flexible Partnership Policy*, Berlin, April 14–15, 2011.

37. Interviewed by author, April 8, 2011.

38. As one national official at NATO headquarters put it. Interviewed by author, April 8, 2011.

5 Back from the Brink: British Military Adaptation and the Struggle for Helmand, 2006–2011

Theo Farrell

Britain's war in Afghanistan has, since 2006, focused on Helmand Province in the south of the country. Under Operation (Op) HERRICK, the codename for Britain's overall military campaign, forces were deployed into Helmand to provide security for the development of governance, services, and infrastructure. The British military soon found themselves drawn into a bloody struggle against insurgents for control of the province. In the Battle of Maiwand in 1880, the British Army suffered ignominious defeat at the hands of a Helmandi army.[1] It came close to doing so again. However, superior skill, firepower, and gritty determination enabled British troops to cling onto their positions across Helmand. Then, from 2007 on, the British military gradually adapted to the demands of the campaign. This chapter explores that painful process.[2]

By 2011, working alongside the US Marine Corps, the British military had managed to secure the key population areas in Helmand and dislodge the Taliban from southern and central Helmand. In short, the British military adapted their way back from the brink of defeat. Less in evidence is adaptation at the strategic level. There is gradual adaptation, in terms of increased force levels and resources for the campaign. But options for adapting Britain's commitment to an increasingly costly campaign have been constrained by the absence of a clear strategy and by alliance politics.

THE CAMPAIGN FOR HELMAND

Until the initial surge of US forces into southern Afghanistan in 2009, responsibility for extending security and protecting Afghan government in Helmand rested on Britain's shoulders. Helmand was and remains the most challenging province in Afghanistan to stabilize, given the scale and reach of the narco-economy, the level of corruption, the degree of local support for the Taliban, and the extent of underdevelopment.[3] Successive British brigades, each on a six-month tour, have formed the International Security Assistance Force (ISAF) task force for Helmand.

This section examines the evolution of the British campaign in Helmand from 2006 to 2011.[4] In the early stages of the campaign, British brigades tried various ways to leverage combat power to destroy the Taliban, or at least dislodge the insurgents from the main populated areas in Helmand. The British managed, just about, to hold insurgents at bay. From late 2007 on, the British military began to focus increasingly on securing and reassuring the population, supporting the development of governance, and working more in partnership with Afghan political authorities and security forces. In short, the British Army and Royal Marines adapted.

Battling the Taliban

The initial British deployment was led by the British Army's elite high-readiness force, 16 Air Assault Brigade. The British government decided to cap the UK task force at 3,150 troops (that is, a half-brigade in strength), both for financial reasons but also because the threat from the Taliban was underestimated. The primary job of 16 Brigade was to provide security and support development in the key ground between and surrounding the provincial capital of Lashkar Gah, and the main economic town of Gereshk. The British task force was prepared and was postured for stabilization operations in Helmand, not combat operations. When initial elements of the British force encountered stiff resistance from the Taliban, it was decided to accelerate the deployment and send an additional 1,500 troops. However, these measures were too little, too late. It still took three months for the task force to deploy in full, and hence it lost important momentum in its break-in battle.[5]

The British made the removal of the highly corrupt governor of Helmand, Sher Mohammed Akhunzada (known as "SMA"), a precondition for their deployment. This was necessary, as SMA's predatory rule was a major cause of local support for the insurgency. However, SMA responded by sending his 3,000-strong militia over to the Taliban side. British planners failed to take account of such an eventuality. To make matters worse, 16 Brigade came under immense pressure from the new provincial governor, Mohammad Daoud, to deploy to northern Helmand, to prevent the towns of Now Zad, Sangin, and

Musa Qaleh from being overrun by insurgents. 16 Brigade had only 600 deployable infantry, who were about to become stretched very thin across Helmand. Appreciating the political imperative to support Governor Daoud, the British task force commander adopted a "platoon house" strategy, sending small packages of between 40 and 100 troops to garrison the northern towns.[6]

These small British garrisons promptly came under almost constant attack from Taliban forces. Now Zad and Musa Qaleh became so difficult to resupply (requiring very dangerous helicopter missions) that 16 Brigade eventually had to abandon them. In its post operation report, 16 Brigade argued that the platoon house strategy had "undoubtedly contributed to the attrition of Taliban forces." However, 16 Brigade also recognized that the focus on fighting had prevented it from supporting development in Helmand.[7]

In October 2006, 3 Commando Brigade took over from 16 Brigade with a different concept of operations (CONOPS). The Royal Marines were also focused on the economic and political center of Helmand, which by then was designated an Afghan Development Zone (ADZ). But the Marines adopted a very different approach that ironically took the focus of their activities outside the ADZ. 3 Brigade had been observing the problems with the Platoon House strategy, which had fixed British forces and prevented them from supporting development activities. The new task force commander, Brigadier Jerry Thomas, declared "reconstruction of the ADZ" to be his main military effort. However, given the dire security situation, Brigadier Thomas determined that "security represents the current pre-eminent line of activity" and that "once our security is in place, development will gather momentum and provide the best tool with which to counter this insurgency."[8] For the security line of operation, Thomas intended to "unfix the North" and "manoeuvre to threaten, disrupt and interdict the enemy."[9] To this end, 3 Brigade created a number of Mobile Operations Groups (MOGs)—250-strong flying columns in forty vehicles (a mix of Vikings and Land Rovers)—that were tasked with seeking out and engaging the Taliban. Conceiving of the "desert as a sea," the Royal Marines expected their MOGs to roam wide and hunt down the enemy. However, the Taliban proved too wily to be drawn into prepared kill zones, and often as not the MOGs ended up engaging the Taliban on their terms. Indeed, this led Royal Marines to coin the term "advance to ambush" to describe MOG operations.[10]

3 Brigade went into Helmand with the right intent. Influence was to be at the center of their CONOPS. The brigade intended to "tread softly," only escalating to violence as required and with due care to minimize collateral damage so

as to "avoid breeding new enemies." 3 Brigade's Operational Design correctly noted that "[a]lmost all tactical engagements (win or lose) favor the enemy. By definition, they demonstrate instability and insecurity thus undermining perceptions of Government of Afghanistan influence, control, and credibility." However, fighting in Helmand *escalated* under 3 Brigade. The total number of engagements with enemy forces went up from 537 under 16 Brigade's tour, to 821 under 3 Brigade's tour.[11]

In its post operation report, 3 Brigade claims to have thrown the Taliban off balance: "The 'dynamic unpredictability' arising from being able to engage the enemy at times and places of our choosing has disrupted him, undermined his will and shattered his unity." It is clear that the Royal Marines were more successful at taking the fight to the enemy than 16 Brigade had been: under 3 Brigade, there was a fourfold increase in the number of engagements initiated by British forces (up to almost 300 engagements). However, as noted earlier, evidence from the field suggests that many of these engagements were actually at a time and place of the enemy's choosing. Moreover, the brigade that followed found Taliban will and unity to be far from "shattered." The larger point is that 3 Brigade never managed to create the security necessary to allow stabilization and development to proceed. The MOGs mostly operated outside the ADZ. This had some effect in keeping the Taliban at bay, resulting in a 45 percent reduction in attacks on British bases guarding district centers. But it also reduced the presence and hence enduring security effect of British forces within the ADZ.[12]

16 and 3 brigades are both elite, high-readiness brigades. With the deployment of 12 Mechanized Brigade in April 2007, the British Army committed one from the regular line. 12 Brigade realized that the mobile operations of 3 Commando Brigade were doing little to secure the populated areas of Helmand. 12 Brigade's observation was that the Royal Marine MOGs "would maneuver around in the desert, hit the crust [that is, the edge of the Green Zone],[13] have a contact [with the Taliban] at range, but have no enduring effect, rarely penetrating into the Green Zone."[14] Unlike 3 Brigade, 12 Brigade focused on securing the ground around and between population centers inside the Green Zone, especially the area between Gereshk (the main economic hub in Helmand) and the newly captured town of Sangin. This involved a series of major "clearance operations," leading to a further escalation in fighting with the number of engagements with enemy forces rising from 821 under 3 Brigade to 1,096 under 12 Brigade.[15] The task force commander, Brigadier John Lorimer, was prepared

for an increase in fighting on 12 Brigade's tour. The Taliban were promising a summer offensive, and so Lorimer was determined that if there was to be an offensive, it would be an ISAF one.[16] Hence, he emphasized core combat skills to be "mastered by everyone, irrespective of rank or cap badge" in predeployment training for his brigade.[17]

At the heart of 12 Brigade's CONOPS was creating security through "persistent presence." Brigadier Lorimer appreciated that security for Afghans meant "compound walls"—that is, solid evidence of security force presence. It was also well understood that areas cleared of Taliban had to be held in order to prevent the insurgents from simply returning. Accordingly, 12 Brigade established a number of patrol bases in the Upper Gereshk Valley. The hope was to backfill areas cleared by British forces with the Afghan National Army (ANA) and Afghan National Police (ANP). However, much like its predecessors, 12 Brigade simply lacked the resources to realize its ambitions. For most of his tour, Brigadier Lorimer had only a battle group and some rifle companies of deployable forces. Most of the additional British forces that came into theater with 12 Brigade were actually "enablers" (such as combat service support).[18]

The ANA were unable to fill the gap in security forces. There were three ANA *kandaks* (battalions) in Helmand; one each in Gereshk and Sangin, and one for training and force generation at Camp Shorabak. These were capable of conducting offensive operations, provided they received sufficient mentoring and logistical support from Task Force Helmand (TFH), but they were poor at holding ground. The ANA view was that this was a job for the ANP. The ANA preferred to return to barracks once offensive operations had been concluded. The ANP were extremely poor quality in Helmand, riddled with tribal factionalism, outright corruption, and drug abuse. Indeed they were far more of a destabilizing factor than a force for stability. Thus, despite their best efforts, 12 Brigade lacked the forces and Afghan partners to provide for enduring presence in territory cleared of Taliban.[19] Accordingly, the Taliban were able to reinfiltrate. The lack of real progress led a frustrated Brigadier Lorimer to reflect that it felt rather like "mowing the lawn."[20]

Protecting the People

The military campaign changed direction with 52 Infantry Brigade in October 2007. The incoming brigade's CONOPS was "clear, hold, build." 52 Brigade's Operational Design conceptualized the campaign center of gravity in terms of the local population instead of the enemy's will and ability to fight. Indeed, for the new task force commander, Brigadier Andrew Mackay, the Taliban body count was a "corrupt measure of success."[21] Accordingly, 52 Brigade's campaign focused on influence operations to win the consent of the population. This resulted in a different approach to "persistent presence." Whereas 12 Brigade rotated units through its Forward Operating Bases (FOBs), 52 Brigade committed units to FOBs for the entire tour, in order to "provide reassurance and provide clear proof that our presence is not transient or temporary."[22] As the brigade plans officer noted, the position of FOBs was important "to establish[ing] a pattern of life locally." He also observed how "locals had something of a 'castle of the hill' mentality," and that by having an enduring presence, "locals would begin to engage the British and provide information."[23]

Thus, "influence operations"—these being activities to engage with the population and win local consent—were absolutely central to 52 Brigade's campaign. Previous brigades had not ignored influence operations. But they had not given it the same prominence in their operational design, nor resourced it to the level as in 52 Brigade. Mackay developed a methodology, called the Tactical Conflict Assessment Framework (TCAF), for targeting and assessing the effectiveness of influence activities. He also created two-man N-KETS ("Non-Kinetic Effects Teams") in each of his battle group HQs, as well as staffing a larger influence team in TFH HQ.[24]

The contrasting approaches of 52 and previous brigades is also evident in their respective approaches to battle. Clearance operations by earlier brigades had involved such extensive use of the force that they often caused the population to flee their homes, in many cases never to return. When 52 Brigade was tasked with retaking Musa Qaleh from the Taliban, Brigadier Mackay was determined that this should be done in a way that minimized the displacement of the local population. For Brigadier Mackay, "[T]he people were the prize," not ground or enemy attrition. Thus he planned a cautious offensive, involving armored forces slowly approaching the town on both flanks weeks in advance of the main assault. The intention was not to capture and kill Taliban forces, but rather to coerce them into leaving the town and thereby avoiding the necessity for a major battle. This approach also gave the local population plenty of warning so they could seek safety in advance of the final assault. In other words, 52 Brigade were prepared to compromise operational security in order to keep the local population on side.[25] The plan largely worked; the Taliban fled Musa Qaleh after a brief fight.[26] Within hours of recapturing the town, 52 Brigade started preplanned stabilization activities. As one journalist on the ground noted: "The residents of Musa Qala voted with their feet soon after the arrival of the Afghan army. Tractors, pick-up trucks and carts started bringing people home."[27]

This new population-centric approach continued under 16 Brigade on its second tour in Helmand, from April to October 2008. This time under the command of Brigadier Mark Carleton-Smith, the brigade's CONOPS was to "go deep not broad." Hence the British task force focused on protecting population centers, and on developing the Afghan government's influence and authority in those areas that realistically could be secured and held. In terms of the enemy, the focus was on undermining Taliban influence rather than fighting their forces. Like 52 Brigade therefore, 16 Brigade strove to achieve an appropriate mix of kinetic and nonkinetic activities. Indeed, Brigadier Carleton-Smith prepared his brigade for a very different campaign than the one conducted by 16 Brigade in 2006: "I wanted to dispel any danger that 16 Brigade might have confused the intensity of the fighting on HERRICK 4 with the nature of the campaign we were seeking to prosecute."[28] Critically, like Mackay, Brigadier Carleton-Smith emphasized the centrally of influence operations over combat operations:

> Our plan for the Taliban was to try to undermine their strategy, rather than merely fight their forces. Their strategy seemed to be more one of influence, intimidation, and the provision of parallel shadow government than of tactical engagement. Therefore the aim was to marginalize their influence in the centres of population.[29]

Thus brigade command staff understood that influence is "everything we do."[30]

To this end, 16 Brigade developed greater integration with the expanding British-led Provincial Reconstruction Team (PRT) in the provincial capital, Lashkar Gah. The brigade planning section (J-5) was physically moved into the PRT building. Moreover, the deputy brigade commander, Colonel Neil Hutton, became one of the three PRT deputies. This second tour by 16 Brigade also coincided with a major increase in PRT capacity, with the civilian staff increasing from forty-five to ninety-four, and military personnel from nineteen to forty-eight. Naturally, civil-military integration was far from perfect, and tensions remained between military and civilian partners. But this was a significant step in the right direction.[31]

A shift in Taliban tactics facilitated the shift to a "softer" way of counterinsurgency (COIN). In 2006–7, Taliban forces suffered considerable attrition in heavy fighting against ISAF forces. British Defence Intelligence put the number of Taliban dead in the thousands (though some British commanders have expressed doubts at such high figures).[32] Accordingly, since early 2008, the Taliban have been less inclined to launch major assaults on district centers and ISAF bases. Thus, when US Marines launched an offensive against the Taliban strongholds in Garmsir District in 2008, the Taliban main force retreated rather than put up a fight. Equally, when 16 Brigade launched an air assault on Taliban villages south of Musa Qaleh, they found that the Taliban had fled.[33] In Kajaki, an Afghan interpreter hired by the British to listen to Taliban communications in 2008 "described almost comical attempts by different commanders to shirk combat and foist the responsibility on other commanders."[34] Essentially the Taliban learned the cost of engaging in direct attacks on ISAF forces. From 2008 on, the Taliban switched to increasing use of Improvised Explosive Devices (IEDs) to target British forces and to defend their own positions. The Taliban did continue to engage in some fight-fires with British forces, including some spectacular attacks (as shall be discussed shortly). Overall, however, we see a more cautious approach from 2008 by a battered Taliban which, while no less deadly to ISAF forces, has been more conductive to stabilization operations because it has resulted in fewer pitched battles in and around district centers.

Securing Central Helmand

Since 2008, the British military effort has focused on securing central Helmand. This has involved a fair amount of fighting as British forces have pushed into Taliban-held territory. Notwithstanding this, the British military have continued to pursue a campaign that is designed to minimize civilian casualties and the destruction of property, and to support the rapid delivery of governance and basic public services.

3 Commando Brigade returned to Helmand in October 2008 with no intention of battling the Taliban. 3 Brigade was clear on its main effort: "The spread of governance allied to the development of capacity at the district level must be viewed as the principal purpose of our security effort and our laydown and tactics should be configured accordingly."[35] However, within days of arriving, the Royal Marines discovered a large Taliban force massing to assault Lashkar Gah.[36] Initial intelligence suggested that up to 1,000 insurgents were moving in to launch a two-pronged attack. In the event, 300 insurgents were involved in a three-pronged attack.[37] The Royal Marines believe the Taliban intended to

decapitate the provincial government and "discredit the British stewardship of Helmand."[38] 3 Brigade had hardly any troops to defend the provincial capital, as the bulk of TFH was deployed in the north (Sangin, Now Zad, Musa Qala, and Kajaki) and the south (Garmsir). Hence, it was left up to Afghan security forces backed by British attack helicopters to fight off the insurgents.[39] Around 150 insurgents were killed. Escaping insurgents were tracked back to where they had come from in central Helmand—namely, the district of Nad-e-Ali, which lies to the west of Lashkar Gah.[40]

This event underlined the extent to which the British, by focusing their efforts on pushing north, had lost control of central Helmand over 2007–8. It led to a complete reassessment of the British campaign plan. The new plan was agreed between the commander of 3 Brigade (then Brigadier Gordon Messenger), the head of the PRT (then Hugh Powell), the Provincial Governor (then Gulab Mangal), and the ANA 205 Corps Commander, that the principal task must be to clear the Taliban from Nad-e-Ali. It was understood that this would require a series of operations stretching over a number of British brigade tours.[41] In late 2008, 3 Brigade launched Op SOND CHARA, a major offensive into Nad-e-Ali. As one senior PRT officer noted, this operation "established a lodgement, a district government and district stabilisation team in Nad-e-Ali from day one, and enough security to develop, with a core of elders, the conditions for later success."[42] 19 Brigade followed through with Op PANCHAI PALANG in June 2009. This operation involved some 3,000 British troops, as well as Danish and ANA forces, to clear the Taliban from the "Babaji Pear," an area on the northern edge of Nad-e-Ali. This operation was one of a series to create the security conditions for elections to take place in central Helmand. It also had a more specific objective, to remove a major threat to Lashkar Gah (the Taliban had been launching rocket attacks from the Babaji Pear). 19 Brigade was concerned to minimize the risk to civilians. Thus the operation was preceded by "messaging" to help civilians evade the fighting and to encourage insurgents to flee.[43]

Following this, 11 Brigade conducted Op TOR SHPAH in December 2009, and Op MOSHTARAK in February 2010, to clear insurgents from the center and north of Nad-e-Ali. Crucially, these operations were also conducted in a way to minimize fighting and civilian casualties. In Op TOR SHPAH, the 1 Grenadier Guards Battle Group in Nad-e-Ali focused on persuading local elders to encourage Taliban to leave their areas in advance of the arrival of British forces. In this way, the grenadiers were able to "talk our way in" to many villages in central

Nad-e-Ali. This became the template for the larger Op MOSHTARAK, which involved a major push by TFH into the Chah-e-Anjir Triangle, a Taliban-held area in north Nad-e-Ali. Thus, much like the approach taken in Op PANCHAI PALANG as well as TOR SHPAH, messaging was used to intimidate Taliban into leaving and to reassure locals that they would not be abandoned afterward. Combined with this were precision strikes against Taliban tactical command and control. This psychological and physical pressure caused the Taliban defense to collapse. The insurgents failed to present organized armed resistance to British and Afghan forces when they air assaulted into Chah-e-Anjir in early February.[44]

By this stage, the British campaign was benefiting from the massive influx of US forces into Helmand (eventually to number almost 20,000) with the 2nd Marine Expeditionary Brigade (2 MEB) in the summer of 2009, replaced by the 1st Marine Expeditionary Force (1 MEF) in April 2010. Incoming US forces released British units from the south of Helmand to support operations in central Helmand. In addition, the major British operations were synchronized with those of the US Marines in order to minimize the Taliban's ability to respond to ISAF's scheme of maneuver in Helmand; thus, Op PANCHAI PALANG was synchronized with Op KHANJAR (the 2 MEB's clearing of the southern districts of Nawa, Garsmir, and Khanishin). Equally, Op MOSHTARAK involved simultaneous offensives by TFH in north Nad-e-Ali, and by the US Marines in Marjah to the south of Nad-e-Ali.

By the spring of 2011, the British had secured central Helmand. The US Marines had secured southern Helmand, and were taking over responsibility from the British for security operations in northern Helmand. Hence, the British Army's 4 Brigade, which replaced 11 Brigade in May, did not plan for or conduct any major "totemic" combat operation. Instead, the emphasis was on "protecting communities and improving freedom of movement," especially for Afghan civilians and officials.[45]

OPERATIONAL IMPERATIVES AND BOTTOM-UP ADAPTATION

After eighteen months of hard fighting, the British military adapted a new approach to the campaign less focused on defeating the insurgents and more focused on protecting and winning over the population. They also concentrated the campaign back on central Helmand, as originally intended.

Operational challenges clearly drove these adaptations. Furthermore, surprise played an important element. Indeed, from the beginning, the British were taken by surprise by the level of armed resistance to their presence in Helmand. For the previous two years, the British military had been operating in the permissive environments of Kabul and Mazar-i-Sharif. There was a limited ISAF presence in Helmand centered on a small US Provincial Reconstruction Team. In fairness, British chiefs were concerned that their understanding of the situation in Helmand was "pretty hazy." Levels of violence were low in Helmand, but intelligence pointed to rising Taliban activity. They understood, as the Chief of the Defence Staff expressed it at the time: "[It] is not the North. It is real bandit country. It is going to be difficult."[46] But it was unclear just how dangerous and difficult it would be.

Two teams of planners—one military and the other civilian—were sent to Helmand in late 2005 to work together on what became the Joint Helmand Plan (JHP). These planners found things to be fairly benign in the province, but they worried about how little they really knew about the situation on the ground. Their concern was about the "uncertainty rather than certainty of threat."[47] Produced in little over a month and briefed to the Cabinet Office in December 2005, the JHP proposed an "inkspot" strategy whereby British forces would focus on creating security in and around the main towns of Lashkar Gah and Gereshk, in order to "create an incubator" for building up Afghan subnational governance. The idea was not for the small British force to try to secure the whole province. The JHP document also contained a strong warning to the British government, that more time had to be spent understanding Helmand before committing the UK task force.[48] This warning was not heeded because there was political momentum behind the British deployment, which was originally scheduled for early 2006.[49] Thus British forces were caught completely off-guard by the level and ferocity of the insurgency when they deployed into northern Helmand, contrary to the injunctions of the JHP.

By the time 52 Brigade arrived, it was clear that the British campaign in Helmand was failing. This drove their search for a new approach; it had become clear that the British could not militarily defeat the insurgency in Helmand. Surprise returns, once again, to play a key role in the second major adaptation to the British campaign. As we saw, 3 Brigade, following on from 52 Brigade, were taken completely off guard by the large Taliban assault on Lashkar Gah. This led to 3 Brigade's operational plan—focusing on security in the north (in Sangin and Musa Qala) and the south (in Garmsir) of Helmand—being "binned" in favor of a new plan that concentrated British military effort on central Helmand.[50]

The process of operational adaptation was primarily bottom-up (that is, led by British task force commanders). British forces came under national and ISAF chains of command. But up to 2009, British task force commanders had considerable latitude to define their own campaign plan and approach to operations. Hence, 16 Brigade were able to depart from Britain's Joint Helmand Plan in 2006, albeit in consultation with superior national command. The follow-on British interagency plan for Helmand, called the Helmand Road Map, also came from the bottom up; it was produced in the field, by 52 Brigade working alongside stabilization advisers in Helmand in late 2007, and then "walked back" to Whitehall for approval.[51] Moreover, the brigades that followed 52 (16 and 3 brigades) promptly drifted off the Helmand Road Map (but crucially, still coordinated closely with the PRT through a new Joint Command Board).[52]

Likewise, the ISAF chain of command was weak. Formally, the British task force came under Regional Command-South (RC-S), which in turn came under ISAF Headquarters. In reality, ISAF lacked a detailed and coherent plan,[53] and RC-S was little more than a "coordinating authority" for the allocation of air and regional assets in support of task force initiatives.[54] The weakness of ISAF command and control was one of the key areas requiring urgent improvement identified by General Stanley McChrystal, when he was appointed Commander of ISAF (COMISAF) by President Barak Obama in the summer of 2009 in order to turn around a failing campaign. Under McChrystal, ISAF developed for the first time a proper campaign plan, and a new corps-level headquarters was created (ISAF Joint Command) to take charge of the running of the campaign and ensure that the plan was driven down through regional commands and implemented by task forces. One year on, the situation had considerably changed. ISAF had a detailed and coherent campaign plan with prioritized objectives for regional commands, and ISAF Joint Command was operating effectively to allocate available resources and ensure regional command compliance with the plan.[55] In addition, with UK 6 Division taking charge of RC-S in October 2009, the command had taken grip of the campaign in the south and imposed a regional scheme of maneuver on national task forces.[56]

At the heart of the McChrystal plan were two key priorities—a focus on securing the population over killing the enemy, and closer partnership with Afghan security forces. This was McChrystal's much lauded "population-centric" counterinsurgency.[57] However, as we have seen, under 52 Brigade, the British had already begun to adopt an approach to operations that was focused on protecting the population and working through Afghan partners—in other words, long before McChrystal appeared on the scene. The effect of McChrystal's COMISAF directives on restraint and on partnering with ANSF was to reinforce the British shift to a more population-focused approach, one that sought to clear the Taliban out of territory without unnecessary big battles and in a way that won over the local population. Hence there were only three civilians killed in Op PANCHAI PALANG, and none in Operation TOR SHPAH or in the British sector of Op MOSHTARAK.

Operational adaptation also extends to changes in how forces are prepared for operations. Here British military adaptation was more sluggish. It was not until the second year of the campaign that proper specialist training was provided for units preparing for deployment to Helmand. In large part this was because the British Army's Operational Training Advisory Group was focused on providing support to the campaign in Iraq. Thus, for the first year, it was "producing the Iraq packages with an Afghan flavour."[58] It took even longer for the British military to update its doctrine for operations in Afghanistan. The US Army and US Marine Corps produced a new COIN doctrine in 2006 that captured and institutionalized the lessons from operations in Iraq from 2004 to 2005.[59] The British Army's new field manual on COIN was published three years later.[60] As it happens, the British Army *did* have a COIN doctrine that was fit for the purpose in 2005, having updated it in 2001. However, because the army was then focusing on peace operations, the updated COIN doctrine was not disseminated or taught. By the time of the Iraq War in 2003 and the Helmand deployment, it was unknown within the army.[61] Thus, for example, 52 Brigade drew on the new US COIN doctrine in developing their CONOPS for Helmand.[62] The British Army decided to rewrite its COIN doctrine in 2005 to incorporate the lessons from Iraq, and had a draft ready in 2007. But publication of this new doctrine was vetoed by the UK military's joint doctrine agency, the Doctrine, Concepts and Development Centre (DCDC), which had in mind a more radical doctrinal rethink. "Intellectual, organisational, and personal dissonance between the joint and army [doctrine] camps prevented progress."[63] It took a further two years for DCDC to produce its new stabilization doctrine and that, in turn, delayed production of the new Army COIN doctrine, as the two had to be synchronized.[64]

THE POLITICS OF STRATEGIC ADAPTATION

The overall picture, therefore, is one of considerable operational adaptation by the British military in response to operational imperatives, especially in terms of the approach and conduct of operations in Helmand. When it comes to strategic adaptation, the record is more mixed. British strategy for Helmand has not adapted for the simple reason that Britain has not had one.[65] The House of Commons Foreign Affairs Committee criticized the government in 2009 on this score, and for allowing "mission creep" to take over.[66] In the absence of a clear national strategy, alliance politics have shaped the British commitment to Afghanistan. This, in turn, has restricted Britain's ability to adapt strategically. However, we do see adaptation by the British state in terms of force levels and resources committed to the campaign, in response to operational imperatives and opportunities afforded by new technologies.

Alliance Politics and British Commitment

Alliance politics appear to have shaped the British commitment to southern Afghanistan in at least three ways. First, NATO credibility was on the line. NATO had taken charge of ISAF in August 2003, in part "get the Alliance back on track" following the transatlantic bust-up over the Iraq War. NATO was to stabilize itself in the process of stabilizing Afghanistan.[67] By 2004, ISAF had expanded anticlockwise to the north and west of Afghanistan, but had yet to expand to the tougher areas in the south and east. The view among British military chiefs was that "[t]he NATO campaign looked completely moribund," with "no appetite" in NATO "for taking the campaign into the South." The wider view in Whitehall at the time was that "there was a clear and pressing need to revivify the NATO campaign in Afghanistan and to try and draw American eyes back to Afghanistan." Accordingly, Prime Minister Blair announced at the NATO Summit in Istanbul in June 2004 that the UK-led Allied Rapid Reaction Corps Headquarters would deploy to Kabul to take command of ISAF, and that UK forces would deploy to Helmand.[68]

Second, Afghanistan provided an opportunity for Britain to redeem its military reputation, especially with the Americans, following the fiasco in Iraq. The British were responsible for securing Basra, Iraq's second largest city. By 2007 it has become obvious to the American military that "the Brits have lost Basra." It was left to the Iraqi Army, with support from hurriedly redeployed US forces, to retake Basra from antigovernment militia. For British military, southern Afghanistan offered a chance to "have a better go."[69] This was not a consideration in the initial decision to deploy to southern Afghanistan. In early 2004, things were going fine for the British in Iraq. But it is an important shaping factor when the British campaign began to run into trouble in Helmand over 2006–7.

The situation was even worse in Basra, and the British military fully appreciated that they could not afford to lose two wars in quick succession.

Third, alliance politics shaped the opening mistake of the British campaign—namely, the "platoon house strategy." The commander of 16 Brigade, Brigadier Ed Butler, decided to deploy small packages of troops to garrison the northern towns of Helmand because he came under immense pressure to do so from Governor Daoud and from the Afghan President, Hamid Karzai. The Taliban were threatening to take control of Sangin, Now Zad, and Musa Qala. Daoud feared that such a move by the Taliban would completely undermine his governorship.[70] The British government anticipated that Daoud would make demands on British forces to do things that were unsustainable. Accordingly, prior to deployment, Butler was advised by the UK Permanent Joint Headquarters (PJHQ) that Daoud "ought to be discouraged from making gestures" that would overstretch the British task force.[71] However, as events unfolded, things began to look different. Butler came to appreciate that as a new provincial governor with no tribal base in Helmand and installed at Britain's insistence, Daoud was highly dependent on British military and financial aid to demonstrate his authority. This was also the view in Whitehall, as noted by Des Browne, who came in as the new Defence Secretary right in the middle of this crisis: "There clearly was an overwhelming view that Engineer Mohammed Daoud's survival as governor was key to the strategic success of what we were trying to do."[72] Hence, Butler agreed to do as requested, following consultation with the Deputy Chief of Joint Operations (who was visiting TFH at the time) and PJHQ back in Britain.[73]

Alliance politics shaped Britain's commitment to Helmand, from the original decision to send British forces to southern Afghanistan, to the decision shortly after arriving in Helmand to garrison the northern towns, to the determination to stay the course at all costs and avoid losing the campaign. This, in turn, reduced options for British strategic adaptation, especially any involving strategic disengagement, as campaign pressures grew.

Civil-Military Relations and Campaign Resources

In terms of resources, Britain did adapt gradually to the campaign with troop increases and new equipment. As noted earlier, 16 Brigade received 1,500 more troops (a 50 percent force uplift) to deal with greater than anticipated armed resistance. Still, this was nowhere near enough. The following four brigade tours saw troop numbers rise progressively from more than 5,000 to

TABLE 1. British Troop Numbers in Afghanistan[74]

Task force Brigade	Tour	Personnel
16 Air Assault Brigade	April–Oct. 2006	3,150 (4,500)
3 Commando Brigade	Oct. 2006–April 2007	5,200
12 Mechanized Brigade	April–Oct. 2007	6,500
52 Infantry Brigade	Oct. 2007–April 2008	7,750
16 Air Assault Brigade	April–Oct. 2008	8,530
3 Commando Brigade	Oct. 2008–April 2009	8,300
19 Light Brigade	April–Oct. 2009	8,300
11 Light Brigade	Oct. 2009–April 2010	9,000
4 Mechanized Brigade	April–Oct. 2010	9,500
16 Air Assault Brigade	Oct. 2010–April 2011	9,500

7,750 (see Table 1). From mid-2008 to late 2009, British numbers stabilized at around 8,300, before creeping up over the course of 2010 to 9,500. Not all of these troops were assigned to the task force in Helmand: a couple of thousand (including one battle-group) were assigned to RC-S and based in Kandahar. At the same time, the British task force was bolstered by an Estonian company and a 750-strong Danish battle-group (deployed in the center of Helmand); the UK task force was renamed Task Force Helmand (TFH) in 2008 to reflect the inclusion of international partners.

Somewhat ironically, civil-military relations were good in the early phase of the campaign, but as pressure grew so did civil-military tensions. The initial force deployment of 3,150 was agreed between the government and military chiefs. Brigadier Butler expressed concerns up the chain of command that he lacked sufficient troops, helicopters, and vehicles. This led the Defence Secretary to seek assurances from military chiefs, which he received, that the task force had sufficient resources for the job. Furthermore, military chiefs reassured ministers that the campaign in Helmand was not dependent on the anticipated military drawn-down from Iraq.[75] Then things got much worse in both theaters, and this simultaneously tied down British forces in Iraq and created demand for more troops in Helmand.[76]

The official record of high-level political and military decisions on Afghanistan is still closed.[77] Thus, any assessment of civil-military relations over this period involves a degree of speculation. It is probably the case that up to 2009, there was broad agreement between military chiefs and ministers on forces and resources for the campaign. The British Army was overstretched waging two military campaigns. Thus, the main constraint on pouring more resources into Helmand would have been military capacity, and not ministerial opposition.

Indeed, ministers appeared to have acted fairly promptly to do what they could to increase resources for the Helmand campaign. Any decision on raising force levels went right up to the Prime Minister. Thus, the decision to send reinforcements was taken by the Prime Minister on recommendation from the Ministry of Defence (MOD) within two months of the British task force deploying into the northern towns in Helmand.[78]

The situation regarding equipment also demonstrates civilian leadership responding to military concerns. As discussed below, ministers took prompt action to authorize the purchase of new armored vehicles and unmanned aerial vehicles (UAVs) for the campaign. Even in the one area that task force commanders most complained about—namely, helicopters—ministers appear to have done what they could.

The British initially deployed to Helmand with six Chinook heavy-lift transport helicopters, four Lynx light-transport helicopters, and eight Apache helicopter gunships. Even before they deployed, 16 Brigade were saying that this fleet had a 20 percent shortfall in required transport helicopter capacity.[79] This was against the original plan that saw British forces based in Lashkar Gah, Gereshk, and Camp Bastion. The situation became infinitely worse when the British deployed into Now Zad, Sangin, and Musa Qala. Not only did this hugely increase the strain on an already overstretched helicopter fleet but flying into these northern towns proved to be extremely hazardous for British helicopters. The MOD, with support from ministers, took fairly prompt action to address this issue. Browne negotiated with the Americans to ensure that they kept a sizable helicopter fleet at Kandahar that the British task force could draw upon. The government bought seven Merlin helicopters, second-hand from Denmark. The MOD had a fleet of eight Chinooks that were equipped for special operations but unused because the United States refused to release the specialist software necessary to operate them. The Defence Secretary promptly "made the decision to strip out all their sophistication and basically turn them into flying buses to get them deployable." But it took time for this new capability to arrive in theater. The Merlins were initially sent to Iraq, and not redeployed to Afghanistan until 2010. Likewise, even though Browne ordered the extra Chinooks to be stripped out and retrofitted for Afghanistan in 2006, that took four years, and so they too did not arrive until 2010.[80] In the meantime, Puma and Sea King helicopters were retrofitted and deployed to provide a medium-lift capacity. However, these were not designed for hot and dusty climates and had difficulty operating in Helmand during the summer months (when insurgent activity tended to be most intense).

Civil-military relations appear to have turned sour under Gordon Brown's tenure as Prime Minister from June 2007 to May 2010. As Chancellor of the Exchequer, Brown had acquired a reputation within the MOD for being indifferent toward the military. In particular, he was responsible for cutting the MOD's helicopter budget in 2004. Thus the stage was set for a showdown.

Between 2008 and 2009, British troop numbers fell in Iraq from just over 4,000 to around 400. This freed up capacity that the British Army wanted to commit to Afghanistan. However, Brown refused the army's request to send 2,000 more troops to support Op PANCHAI PALANG, the major offensive by 19 Brigade in central Helmand. Instead, the Prime Minister agreed to a temporary force uplift of 500 troops. This infuriated the Chief of the General Staff, Sir Richard Dannatt, who, one month before his retirement, went public with this concerns.[81] It was unprecedented for a serving military chief to openly criticize the government in this way. When Dannatt went on a final visit to British forces in southern Afghanistan in July 2009, he was flown around Helmand Province in American helicopters. This was misreported as indicating that the key issue was the shortage of available British helicopters.[82] Actually, Dannatt was more concerned about the "lack of energy and investment" in developing capabilities for countering IEDs (C-IED), which by then were the main killer of British troops.[83] Dannatt returned from Helmand with a "shopping list" for the government, at the top of which were "more boots on the ground" and more investment in C-IED.[84] (As discussed below, action was taken by the MOD on the latter issue a few months later.) Labor ministers reacted with fury, criticizing the army chief for "playing politics."[85] The row between the army and the government spilled over into Westminster, with the Defence Committee rushing out a highly critical report on the helicopter shortage.[86] Shortly thereafter, the chair of the Defence Committee grilled the Prime Minister when he appeared before the Commons Liaison Committee on July 16.[87] There was one benefit from this public spat. The additional 500 troops that had been deployed to support Op PANCHAI PALANG and were due to be withdrawn following the operation remained as a longer-term increase, raising British force levels to 9,500.

Thus, the story on civil-military relations is mixed. As Defence Secretary, Des Browne exercised civilian leadership in responding to military requests for additional resources. As noted earlier, this was crucial in giving 52 Brigade the confidence and capability to adapt a new way of COIN. However, later on, when Iraq freed up extra military resources, Prime Minister Gordon Brown exercised civilian leadership in placing a block on a "mini-surge" of British forces to Afghanistan. This, in turn, hindered further operational adaptation in Helmand.

New Technology and the Counter-IED Fight

New technology has been an important driver of military adaptation by the British. Essentially, the growing threat from IEDs led the British to hurriedly acquire new "protected mobility" and surveillance capabilities. Between 2006 and 2008, there was a fourfold increase in IED attacks in southern Afghanistan. This figure increased a further twofold between 2008 and 2010.[88] The acquisition of new technologies to counter this threat is an area in which Britain did strategically adapt, in terms of resourcing the campaign.

The British Army went into Helmand with lightly armored Snatch Land Rovers. By late 2006, a company of armored infantry were deployed, equipped with Warrior infantry fighting vehicles.[89] However, these heavy tracked vehicles, which to the untrained eye look like tanks, are not ideal for Helmand. They can cause considerable damage to routes, and the flat bottoms are vulnerable to mines. Around this time, the US military was acquiring new vehicles called MRAPs (Mine-Resistant Ambushed Protected), with V-shaped hulls and other features especially designed to combat the threat from mines and IEDs. Between 2006 and 2008, the Defence Secretary announced the acquisition of almost one thousand MRAPs in three packages. Called Mastiff and Ridgeback, these were British variants on the American Cougar MRAP.[90]

Also crucial to countering IEDs has been the enhancement of ISTAR (Intelligence, Surveillance, Target Acquisition and Reconnaissance) capability. ISTAR assets enable British forces to track and target those responsible for producing and planting IEDs. Britain acquired three types of unmanned aerial vehicle (UAV)—the Desert Hawk, Hermes 450, and Reaper—all as Urgent Operational Requirements (UORs), and rushed them into service in 2007.[91] As with MRAPs, these were also bought off the shelf (Desert Hawk and Reaper are US systems, and Hermes 450 is based on an Israeli UAV).[92] In addition to these aerial systems, British forces developed ground-based systems, called BASE ISTAR, for protection of fixed sites. Base ISTAR started off as a series of tracking and validating systems used at the main gates of Forward Operating Bases (FOBs) in Iraq and Afghanistan. This is a prime example of tactical adaptation in the field resulting in more significant capability development. Base ISTAR is now contained in the Cortex program, itself a collection of camera, radar, and ground sensor systems. Cortex has been deployed to some twenty British FOBs in Afghanistan.[93]

British ISTAR development occurs under the umbrella of a far larger Network Enabled Capability (NEC) program.[94] In a clear example of strategic adaptation, this entire program has been refocused on the counter-IED fight. In late 2009, the senior officer response for NEC (and Deputy Chief of the Defence Staff for Capability), Vice Air Marshall Carl Dixon, instigated a review of NEC support to operations in Afghanistan. This was in the context of growing ISAF casualties from IED attacks: more coalition troops were killed by IEDs in the first half of 2009 (sixty-five) than in all of 2008 (fifty-five).[95] As a consequence of the Dixon review, the management and main effort of the NEC program was changed to concentrate on supporting the counter-IED effort.[96]

Most of these new technologies for protected mobility and counter-IED were acquired under the Urgent Operational Requirement (UOR) scheme. UORs are designed to rapidly fill urgent equipment shortfalls in operations, and are paid for out of a special Treasury fund and not the regular MOD budget. The British soldier and marine also received around twenty enhancements (rifle upgrade, new body armor, night-goggles, and so forth) under the UOR scheme, and existing platforms (such as Warrior) were upgraded. Much of this new equipment and many of the enhancements were acquired and fielded within a year. Mastiff acquisition was especially quick, taking just five months, because the US military agreed to interrupt their own Cougar production schedule to divert MRAPs to meet Britain's UOR. By 2009, the MOD had spent £4.2 billion on UORs, with almost 40 percent devoted to protected mobility.[97] The scale of the UOR scheme not only demonstrates strategic adaptability; it also in turn enabled operational adaptability. Improved protected mobility has given the task force greater ability to establish presence. Even more significant have been improvements in ISTAR, which is seen as a "key-enabler" of operations. 3 Brigade relied on ISTAR to find the insurgents attacking Lashkar Gah, and then to follow the survivors back to Nad-e-Ali District and pinpoint where they had come from. In the final clear of Nad-e-Ali a year later, 11 Brigade relied heavily on ISTAR to identify and take out the local Taliban leaders in advance of the air assault, and thereby successfully removed the risk of a big battle (as the Taliban defense collapsed).

CONCLUSION

Britain's military campaign in Helmand started off badly. The British military failed to anticipate that their deployment in Helmand would "stir up a hornets' nest," and hence went in unprepared.[98] For the first two years security got worse, for locals and internationals, in Helmand. The British military had two "adaptive moments" that turned the campaign around. The first happened in late 2007, when 52 Brigade developed and resourced a population-centric approach to operations. The second occurred a year later, when 3 Brigade focused the British campaign back on central Helmand.

Both adaptive moments were triggered by the prospect of campaign failure. With extra resources and arriving at the end of fighting session, 52 Brigade had a good opportunity to experiment with a new approach to operations. The Taliban assault on Lashkar Gah in late 2008 revealed just how perilous remained the British hold on the province. This was to be a knockout blow against the British and Afghan government. Thereafter, by degrees, the British refocused on central Helmand and pushed the Taliban out. The massive influx of US Marines and growing ANA capabilities were crucial to turning things around in Helmand. But the British military were also able to come back from the brink of defeat because they adapted to the demands of this most challenging campaign.

Some critics argue that the British military were too slow in adapting to operational imperatives in Helmand, and that even after 2007 the British were inclined to spread themselves too thinly, undertake big clearance operations, and rely too much on firepower (artillery and airstrikes) when they got into trouble.[99] It took the British two years to adapt a new approach to operations. There really is no benchmark to say whether this was too long; it is entirely a matter of judgment. On balance, it does not seem excessively long, especially given the immense pressure on commanders and the paucity of resources. Moreover, it might reasonably be argued that the first eighteen months of fighting were unavoidable in that it was necessary to fight the Taliban to a standstill, and forced them to switch tactics (to the use of IEDs) that while still deadly were less destabilizing.

In any case, from late 2007 on, the British military did begin to adapt, by developing new capabilities for influence operations, improving partnership with the PRT, and refocusing the campaign on securing the population and working with Afghans. Training and doctrine lagged behind adaptation in tactics and approach but have since caught up. It is true that in 2007–8 the British forces undertook a number of major operations that drew them out of central Helmand. 52 Brigade went north to retake Musa Qaleh, and 16 Brigade moved large turbines up to Kajaki Dam, traveling in force up through neighboring Kandahar Province. But from late 2008, the main British effort returned to central Helmand with major operations by 3, 19, and 11 brigades to remove the Taliban threat to Lashkar Gah. Furthermore, all these operations were undertaken successfully in a way so as to minimize the risk to civilians.

Some critics might also argue that the refocusing of British effort in central Helmand was simply a form of tactical retreat, with the over-stretched British handing over the south and north of the province to more powerful US Marine forces. Certainly, the US Marines used more firepower and more aggressively pushed the Taliban lines back away from district centers in Garmsir and Nawa in the south in late 2009, and Musa Qala and Sangin in the north in 2010–11. There does appear to have been an Anglo-American difference, if not disagreement, in COIN tactics, with the British preferring to focus on securing district centers and using development assistance to project influence into contested territory, and the Americans preferring to seize contested territory by force.[100] Nonetheless, the mass arrival of US Marine forces did enable the second adaptive moment in the British campaign. Free to concentrate on central Helmand, the British task force finally achieved significant progress on the ground.

Britain's record on strategic adaptation is more mixed: very poor on strategy, a bit better on force levels, and pretty good on equipment. Alliance politics—the need to bolster NATO unity and purpose, recover Britain's military reputation with America, and support the Afghan government in Helmand—have been the major strategic drivers for Britain's war effort. Britain has taken advantage of new technologies to improve protected mobility and ISTAR for Afghanistan. Sorting out the shortage of helicopters has taken longer, in spite of prompt ministerial action. Overall, in terms of resources, civil-military harmony gave way to increasing tension as operational pressures mounted. As noted, the massive influx of US military resources into Helmand, especially in 2010, took the pressure off this issue, combined with the arrival of additional British helicopters and enhanced C-IED capabilities.

Ultimately, it remains to be seen if Britain has adapted enough. Improved approach to operations, a decent force size, and better equipment may not be sufficient, in the absence of a strategy for success.

NOTES

1. In modern Afghanistan, Maiwand District is in neighboring Kandahar Province. The Afghan army that defeated the British was composed mostly of Alizai tribesmen from Helmand Province. Rob Johnson, *The Afghan Way of War* (London: Hurst, 2011), 129–38; Mike Martin, *A Brief History of Helmand* (Warminster: British Army Afghan COIN Centre, 2001), 18-19.

2. This chapter cites Task Force Helmand post operation reports that are held by the British Army Land Warfare Centre. The reports are identified by the number of the HERRICK tour; hence POR H7 = post operational report for HERRICK 7.

3. Theo Farrell and Stuart Gordon, "COIN Machine: The British Military in Afghanistan," *Orbis* 53, no. 4 (2009): 667–69.

4. This section draws on material published in Theo Farrell, "Improving in War: Military Adaptation and the British in Helmand Province, Afghanistan, 2006–2009," *Journal of Strategic Studies* 33, no. 4 (2010): 567–94.

5. Patrick Bishop, *3 Para: Afghanistan, Summer 2006* (London: HarperPress, 2007), 34–36; Sean Rayment, *Into the Killing Zone* (London: Constable, 2008), 37–40; Col. Stuart Tootal, *Danger Close: Commanding 3 PARA in Afghanistan* (London: John Murray, 2009), 23–24, 47.

6. Michael Clarke and Valentina Soria, "Charging Up the Valley: British Decisions in Afghanistan," *RUSI Journal* 156, no. 4 (2011): 80–89.

7. POR H4, October 19, 2006.

8. POR H5, April 11, 2007, 2-0-4, para. 2 and 3.

9. Ibid., 2-0-4, para. 14.

10. Ewen Southby-Tailyour, *Helmand, Afghanistan* (St. Ives: Ebury Press, 2008), 77–78.

11. POR H5, 2-0-3.

12. Ibid., 2-0-3, 2-4-1, 2-4-2.

13. The Green Zone is the belt of fertile ground, between one and five miles wide, that runs the length of the Helmand River all the way from Kajaki in the North to Khanishin in the South. The population in Helmand is concentrated in this area.

14. POR H6, October 19, 2007, 5.

15. Ibid., 6, 2.

16. Author interview with Brig. John Lorimer, Permanent Joint Headquarters, Northwood, London, 29 June 2010. Lorimer was CO of 12 Brigade on HERRICK 6.

17. POR H5, 3.

18. Author interview with Brig. Lorimer.

19. Ibid.

20. Stephen Grey, *Operation Snakebite* (London: Viking, 2009), 61–65.

21. Commander British Forces, Op HERRICK 7, "Counterinsurgency in Helmand, Task Force Operational Design," TFH/COMD/DO7, January 1, 2008, 2.

22. Annex A to COMD/DO 7, DTD, January 1, 2008, para. 8.

23. Author interview with Maj. Geoff Minton, 52 Brigade HQ, Redford Cavalry Barracks, Edinburgh, June 29, 2009.

24. For discussion on NKET and TCAF, see Farrell, "Improving in War," 579–80.

25. Author interview with staff planning officer, 52 Bde HQ, Redford Cavalry Barracks, Edinburgh, June 29, 2009. Toward the end of their tour, 12 Brigade were also warning locals of impending operations to give them an opportunity to flee before battle. Discussion with staff officer, 52 Brigade, Joint Services Command and Staff College, Shrivenham, March 9, 2010.

26. Tom Coghlan, "Taliban Flee as Troops Retake Musa Qala," *Telegraph*, December 10, 2007; Jason Burke and Richard Norton-Taylor, "Allies Move into Town Held by Taliban," *Guardian*, December 10, 2007.

27. Grey, *Operation Snakebite*, 280.

28. Post Operation Interview (POI) with Brig. M. A. P. Carleton-Smith, Comd 16 AA Bde, Op. HERRICK 8 (April–October 2008), November 24, 2008, Colchester, 2.

29. POI, Carleton-Smith, 6.

30. Author interview with staff officer, 16 Brigade HQ, Merville Barracks, Colchester, June 16, 2009.

31. Civil-military integration in theater was supposed to be mirrored and supported by a more integrated civil-military approach back in the United Kingdom. However, cultural and bureaucratic differences have prevented the development of this so-called Comprehensive Approach in British government to overseas operations. See Farrell and Gordon, "COIN Machine," 23–24.

32. Author interview with staff officer, Defence Intelligence, Ministry of Defence (MOD), London, November 2008.

33. Sam Kiley, *Desperate Glory: At War in Helmand with Britain's 16 Air Assault Brigade* (London: Bloomsbury, 2009), 33–34, 137.

34. Tom Coghlan, "The Taliban in Helmand: An Oral History," in Antonio Giustozzi, ed., *Decoding the New Taliban* (London: Hurst, 2009), 145.

35. POR, H9, 3.

36. Author notes from Collective Debrief of 3 Commando Brigade, at RM Barracks, Stonehouse, Plymouth, July 7, 2009.

37. Author interview with senior staff officer, 3 Commando Brigade, MOD, London, July 1, 2010.

38. POR, H9, 18.

39. For a dramatic account of the Taliban attack, see Ewen Southby-Tailyour, *3 Commando Brigade: Helmand Assault* (London: Ebury Press, 2010), 55–66.

40. Ibid., 65.

41. Author interview with Maj. Gen. Gordon Messenger, MOD, London, May 11, 2010; author telephone interview with Hugh Powell, London, April 15, 2010. Messenger was commander (Comd) 3 Cdo Bde on HERRICK 9, and Powell was head of PRT.

42. Author interview with Jon Moss, thematic head subnational government, Helmand PRT, Lashkar Gah, Helmand, May 29, 2010.

43. Author interview with Brig. Tim Radford, MOD, London, March 10, 2011. Radford was Comd 19 Bde on HERRICK 10.

44. For assessment of these operations based on extensive field research, see Theo Farrell, *Appraising Moshtarak: The Campaign in Nad-e-Ali District, Helmand* (London: RUSI, June 2010): available at http://www.rusi.org/downloads/assets/Appraising_Moshtarak.pdf, accessed July 6, 2011.

45. Author interview with Brig. Richard Felton, TFH HQ, Lashkar Gah, Helmand, May 27, 2010. Felton was Comd 4 Bde on HERRICK 12.

46. All quotations from testimony by Air Chief Marshall (Rtd) Lord Stirrup, former Chief of the Defence Staff, to the House of Commons Defence Committee, May 4, 2011. House of Commons Defence Committee, *Operations in Afghanistan*, Fourth Report of Session, 2010–12, HC 554 (London: TSO, 2011), Ev 140 at Q611, Q630, and Q655.

47. Author interview with Maj. Gen. Messenger. Messenger led the military team of planners.

48. Author interview with Minna Jarvenpaa, London, February 16, 2012. Jarvenpaa was a lead civilian planner for the JHP.

49. Jack Fairweather, *A War of Choice: The British in Iraq, 2003–9* (London: Jonathan Cape, 2011), 232–34.

50. Author interview with military planner, 3 Commando Brigade, MOD, London, July 1, 2010.

51. Farrell and Gordon, "COIN Machine," 672.

52. Author notes from collective debrief of 3 Commando Brigade at RM Barracks, Stonehouse, Plymouth, July 7, 2009.

53. Author interview with very senior former ISAF officer, 2007–9, London, April 27, 2011.

54. Author interview with Brig. Mark Carleton-Smith, MOD, London, May 20, 2010. Carleton-Smith was Comd 16 Bde on HERRICK 8.

55. ISAF Joint Command study (classified), October 2010.

56. Farrell, *Appraising Moshtarak*.

57. General Stanley McChrystal, COMISAF/CDR USFOR-A, "Commander's Initial Assessment," August 30, 2009 (unclassified).

58. POI, Mackay, 6.

59. US Army/US Marine Corps, *Counterinsurgency Field Manual* (Chicago: Chicago University Press, 2007).

60. Army Field Manual *Countering Insurgency*, Vol. 1—Part 10: January 2010, AC 71876.

61. Colonel Alex Alderson, "The Validity of British Army Counterinsurgency Doctrine after the War in Iraq, 2003–2009," Ph.D. dissertation, Cranfield University, 2009.

62. Interview with Maj. Gen. Andrew Mackay, former Comd 52 Brigade, on "Afghanistan: The Battle for Helmand," broadcast on BBC2, July 3, 2011.

63. Email correspondence with senior army officer, Land Warfare Centre, October 25, 2011.

64. Development, Concepts and Doctrine Centre, *Security and Stabilisation: The Military Contribution*, Joint Doctrine Publication 3–40, November 2009. Based on numerous discussions with officers at the Land Warfare Centre and the Development, Concepts and Doctrine Centre in 2009. Farrell was an adviser to the production of both AFM *Countering Insurgency* and JDP 3–40.

65. David Betz and Anthony Cormack, "Wars amongst the People: Iraq, Afghanistan and British Strategy," *Orbis* (Spring 2009); Hew Strachan, "The Strategic Gap in British Defence Policy," *Survival* 51, no. 4 (2009): 49–70; Paul Newton, Paul Colley, and Andrew Sharpe, "Reclaiming the Art of British Strategic Thinking," *RUSI Journal* 155, no. 1 (2010).

66. House of Commons, Foreign Affairs Committee, *Global Security: Pakistan and Afghanistan*, Eighth Report of Session, 2008–9, HC 302 (London: TSO, August 2, 2009), para. 225.

67. Tim Bird and Alex Marshall, *Afghanistan: How the West Lost Its Way* (New Haven: Yale University Press, 2011), 116–17.

68. Testimony of Gen. (Rtd) Sir Robert Fry, Deputy Chief of the Defence Staff (Commitments) to the House of Commons Defence Committee, February 8, 2011. House of Commons, *Operations in Afghanistan*, Ev 86, at Q397 and Q398.

69. Frank Ledwidge, *Losing Small Wars: British Military Failure in Iraq and Afghanistan* (New Haven: Yale University Press, 2011), 46–59.

70. Interview with Mohammad Daoud, former provincial governor of Helmand, on "Afghanistan: The Battle for Helmand," broadcast on BBC2, July 3, 2011.

71. Testimony by Lord Reid (former Secretary of State for Defence) before the House of Commons Defence Committee. House of Commons, *Operations in Afghanistan*, Ev 90, at Q415 (see also Q420).

72. Testimony by Lord Browne (former Secretary of State for Defence) before the House of Commons Defence Committee. House of Commons, *Operations in Afghanistan*, Ev 122 at Q574.

73. Interview with Brig. (Rtd) Ed Butler, former Comd 16 Air Assault Brigade, on "Afghanistan: The Battle for Helmand," broadcast on BBC2, July 3, 2011.

74. Data from ISAF Troops Placemat archive: available at http://www.nato.int/isaf/docu/epub/pdf/placemat.html.

75. Testimony by Lord Reid, Ev 98 at Q455 and Q456.

76. Testimony by Gen. Fry, Ev 96 at Q446. Britain had 5,500 troops in Iraq in 2007, and more than 4,000 troops throughout in 2008–9.

77. Even the House of Commons Defence Committee were not given access to the

minutes of the Chiefs of Staff committee minutes for their inquiry into operations in Afghanistan. House of Commons, *Operations in Afghanistan*, para. 15.

78. Testimony by Lord Browne, Ev 122, at Q577.

79. Testimony by Brig. (Rtd) Ed Butler (former Commander 16 Brigade) before the House of Commons Defence Committee. House of Commons, *Operations in Afghanistan*, Ev 108 at Q493.

80. Testimony by Lord Browne, Ev 124 at Q587.

81. Jonathan Oliver and Michael Smith, "Labour Clashes with Army as Afghan Death Toll Mounts," *Times*, July 12, 2009.

82. Michael Evans and Philip Webster, "British Army Chief Forced to Use U.S. Helicopter in Afghanistan," *Times*, July 15, 2009.

83. Author interview with Lord Dannatt, London, February 16, 2012. General Richard Dannatt was Chief of the General Staff, 2006–9.

84. Michael Evans and Nico Hines, "General Dannatt Forces Afghanistan Shopping List on Gordon Brown," *Times*, July 17, 2009.

85. Francis Elliot and Michael Evans, "General Sir Richard Dannatt Told to Keep Out of Helicopter Politics," *Times*, July 16, 2009.

86. House of Commons Defence Committee, *Helicopter Capability*, Eleventh Report of Session, 2008–9, HC 434 (London: TSO, July 16, 2009).

87. Nick Allen and Rosa Prince, "Gordon Brown Faces Grilling over Afghanistan Helicopters," *Telegraph*, July 16, 2009.

88. Clarke and Soria, "Charging Up the Valley," 81.

89. Author interview with Brig. Lorimer.

90. Hansard, July 24, 2006, Column 758W, December 12, 2007, Column 332, October 29, 2008, and Columns 28WS-30WS.

91. House of Commons Defence Committee, *The Contribution of Unmanned Aerial Vehicles to ISTAR Capability*, Thirteenth Report of Session, 2007–8, HC 535 (London: TSO, August 5, 2008).

92. House of Commons Defence Committee, *The Contribution of ISTAR to Operations*, Eighth Report of Session, 2009–10, HC 225 (London: TSO, March 25, 2010), paras. 15–17.

93. Author interview with senior officer, DEC, MOD, London, July 1, 2010.

94. Britain's NEC program is inspired by US military transformation and, in particular, the US concepts and capabilities for Network-Centric Warfare. See Theo Farrell, Sten Rynning, and Terry Terriff, *Transforming Military Power since the Cold War: Britain, France and the United States, 1991–2012* (Cambridge: Cambridge University Press, 2012).

95. Sheila Bird and Clive Fairweather, "IEDs and Military Fatalities in Iraq and Afghanistan," *RUSI Journal* 154, no. 4 (2009): 33.

96. Author interview with very senior officer, MOD, London, July 27, 2010.

97. National Audit Office, *Support to High Intensity Operations*, Report by the Comp-

troller and Auditor General, HC 508 Session 2008–9 (London: TSO, May 14, 2009), para. 5; author interview with military officer, Directive of Joint Capability (DEC), London, MOD, February 6, 2009.

98. House of Commons, *Operations in Afghanistan*, para. 36.

99. Robert Egnell, "Lessons from Helmand, Afghanistan: What Now for British Counterinsurgency?" *International Affairs* 87, no. 2 (2011): 271–97; Ledwidge, *Losing Small Wars*; Anthony King, "Understanding the Helmand Campaign: British Military Operations in Afghanistan," *International Affairs* 86, no. 2 (2010): 311–22; Warren Chin, "Colonial Warfare in a Post-Conflict State: British Operations in Helmand Province, Afghanistan," *Defence Studies* 10 (2010); Stephen Grey, "Cracking On in Helmand," *Prospect* (September 2009): 47–51.

100. Rajiv Chandrasekaran, *Little America: The War within the War for Afghanistan* (London: Bloomsbury, 2012); Rajiv Chandrasekaran, "U.S. Marines, British Advisors at Odds," *Washington Post*, September 4, 2010; Thomas Harding and Ben Farmer, "US Forces 'Ignore British Advice' in Sangin Handover," *Telegraph*, September 20, 2010.

6 The Military Metier: Second Order Adaptation and the Danish Experience in Task Force Helmand

Mikkel Vedby Rasmussen

The expeditionary forces of a small country fight the wars its more powerful allies want to fight and fight them in ways defined by those larger allies. This salient fact for a small country is demonstrated by the Danish experience in Afghanistan, an experience that has been defined by the Danish Armed Forces' wish to realize the army's transformation in the Helmand theater of operations. Thus striving to be part of the North Atlantic Treaty Organization's (NATO's) innovation agenda, the Danes adapted to the organizational environment created by the British in Task Force Helmand as well as the lessons learned in military operations. The military operations conducted by the Danes were defined by the operational framework set by the British and American forces, who were big enough to adapt directly to the operational environment defined by the struggle with the Taliban for control of Afghan society. The lessons on how to improve military performance and/or strategic effect derived from operational and/or strategic challenges are described as adaptation in this volume. However, the Danes were learning another set of lessons—and in fact it was these lessons learned from their Great Power allies the Danish soldiers wanted to learn. Thus this chapter adds another "layer of lessons." The lessons derived from the organizational setup created by larger allies can, following Niklas Luhmann, be defined as "second order adaptation."

This chapter focuses on Danish second-order adaptation analyzing how the Danish Army sought to use the Afghanistan operation as a way to further the transformation of the army from a force dedicated to national defense to an expeditionary force. To the Danes, in other words, adaptation was a point in and of itself. While the Danish military was quite open about its ambition to use the Afghanistan deployment to prove itself in battle, the Ministry of Defence and the Ministry of Foreign affairs promoted a more comprehensive approach to military force, defining the fighting in Helmand as part of a civilian effort to rebuild and develop Afghan capacities. The civil servants and the diplomats did their best to make this "comprehensive approach" a Danish selling point in NATO. While this had considerable traction in Brussels, it had little relevance for the way Danish operations were actually conducted in Helmand. The result was an ever-increasing distance between rhetoric and reality that hampered the effectiveness of the Danish contribution (civilian as well as military) in Helmand.

Danish forces operated in Helmand from 2006 as part of the British Task Force Helmand. Initially, Denmark deployed a task force consisting of one Husar company with supporting units. In 2007, the withdrawal from Iraq enabled the Danish Armed Forces to deploy a battle group with one mechanized infantry battalion at its core supported by tanks and various other support troops. Totaling 650 troops, the Danish force was deployed along the Helmand River north of Gereshk. From 2007 to 2011, the mission of the Danish battle group was to prevent Taliban infiltration of the Upper Gereshk Valley in order to secure the entrenchment of the central government's power in the Gereshk area itself. In cooperation with British troops, the Danish troops have done so from bases along the river. In 2011, the Danish troops were withdrawn to a Patrol Baseline on the outskirts of Gereshk itself, anticipating the gradual withdrawal of Danish troops, which are expected to leave Afghan units in control of the river valley. The year 2012 will be the last with a Danish Army Battle Group in Helmand, and the remaining Danish troops are to be committed to training and mentoring activities until 2015. After this, all Danish troops are expected to be withdrawn with the rest of the British Task Force.

SMALL STATE ADAPTATION

This chapter describes Danish military adaptation in Helmand from 2006 to 2011. This volume defines military adaptation as a "change to strategy, force generation, and/or military plans and operations, which is undertaken in response to operational challenges and campaign pressures."[1] In other words, Farrell, Osinga, and Russell define adaptation as a military force's response to its environment and how it learns to improve these responses when faced with their effects. A military force operates in a given theater of operations, which is defined by, at least, two environments: a political environment, which influences the military force in terms of international or domestic politics and strategic imperatives, and an operational environment, which provides the military actor with certain experiences in the field that, at best, prompt changes in tactics, techniques, or technologies. The military organization is to adapt its performance to these environments in order to achieve the desired effect. This is the basic adaptive framework used in this volume and that describes adaptive pressures, adaptive events, and organizational responses in the Afghanistan campaign. When used on a small state like Denmark, which deployed only about 650 soldiers in the Helmand theater of operations, one has to unpack the notion of the military actor and the environment to which it adapts. As I will demonstrate in this chapter, the adaptation perspective allows one to analyze the armed forces as organizations as well as fighting entities.

In the case of small military actors the distinction between the organizational and operational levels makes a significant difference. In operational terms, first one and then two Danish Husar companies were stationed in the northeastern part of Helmand province in the first year of operations. From 2007 a battalion was deployed in central Helmand, primarily in the Gereshk area. In the summer of 2006 and the fall of 2007, particularly the Husars and the Royal Life Guard engaged in heavy fighting comparable to what British and American units have experienced in Helmand. During these operations the Danes' environment was defined by the engagement with the Taliban, and organizational learning was shaped by these experiences. When these experiences are to be adapted by the Danish Army, however, it becomes clear that the Danish forces exist in an environment defined by the British Task Force Helmand—in which the Danish troop contribution (only about 5 percent of the task force's total strength) is very much the junior partner. Another layer is added to that environment when one takes into consideration how the British task force itself became the junior partner to the Americans when the US Marines arrived in force in Helmand in 2009. At least until 2009, the British were able to define Helmand as their theater of operations and to a considerable extent define strategy on their own terms because of the weak International Security Assistance Force (ISAF) chain of command until General McChrystal established a proper campaign plan.[2] Hence to the British the adaptive environment was constituted by the Taliban and British commanders adapted to the dialectic of war described by Clausewitz. The Danes adapted to an environment defined by its larger coalition partner because, being the junior partner, the Danish commanders were not able to set strategy and shape operations independently of their British superiors.

I will term this "second order adaptation" following Niklas Luhmann's notion of second-order observations, which describes the way in which social agents relate not only to other social agents in an environment but also to the environment itself.³ In other words, whereas British troops adapted to an environment defined by the confrontation between themselves and the Taliban, the Danes adapted to the Taliban as well as to the way the British adapted to the Taliban. Second-order adaptation is thus the adaptation of adaptation.

Studying adaptation from a first- as well as a second-order perspective means that one has to be very precise about what adaptation actually entails. Jean Piaget defined adaptation in terms of "assimilation" and "accommodation."⁴ Piaget developed these terms to study the cognitive evolution of the child, and is useful as a tool to describe how responses from the environment make an actor adapt. As such, Piaget's notion of adaptation can be used in Camp Bastion as well as in a kindergarten. The reason for this is that Piaget bases his notion of adaptation on the concept of a "schema": a concept that refers to a general cognitive process rather than a specific childhood experience. A schema is a routine for successfully manipulating one's environment. A schema is based on experience—a child learns to pat the family dog and is rewarded by emotional experience and parental recognition. Military operations are similarly based on experience—the way to drive in a convoy with exactly the right distance between the cars, for instance. This captures Farrell's observation that "military organisations are both rational and routine-bound." Through trial and error, military organizations have learned how to do things and have established schemas learned at Staff College's and boot camps.⁵ Actors use schemas to adapt to their environment. If the child has established the schema of patting the family dog and then uses this schema to pat a strange dog in the street, the child may experience that the dog in the street barks back at the child. Frightened by this response, the child *assimilates* the new experience with dogs into the schema—"patting the family dog is ok, patting strange dogs should be avoided." In the same manner, Western forces in Afghanistan learned to drive differently on and off the roads in order to avoid Improvised Explosive Devices (IED)—that is, the driving schema was changed.

Not all new experiences fit existing schemas, however. Piaget points to a second type of adaptation—*accommodation*—that occurs when experiences cannot be fitted into an existing schema but constitute something new for which a new schema has to be made. Teaching is often about provoking a "perturbation" in the student in order for the student to accommodate to a new schema.⁶

Many university seminars are conducted on this model of learning. The same type of learning occurs on the battlefield, however, when the enemy realizes the schema you are operating by and uses that knowledge against you. Hence, having established a new schema for driving in order to avoid IEDs, the Taliban will use IEDs differently, perhaps supplemented by suicide bombers, in order to hit the Western forces. In turn, the Western forces have to learn to do things a new way or conduct another type of operation in order to achieve the same goal. John Boyd's OODA-loop describes this type of accommodation.[7]

By focusing on second- as well as first-order adaptation and by operationalizing the concept of adaptation in terms of Piaget's distinction between assimilation and accommodation, it is possible to avoid the pitfalls many analyses of small states' contribution to military coalitions fall into. That is either to overemphasize the scope of choice of a small state with a small contingent in a military collation, or to reduce strategy, operations, and tactics to political or perhaps even symbolic actions. On the first count a number of Danish authors have emphasized how the end of the Cold War removed the constrains of an outside threat (that is, the Soviet Union), at the same time as international institutions like the European Union (EU) and NATO created a framework of action that has given small European countries like Denmark a greater scope to be a "strategic actor."[8] Such geopolitical explanations highlight important elements of the new willingness of small European countries, like Denmark, to deploy military forces. This level of analysis attempts to affirm the feasibility of the studies—but they neither explain why Denmark embarked on a much more "militarist" foreign policy than, for example, Norway, nor do they investigate the role a small country and its armed forces play once committed to a given operation. The geopolitical focus makes the actual operations an aside. Oddly, this is also the case in some studies focusing on the making of small state doctrine. Kjell Inge Bjerga and Torunn Laugen Haaland thus argue that small countries' "need to adjust and learn from operational experiences is therefore less pressing than that of a great power, and their military doctrines can develop fairly independently of operational experiences."[9] While this statement captures the fact that small states are dependent on first-order adaptation from larger countries, it disregards second-order adaptation. Small countries need to learn and adapt just as much as big countries—they just do it differently, learning a different type of lesson. This chapter tells one such story of small state adaptation.

First, this chapter describes how the Danish troops arrived in Helmand with the ambition to translate military innovation into a concrete deployment of a battalion-size battle group. This demonstrates a disconnect between the rhetoric on comprehensive planning from the Ministries of Defence and Foreign Affairs, and a practice in which little attempt was made to conduct full-scale counterinsurgency (COIN) operations by integrating civilian and military efforts. The first phase of the Danish engagement in Helmand was defined by assimilation to the British Task Force Helmand. Second, the chapter deals with how the Danish military sought to adapt an existing schema of mechanized warfare to the battles in the Green Zone. This proved possible by second-order adaptation to British mechanized units and because the military and civilian engagement in Helmand was regarded as two distinct efforts. Third, the end game in the Helmand campaign—in which Britain and the US have ambitions of drawing down the commitment by 2015—demonstrates how Denmark adapted to the strategies set by the Great Powers, a clear case of second-order adaptation.

ARRIVING IN HELMAND

In June 2006 the first contingent of Danish soldiers arrived in Afghanistan as part of the British-led Task Force Helmand. The DANCON (Danish Contingent) was constituted of one company of reconnaissance troops with support elements stationed at the British main base at Camp Bastion in central Helmand.[10] Equipped with Eagle patrol vehicles with 12.7 machine guns, the unit was one of two reconnaissance companies of the Husars Regiment that were held in a higher state of readiness and had thus served as the "first-in" units for Danish Army deployments in the Balkans and in Iraq. The unit was based on the island of Bornholm in the Baltic, and the soldiers' feeling of regional identity was so strong that they flew the island's semiofficial flag (a red flag with a green cross, instead of the white cross of the Danish national flag) from their vehicles. As that flag unfolded in the desert wind, it is worth considering how the military leadership in Copenhagen considered this first company the first unit of a much larger Danish commitment to the operations in Helmand.

When it became clear that NATO was going to take over the mission in Afghanistan and as the Danish government committed itself to take part, the then Chief of Defence, General Jesper Helsø, was determined to use the opportunity to deploy a battalion-size battle group. The general's ambition was a case of accommodation. The 2004–9 defense budget had the stated purpose of producing transformation in the way the armed forces were organized and in the roles and operations conducted by the armed forces.[11] In spite of Danish involvement in the Balkan wars of the 1990s, as peace-keepers in Croatia as well as peace-enforcers in Kosovo, the armed forces were basically an organization designed to produce a conscript force to defend the national territory in case of invasion. The 2004–9 budget was to change that. Inspired by the agenda of reform set by the NATO Response Force,[12] the armed forces were to reconfigure its professional ethos from a national defense force to an intervention force. In Danish military parlance the forces had to be *helstøbte*,[13] which was a translation of the "essential operational capabilities" defined by NATO's MC317/1 (Military Committee Document 317/1).[14] The armed forces thus embraced NATO's transformation agenda in order to use this agenda to redefine its core mission.

In Piaget's terms, the 2004–9 budget had defined a new schema for the Danish Armed Forces. This schema was established by adaptation to the new NATO agenda on military transformation that came into being when NATO embraced the American Revolution in Military Affairs (RMA) agenda at the 2002 Prague Summit.[15] Defining the Danish Armed Forces in a NATO context, the Danish military came to think of their profession not in terms of the old schema of national defense but in terms of a new schema that had a capability-based approach to armed force and put a premium on using those capabilities in international operations. The schema thus redefined the organization as well as the professional ethos of the Danish Armed Forces. By 2006 this process was so far under way that the Chief of Defence wanted not only to demonstrate that his forces could actually produce the number of troops needed for a battalion scale deployment but, perhaps more important, also to prove that the new organizational setup could actually deliver essential operational capabilities and stomach the realities of war. General Helsø wanted a mission that "belonged to the military métier."[16]

The capabilities schema, which the armed forces embraced, defined the operations in Helmand in military terms. The armed forces had focused on producing essential operational capabilities leading them to focus on what could be generated by the military's own resources rather than on a whole-of-government approach to security issues. General Helsø was well aware that in Helmand the civilian element was going to be a decisive factor, but true to the schema he was very clear that nation-building was not part of the army's "métier." Actually, the general did not believe that the Danish government had the capability to take over a Provincial Reconstruction Team (PRT) when it deployed to Helmand.[17] Believing that the civil-military cooperation (CIMIC) inherent in the

PRT concept was beyond the resources of both the Danish Armed Forces and the Development Agency in the Ministry of Foreign Affairs, the armed forces simply chose to ignore the civilian element of the operation.

This approach would probably have surprised diplomats in NATO and elsewhere that had heard officials from the Danish Ministry of Defence and Ministry of Foreign affairs promote a "comprehensive approach" since 2005.[18] That year Denmark attempted to put the comprehensive planning of military and civilian operational capabilities on NATO's agenda with a nonpaper circulated among the NATO-member states followed up by a seminar in Copenhagen. Leading officials from the ministries of defense and foreign affairs wrote about "Improving civil-military co-operation the Danish way" in the *NATO Review* in the summer of 2005.[19] In fact, there seem to have been two Danish ways. One schema for operations was promoted by the civilian part of the Ministry of Defence in collusion with the Ministry of Foreign Affairs—this schema was about the comprehensive planning of civil-military cooperation. Another schema was, as mentioned, adopted by the armed forces embracing the "military métier" and explicitly rejecting an integration of civilian and military means to further operational ends. The first schema was focused on political ends and the second schema focused on military means, without proponents of either making a serious effort to integrate the two. The civilian officials thus promoted an agenda that gave some diplomatic clout in NATO but that had little connection to the way in which Denmark was actually planning for operations in Afghanistan.

Neither the armed forces nor their civilian leadership had an interest in resolving this discrepancy between rhetoric and reality, however. The rhetoric served a diplomatic purpose in and off itself in a diplomatic context, while the low level of ambition on actual civilian-military partnership served the bureaucratic interests of the army as well as the Danish Development Agency in the field. Applying one schema in Copenhagen and another in Helmand served everybody's interest. The only organizational entity, which found itself in trouble because of this constructive ambiguity, was the Ministry of Defence, which could neither focus exclusively on the diplomatic rhetoric, as the diplomats in the Ministry of Foreign Affairs were able to, nor focus exclusively on the operational realities, as the army officers were able to. The Ministry of Defence ended up applying the "comprehensive approach" schema to operations that were not arranged according to that schema by the operational commanders. Rynning and Ringsmose point out that one reason why it was difficult for the

Danish Ministry of Defence to actually realize the comprehensive planning into comprehensive action was in part that the civilian side of the civilian-military (CIVMIL) equation was expected to come mainly from nongovernmental organizations (NGOs). The Danish approach was in many ways not a whole-of-government approach but an NGO-military approach. This followed from a tradition of extensive NGO involvement in the formulation and implementation of development policies, but this schema did not adapt well to the Afghan theater. Because of the security situation the conventional schema for aid distribution and capacity building could not be applied in Afghanistan. This made it difficult for the military to identify reliable partners that could be placed in the chain of command and impossible for the NGOs to cooperate directly with the military, exactly because being part of a military effort went against their code of conduct. Having defined CIMIC as cooperation between the armed forces and NGOs, the Ministry of Defence and the Ministry of Foreign Affairs in effect wrote themselves out of their own policy, thus making it very difficult to actually have a comprehensive approach to security, reconstruction, and development operations in the Danish area of responsibility.[20]

Having invested considerable prestige into the comprehensive approach, the Ministry of Defence had come to believe its own rhetoric to be true. In 2010, Minister of Defence Gitte Lillelund Bech thus concluded that "we have embraced the comprehensive approach at the tactical and ministerial level for some years."[21] Tellingly, left out in the minister's statement was the level that mattered most in actually implementing a comprehensive approach—the operational level.

OPERATING IN HELMAND

In August 2006 the Husars from Bornholm were under siege in the district center in the town of Musa Qala in northern Helmand. The Danish unit was no more suited for that operation than the similar unit from 3 Para, which the Danes relieved. However, they were implementing the "platoon house strategy" currently applied by the British commanders in Helmand.[22] The Danish commanders accepted the assignment to hold a fixed position, but they would most certainly have preferred the mobile operations they were trained for. When 3 Commando Brigade took over operations from 16 Air Assault Brigade in October 2006, the new leadership gave up on the "platoon house strategy," opting for maneuvers that set up Mobile Operations Groups, which suited the Husars just fine.[23] Danish commanders felt that the British Task Force appreciated their unit, which at this point was "heavier" than most other units in theater, but were also aware that they were being appreciated for how they fit into the British concept of operations.[24] Contributing 250 to the overall deployment of 3,000–4,000 soldiers, the Danes did not count for much at the frontline, and even less when it came to the operational management of the operations. Being very much the junior partner in Task Force Helmand, the Danes had to adapt to the British concept of operations. This "second-order adaptation" meant that they might have to fit into a British plan, even if they disagreed with the overall strategic outlook. As Jakobsen and Thruelsen note, guidance from ISAF command was not as important in shaping Danish operations as was the guidance from the Danish officers' immediate superiors at Task Force Helmand.[25] The deployment to Musa Qala was one example of that. In practical terms, this was emphasized by the fact that the Danish troops rotated in the summer, while the British rotated in the fall. The Danish contingent was fully operational at the moment new British troops arrived and had to adapt to the theater themselves. When the British had settled in, the Danes would go into periods of leave, thereby reducing the number of troops available for operations.

Some Danish officers complained that they were met with "British arrogance" by a task force headquarters not really disposed to taking advice from a subordinate command, let alone a foreign one. The British maintained a hands-on approach toward the Danish units, regarded as any other British army unit in the task force's area when it came to operational command and control. Thus British commanders repeatedly overruled Danish concerns on adverse effects from British operations outside the Danish area of operations and deployed Danish units outside that area to support British operations—with negative results for the security situation in the Danish area.[26] Anglo-Danish cooperation seems to have been at its best when the British officers were able to assimilate the Danish units under their command. The End of Tour Report from the first unit in Helmand explained how the Danes did not have official access to British intelligence and computer systems. The report described how the Danish officers had asked for access as informal favors from their British colleagues. With this background, the Danish commander advised succeeding units to make an effort to be integrated into the British chain of command.[27] Aware of this, Danish officers made an effort to explain their capabilities and outlook to their British superiors—one Danish commander termed this "PsyOps."[28] The purpose of such PsyOps was to ensure that the British assimilated the Danish troops, while the Danes on the other hand had to accommodate to the environment defined by the British task force. This accommodation seems to have been most successful when the Danish units were of a similar kind to those dominating the British Task Force.

Anthony King has argued that the particular culture of the British regiments assigned to Task Force Helmand goes a long way in explaining the strategic and operational decisions taken by the commanders.[29] The Danish units had similar cultural schemas for how the "military métier" was to be conducted. The Husars from Bornholm had embraced the ideas of patrols in the desert operating as "warriors, developers and diplomats," as one of their commanding officers put it.[30] The Royal Life Guards had another approach. The guards regiment was trained as mechanized infantry and formed the bulk of the battalion that was finally deployed to Helmand in 2007 in order to realize the Chief of Defence's vision of a Danish battle group which could demonstrate how far military innovation had changed the Danish Army. The Royal Life Guards deployed in the summer of 2007. 12 Mechanised Brigade had taken over Task Force Helmand in April, and the two mechanized formations had very similar schemas for operating. 12 Brigade wanted to establish patrol bases in the population centers of the Green Zone along the Helmand River in order to demonstrate "enduring presence."[31] Perhaps most important from the perspective of a mechanized brigade, the bases were to serve as the jumping off points for the more logistically challenging mechanized operations. The Danes were tasked to set up a number of bases in the Green Zone in the fall of 2007.

The Royal Life Guards assimilated easily to British operations focusing on establishing "enduring presence" by maneuver, which was carried on by 52 Infantry Brigade, which took over from 12 Brigade in October 2007. According to the guards' schema, however, an important element of mechanized warfare was missing: tanks. The army went to Helmand in order to unfold a schema. The main objective was to adapt to a schema for doing conventional warfare realizing the full potential of the "military métier," in General Helsø's term, rather than to adapt to a counterinsurgency approach. Furthermore, the Danish Army used tanks with great effect and international recognition in Bosnia.[32] Here was another experience that shaped a new schema. General Helsø thus explicitly refers to the reasons for deploying tanks in Bosnia as a template for the deployment of tanks in Helmand.[33] The Canadian use of tanks in Kandahar was mentioned as an argument, but the key reference was the army's use of tanks in Bosnia. As opposed to Canada, deploying tanks was *not* an example of accommodating to a new schema that came about fighting the Taliban;[34]

deploying tanks was assimilating the mission to fit the schema of mechanized warfare. This is not to say that the army denied that this was a counterinsurgency operation. On the contrary, the army and the defense leadership were very conscious of the demands of counterinsurgency operations. That was the reason why General Helsø had explicitly rejected the establishment of a Danish PRT and the actual realization of comprehensive planning for the Helmand campaign. The army wanted to supply the military element of the campaign, while leaving it for others to do reconstruction and development.

In the first stage of deployment the smaller Danish unit and the British strategy of Mobile Operation Groups produced a schema that, at first glance, fitted the COIN-schema well but was actually a conventional military operation. This is illustrated in the fact that the Civil-Military Cooperation Unit (CIMIC-unit) of the 3rd Danish rotation left Helmand having used only a third of the funds appropriated for development and CIMIC-work.[35] The next units deployed were battalion-size battle groups, and in such numbers the military focus was more explicit. The army applied the schema of mechanized warfare.

If Danish operations in Helmand had actually followed the schema of the comprehensive approach, the Minister of Defence should not have approved the request for deployment of tanks. Minister Søren Gade publicly denounced the idea at first, but in the end the schema of operations followed by 12 Mechanised Brigade and the Royal Life Guards dictated tanks; in November 2007 tanks were indeed transported to Helmand.[36] The debate about the deployment of tanks by the Danish forces in the Green Zone illustrates the strengths and weaknesses of the Danish approach to adaptation. The army was strong on assimilation. It had a strong schema for how to conduct operations at the tactical level and was rigorous in its assessment of its needs, especially in terms of materiel and planning. Within the army, the debate on tank deployments in 2007 was widely regarded as an example of how army professionals stood up for was what right in the face of political opposition. The term most often used to describe a professional military assessment in Danish is *militær faglig*, which is perhaps best translated as "the military vocation." The word *faglig* not only refers to a type of professional knowledge, which anyone schooled or experienced in military matters might possess, but the word also conflates knowledge and profession in the same way as one would say that a carpenter knows more about fashioning furniture in wood than a noncarpenter. In some ways this notion of the military as a vocation follows Samuel Huntington's notion of the military as a distinct profession with a distinct skill-set.[37] Military knowledge is thus something that only military officers have, and one of the reasons why academics like Peter Viggo Jakobsen was scolded for his opposition to deploying tanks to Helmand.[38]

Adaptation is about the way an organization defines its relationship with its environment. The organizational approach to human resources goes a long way to explain how this relationship is carried out in practice. Alexandra Michel describes how two investment banks organized their personnel policies differently in order to prepare their employees to deal with uncertainty and risk. One bank emphasized risk reduction, another encouraged its employees to embrace risk. In the first bank, bankers are taught to be "superstars." These superstars have elaborate titles and clearly defined roles in a hierarchical organization. The employees know their place and their role with the purpose of creating security and building up their reputation in the meeting with clients and other actors external to the organization that read the bankers' business cards and understand that they are experts. The second bank eschews titles in favor of a project-based, ad hoc organization. This organization is less hierarchical, and therefore employees do not have fixed positions and portfolios. Instead, the bankers work on different projects and are thus generalists rather than experts. In this bank, the employees regard themselves as investigators of reality, while employees in the first bank find that they define reality.[39] Michel's point is that both banks do well, and from the stockholders' perspective there is little difference between the two very different HR-approaches. In operational terms, the two banks approach adaptation very differently.

Danish officers are taught to be superstars in the manner of the employees in Michel's first bank. The notion of *militær faglig* makes them the only national experts in their profession. Thus Danish officers are the "Danish superstars." They are the kingdom's experts on military matters, and they regard themselves as a closely knit community of practice. Officers know one another personally, which is one reason why information as well as experience is often transmitted via personal networks.[40] As a vocational expert, one commanding officer has a fairly easy time transmitting his experience to the next officer taking his place in Helmand. The army tries to facilitate such exchanges by means of an "experience seminar" in which officers who have just returned from Helmand transmit what they have learned to those officers about to leave for Helmand. While the transmission of Danish experiences is thus institutionalized, personal networks are utilized to get experiences from other nations.[41] The point is that as the lessons learned networks are focused on experience, this necessarily makes the whole process focused on the operations of platforms.[42]

As head of the Army Combat School and responsible for much of the lessons learned work done in the army, Colonel Lars Møller is understandably proud about the speed with which the army was able to turn experiences in Helmand into training for the soldiers going to be deployed there. However, the colonel also notes that learning within the army is focused on the tactical level. "An officer," Colonel Møller notes, "is so deep in day-to-day operations, that heavy thinking is done off-duty."[43] A report from the Danish Institute for Military Studies supports the colonel's assertion and concludes that compared with Dutch and British practices there is little interaction between the tactical and strategic level in the Danish Army in particular and the armed forces in general. The Danish Armed Forces have no lessons learned procedures that systematically create a conceptual link between Danish tactics and the Danish—or indeed coalition—strategic planning.[44]

By using their vocational networks and with the aid of the Army Combat School, the army developed new Tactics, Techniques and Procedures (TTPs). The Danish troops became better and better at patrolling the Green Zone and using tanks in the peculiar environment of the Helmand River Valley while trying to avoid IEDs. These operational adaptations, however, did not translate into doctrinal change. The Danish Army prided itself on being agile in the face of new operational circumstances at the same time as the officers were vocationally committed to the notion that while operational circumstances might change, the nature of warfare remained the same. Army doctrine was thus resolutely conservative. The troops fighting in Helmand were fighting according to a doctrine set down in 1943 and revised in the following years into a field manual that has changed surprisingly little in seventy years. The field manual is an important instrument in defining the vocational nature of being a Danish Army officer. Captains are taught how to conduct operations according to the field manual at the Army General Staff Course, and at their final exam the commander of Army Operational Command himself often examines them.[45] The staff course serves to produce superstars on the model described by Michel. In this context, doctrine becomes a standard that officers are measured by, rather than a tool to conduct operations. The Danish Army was thus quite willing to learn how to conduct COIN-operations on the TTP level because it was part of the vocational commitment by the officers that they should be able to adapt to changing circumstances; but at the same time the vocational identity dictated a rejection of adopting a particular doctrine for counterinsurgency as the US Army and Marine Corps did when they adopted the Counterinsurgency Field Manual in 2006.[46] In 2011 the army was reportedly working on a subsection of the field manual to deal with operations by a battle group like the one operating in Afghanistan. While this would undoubtedly be important to institutionalized lessons learned in the years operating in Helmand, this addition to the field manual did not constitute a COIN doctrine. Declining to do doctrinal work on its own, the army relied on NATO doctrine and TTPs that often defined the operational concepts of using certain platforms or fighting in certain terrains, but seldom served a doctrine's real purpose by connecting strategic and operational considerations.

Because of the rejection of doctrinal development on vocational grounds by 2009, the Danish troops had become better and better at fighting in Helmand at the same time as they moved further and further away from the possibility of winning. Adapting the schema of mechanized warfare to the operational environment in Helmand had brought no decisive change in the security situation in the Danish area of operations around Gereshk. One might even argue that the Danish troops had adapted so perfectly to the environment that they were able to keep operations going indefinitely. However, they were not able to change their environment in a way that led to the defeat of the Taliban. Officers who, in Møller's phrase, actually did some strategic thinking in their time off had at that point realized that refining tactics would not make the strategic situation more favorable, but "we needed someone with power to say it," noted Colonel Møller. With his *Commander's Assessment* General Stanley McChrystal had the power to identify the strategic dead-end that the Afghanistan campaign had entered.[47]

LEAVING HELMAND

By 2009 the Danish contingent had suffered twenty-six casualties (by November 2011, forty-two Danish soldiers had died in action). This made Denmark one of the nations that sacrificed the most soldiers per capita in the Afghanistan campaign. It is worth noting that while all of the Danish dead in 2007 died in one of the fire fights that brought the Royal Life Guards in control of the Upper Gereshk Valley, the majority of subsequent casualties were the result of IED attacks. This change in the Taliban's mode of attack meant a significant increase in the number of wounded. In 2009, forty-seven Danish soldiers were wounded in various degrees, while only eight were wounded in 2007.[48] The fact that the soldiers were losing their lives while on the defensive put a special responsibility on the politicians to explain why Danish soldiers were dying; the wounded returning to Copenhagen were a constant reminder of how little security DANCON was able to maintain in its area. In the absence of the gains of territory that one could point to in 2007, the political discourse increasingly focused on the value of fighting the war in and of itself. In a sense this argument mirrored General Helsø's ambition to embrace the "military métier." However tragic the death of the individual soldier, the sacrifice he or she made became a monument to Danish grit and commitment. To the right-wing government in power from 2001 to 2011, erecting a monument for Danish values in the public debate was an important goal in itself.[49] Focusing on the values that the individual soldiers upheld by volunteering for service in Afghanistan and affirmed by dying there, the vocational nature of Danish military life came into focus. The strategic reasons for sacrificing the soldiers' lives mattered less.

Perhaps the focus on the virtue of the individual soldier reflected the fact that overall strategic direction and lack of progress was to a large extent beyond Danish control. The Danish Armed Forces adapted to the conditions for operations in the Helmand theater in the sense that they assimilated to the British strategy and accommodated to the development in tactics from the Taliban. The strategic direction of the campaign (that is, the changing of the environment itself in a direction more beneficial to ISAF) was something in which the Danes showed little interest. One reason for this was geopolitical. The Danish government could not reasonably expect to be able to change the direction of the war. Another reason was the way in which the vocational culture of the Danish Armed Forces made Danish officers focus on tactics, techniques, and procedures rather than on doctrines and strategy. For a small country like Denmark, adaptation on the strategic level was Second Order Adaptation. To Danish planners, the immediate environment, which they had to adapt their plans to, was not defined by the engagement with the enemy (the Taliban) but defined by the British and, increasingly, the Americans. This created a dilemma for the Danish government when it came to changes in strategy. On the one hand the Danish government had to be loyal to ISAF strategy, as defined by the British and the Americans, but on the other hand the government had to be flexible enough to be able to change strategy when the British and the Americans decided it was time to do so. The Danish government had to get its timing exactly right but rarely managed to do so.

The diplomats' focus on the comprehensive approach must be regarded in light of this dilemma. The best, and perhaps only, way in which the government in Copenhagen could influence strategy was to push for certain principles on how strategy was done (that is, the comprehensive approach), rather than on specific strategic priorities. If these principles suited an Anglo-American agenda, that was all the better for the Danes, who then could entrench cooperation on the operational level on the strategic level as well. The problem was of course that since the Danish government was actually not able to do comprehensive planning and action on its own, the discrepancy between rhetoric and reality became so wide at times that the adaptation on the diplomatic scene did not mirror the adaptation in Helmand. A small country needs to coordinate its policy better than most in order to achieve anything substantial,[50] and this created a special challenge for the Danish government that it had great difficulties in meeting.

The fact that the governments' policies in Afghanistan were supported by a large majority in Parliament, which included the main opposition parties, made support for the war seem much more solid from the outside than it actually was. In the winter of 2007, Chief of Defence General Helsø "felt a drop in support for the mission, also within the Armed Forces themselves—people were asking 'are we doing the right thing?'"[51] According to an opinion poll of October 27, 2007, some 52 percent of the population wanted the troops withdrawn from Helmand. The lack of confidence in the mission was mirrored in a record high number of soldiers that left service during training because they did not want to deploy to Afghanistan. The disjoint between rhetoric and realities that served the government well in terms of foreign policy was beginning to have domestic repercussions. This was especially a problem for the opposition, since the Social Democratic Party and its Social Liberal partner had an electorate more skeptical toward the use of armed force than the right-wing government. The opposition thus demanded a coherent Danish strategy for Afghanistan that was not only to define the end-state of the mission but also to identify benchmarks that could show progress toward those end goals. The approach was clearly inspired by the exit-strategy from Iraq presented by the Democrats in the US Congress at that time.

The Ministry of Defence was quite happy to comply. "It was obvious that the conflict would not be won as a military victory," then Permanent Under-Secretary for Defence Bjørn Bisserup concluded. If the military side could not win, it was important for the Ministry of Defence to define to what extent the military had the responsibility for a potential defeat. As ever focused on the tactical level, the Ministry of Defence and the uniformed military clearly believed that by defining civilian and military goals it would be obvious that the soldiers had done their bit successfully; never mind that the civilian efforts, which the military also recognized as crucial for success of the overall mission, had faltered. Thus it was important, from the Ministry of Defence's point of view, to make clear that the civilian side, especially the Ministry of Foreign Affairs, had a responsibility to do its part.[52] The Ministry of Foreign Affairs was tasked with producing the strategy and produced a first draft so vague that it seemed to confirm the fears in the Ministry of Defence that the diplomats would take no responsibility for the mission in Afghanistan. The opposition parties rejected the draft and sent the diplomats back to write a more specific plan with more emphasis on the military side of the strategy.[53] This pleased General Helsø, who still wanted a clear distinction between the military operations under his command and the civilian reconstruction effort led by the Ministry of Foreign Affairs.[54]

The result was a "Helmand Plan" and an "Afghanistan Strategy."[55] These strategy documents were closely coordinated with London; one might indeed argue that the British plan made it possible for the junior partner in Copenhagen to make a plan in the first place. To the British, as well as Denmark and other states allied to Britain. this constitutes an important lesson. Planning by the nation dominating the task force is a prerequisite for the smaller coalition partners being able to plan their own engagement. First produced in 2008 and redefined once a year, the Danish planning defined the mission as a process toward the Afghan security forces gaining the capability to take over the fighting. This approach had much more in common with the investigative approach to a risky and unpredictable environment identified by Michel in bank number two than with the superstar approach of bank number one, previously used by the Danish Armed Forces. By accepting the need for a strategy that linked strategy, operations, and tactics, the military establishment had tacitly accepted that military officers were not the only authority on how one should measure progress in the operations in Helmand.

For the armed forces it was now important to be able to produce "measurable progress," in the words of Bjørn Bisserup.[56] Defining the end toward which operations were to progress, however, was still a matter for the Great Powers. In November 2010, NATO defined the mission in Afghanistan in terms of "Afghanization" at the Lisbon Summit, and Prime Minister Cameron stated that British troops would only have an advisory capacity from 2015 onward.

This defined the overall ends of the alliance and the priorities of Task Force Helmand. On that basis the parliamentary majority behind the deployment to Helmand and the government was able to agree on a new Helmand Plan in the spring of 2011 that described a gradual withdrawal of the battle group in favor of units training the Afghan Army. At the same time Special Forces units and air force personnel were planned to be deployed at various times until the complete withdrawal in 2015. The military contribution was thus shifted from entirely military operations firmly within the "military métier," as defined by General Helsø, toward a more varied contribution of small units more integrated in the British Task Force and in closer cooperation with the civilian side of operations.

CONCLUSION

In the Helmand theater of operations the Danish Army proved itself. General Helsø wanted an operation that showed that the army could fight within the "military métier." General Helsø and his colleagues in the army had a very particular understanding of what constituted the "military métier," an understanding that was defined by a strong vocational identity that defined military operations in terms of mechanized warfare and doctrine in terms of the enduring nature of armed conflict. Neither this preference for mechanized warfare nor the inflexibility on doctrine made the Danish Army particularly well prepared for counterinsurgency operations, nor did this conceptual background make it easy for the army to adapt to realities on the ground in Helmand. While the army swiftly learned how to adapt tactics, techniques, and procedures, it preferred not to let these operational lessons influence either its force posture or its doctrine. Instead, the realities of the Helmand theater were adapted to the "Danish Way of Warfare," rather than the other way round. The armed forces were quite candid in stating their commitment to the "military métier." However, that stand did not fit well with the way in which the civilian part of the effort was conceived. The civil servants wanted the soldiers to perform a supporting role in a narrative on comprehensive planning and action that placed civilian NGO's in the center of the effort to create a stable Afghan society. This created two conflicting perspectives on operations in Helmand, with conflicting priorities and notions of success. From a military perspective, conducting operations with the British on an equal level was proof that the army had successfully transformed to an expeditionary corps, and that constituted success on an organizational level. In operational terms, the army was intensely proud of the battles against the Taliban in the fall of 2007, and these victories mattered to the organization even if they had not produced a durable stability in the area. From a civilian perspective, however, the military campaign was instrumental only in providing the security for development projects. For that reason the civilian side never took the integration between civilian and military efforts seriously, but saw them as preconditions for each other carried out independently in a comprehensive plan rather than as a continuous, comprehensive effort.

What the military and the civilian perspectives had in common was "second order adaptation." The Danish troops were adapting to the British, which they regarded as a benchmark for military transformation, and the civilian side adapted the mission to United Nations (UN) standards for development and Danish development policies. In neither case were the operational results in Helmand as important as the ways in which the operation benefited organizational goals set in Copenhagen. As the Danish military presence in Helmand is being drawn down, the military perspective on the campaign is rapidly being forgotten. Without the news of military engagements to rock the boat, the development perspective is increasingly dominating the official version of events. This narrative is obviously being reinforced by the focus on training and mentoring of Afghan police and military units. As the Danish soldiers are withdrawn to fixed positions around Gereshk, leaving offensive operations to the Americans, the British, and the Afghans, and as the military role is taken over by officers and noncommissioned officers (NCOs) responsible for training the military commitment fit the comprehensive approach narrative of the Ministry of Defence and Ministry of Foreign Affairs. Unfortunately, this approach reinforces the disconnect between the military campaign and the civilian effort. The military can draw down without having to discuss the meager strategic impact of the campaign, while the civilian effort can focus on goals so far in the future that their impact is not yet up for discussion. Thus the Danish engagement in Afghanistan will probably end in 2015 without a clear and open assessment of what was achieved and what has been a failure. On the contrary, the rhetoric of the comprehensive approach and the reality of military operations will be as far apart as ever, but because the mission will end in a predominantly civilian effort, the Danish government might end up believing in its own rhetoric. Denmark will thus sign up for missions in times to come believing itself ready to take on complicated civilian-military integration, while in fact civilian and military capabilities remain disintegrated.

ACKNOWLEDGMENTS

I would like to thank Theo Farrell for his constructive comments on earlier drafts of this chapter. I would also like to thank my colleagues at the Centre for Military Studies for their instructive comments on the draft for this chapter, as well as Peter Viggo Jakobsen for his comments. Any mistakes are of course my responsibility.

NOTES

1. Theo Farrell, "Introduction: Military Adaptation in War," this volume. Compare Farrell, "Improving in War: Military Adaptation and the British in Helmand Province, Afghanistan, 2006–2009," *Journal of Strategic Studies* 33 (August 2010): 569.

2. Theo Farrell, "Back from the Brink: British Military Adaptation and the Struggle for Helmand, 2006–2011," this volume.

3. Niklas Luhmann, *Social Systems* (Stanford: Stanford University Press, 1995), 48.

4. Jean Piaget, *The Psychology of Intelligence* (London: Routledge and Kegan Paul, 1950).

5. Farrell, "Improving in War," 570.

6. E. von Glaserfeld, *Radical Constructivism: A Way of Knowing and Learning* (Washington, DC: Falmer, 1995), 68.

7. Frans Osinga, *Science, Strategy and War: The Strategic Theory of John Boyd* (London: Routledge, 2006).

8. Bertel Heurlin, "Actual and Future Danish Defence and Security Policy," in Heurlin, ed., *Global, Regional and National Security* (Copenhagen: Danish Institute for International Affairs, 2001), 197–219; Sten Rynning, "Denmark as a Strategic Actor?," in Per Carlsen and Hans Mourtizen, eds., *Danish Foreign Policy Yearbook 2003* (Copenhagen: Danish Institute for International Studies, 2003), 23–56; Anders Wivel, "The Security Challenge of Small EU Member States: Interests, Identity and the Development of the EU as a Security Actor," *Journal of Common Market Studies* 43 (2005): 393–412.

9. Kjell Inge Bjerga and Torunn Laugen Haaland, "Development of Military Doctrine: The Particular Case of Small States," *Journal of Strategic Studies* 33 (August 2010): 507.

10. Lars Ulslev Johannesen, *De Danske Tigre [The Danish Tigers]* (Copenhagen: Gyldendal, 2009).

11. Danish Defence Agreement 2005–9, Ministry of Defence, June 10, 2004.

12. *The Security Policy Conditions for Danish Defence* (2003), Executive Summary of the Report, August 2003, Royal Danish Ministry of Foreign Affairs [www.um.dk] (September 30, 2003).

13. Danish Defence Agreement 2005–9.

14. John R. Deni, "The NATO Rapid Deployment Corps: Alliance Doctrine and Force Structure," *Contemporary Security Policy* 25 (2004): 508–11.

15. On the NATO transformation agenda, see Theo Farrell and Sten Rynning, "NATO's Transformation Gaps: Transatlantic Differences and the War in Afghanistan," *Journal of Strategic Studies* 33 (2010): 673–99; Bertel Heurlin and Mikkel Vedby Rasmussen, eds., *Challenges and Capabilities: NATO in the 21st Century* (Copenhagen: Danish Institute for International Studies, 2003).

16. Interview with General Jesper Helsø, April 8, 2010. Statements from interviews are translated to English by the author.

17. Ibid.

18. Peter Viggo Jakobsen, *NATO's Comprehensive Approach to Crisis Response Operations: A Work in Slow Progress*, DIIS Report 2008/15 (Copenhagen: Danish Institute for International Studies, 2008), 7.

19. Kristian Fischer and Jan Top Christensen, "Improving Civil-Military Cooperation the Danish Way," *NATO Review*, Summer 2005, accessed March 29, 2011.

20. Jens Ringsmose and Sten Rynning, "The Impeccable Ally? Denmark, NATO, and the Uncertain Future of Top Tier Membership," in Nanna Hvidt and Hans Mourtizen, eds., *Danish Foreign Policy Yearbook 2008* (Copenhagen: Danish Institute for International Studies, 2008), 76–77.

21. Minister of Defence Gitte Lillelund Bech, *Speaking Notes for Intervention at the Conference "Defeat or Sustainable Exit? The Comprehensive Approach and Military Capacity Building in Afghanistan,"* Centre for Military Studies, University of Copenhagen, and the Royal Danish Defence College, Copenhagen, December 6–7, 2010.

22. Farrell, "Improving in War."

23. Ibid., 576.

24. Interview with Colonel H. C. Mathiesen, May 26, 2010.

25. Peter Viggo Jakobsen and Peter Dahl Thruelsen, "Clear, Hold, Train: Denmark's Military Operations in Helmand 2006–2010," in Nanna Hvidt and Hans Mouritzen, eds., *Danish Foreign Policy Yearbook 2011* (Copenhagen: Danish Institute for International Studies, 2011), 79.

26. Ibid., 99.

27. *DANCON/RC(S) ISAF, hold 1, End of Tour Report*, 11.

28. Interview with Colonel H. C. Mathiesen, May 26, 2010.

29. Anthony King, "Understanding the Helmand Campaign: British Military Operations in Afghanistan," *International Affairs* 86 (2010): 311–32.

30. Interview with Colonel H. C. Mathiesen, May 26, 2010.

31. Farrell, "Improving in War," 577.

32. Trevor Findlay, *The Use of Force in UN Peace Operations* (Oxford: Oxford University Press, 2002), 230–31.

33. Interview with General Jesper Helsø, April 8, 2010.

34. See Stephen Saideman's contribution to this volume for an account of the Canadian deployment of tanks to Kandahar.

35. *End of Tour Report for DANCON/RC(S)/ISAF*, hold 3, bilag 5.

36. Mikkel Vedby Rasmussen, *Den Gode Krig?* [*The Good War?*] (Copenhagen: Gyldendal, 2011).

37. Samuel Huntington, *The Soldier and the State* (Cambridge: Harvard University Press, 1981).

38. Peter Viggo Jakobsen, "Kronik: Slaget om kampvognene" ["Op Ed: The Battle for the Tanks"], *Jyllandsposten*, 11. Juni 2007. See also the response to Jakobsen: Bo Grünberger, "Slaget om skrivebordet" ["Desk Battle"], *Jyllandsposten*, 15. Juni 2007.

39. Alexandra Michel, "A Distributed Cognition Perspective on Newcomers' Change Processes: The Management of Cognitive Uncertainty in Two Investment Banks," *Administrative Science Quarterly* 52 (2007): 518. See also Michel and Stanton Wortan, *Bullish on Uncertainty: How Organizational Cultures Transform Participants* (Cambridge: Cambridge University Press, 2008).

40. Kristian Søby Kristensen and Esben Salling Larsen, *At lære for at vinde. Om dansk militær erfaringsudnyttelse i internationale operationer* [*To Learn in Order to Win: On Danish Military Lessons Learned in International Operations*] (Copenhagen: Danish Institute for Military Studies, 2010), 15.

41. Ibid., 16.

42. Ibid.

43. Interview with Colonel Lars Møller, June 1, 2010.

44. Kristensen and Larsen, *At lære for at vinde*, 25.

45. Interview with senior military official, October 2011.

46. *Counterinsurgency Field Manual No. 3–24*, Marine Corps Warfighting Publication No. 3–33.5 (Chicago: University of Chicago Press, 2007).

47. Commander NATO International Security Assistance Force, Afghanistan, US Forces, Afghanistan, *Commander's Initial Assessment*, August 30, 2009.

48. Rasmussen, *Den Gode Krig?*, 96–97.

49. Ibid., 98–102.

50. Henrik Breitenbauch, *Kompas og kontrakt* [*Compass and Contract*] (Copenhagen: Danish Institute for Military Studies, 2008).

51. Interview with General Jesper Helsø, April 8, 2010.

52. Interview with Lieutenant General Bjørn Bisserup, June 7, 2010.

53. *Politiken*, "Partier forkaster Afghanistan-plan" ["Parliament Rejects Afghanistan Plan"], April 25, 2008.

54. Interview with General Helsø, April 8, 2010.

55. *The Danish Helmand Plan 2011–2012: A Report on the Danish Engagement in Afghanistan in 2010* (Copenhagen: Ministry of Foreign Affairs, 2011).

56. Interview with Lieutenant General Bjørn Bisserup, June 7, 2010.

7

Soft Power, the Hard Way: Adaptation by the Netherlands' Task Force Uruzgan

Martijn Kitzen, Sebastiaan Rietjens, and Frans Osinga

INTRODUCTION

Since the end of the Cold War the Netherlands' Armed Forces have gone through a turbulent period of continuous reduction, reorganization, and professionalization. The ability to conduct expeditionary operations was a key theme in this transformation process. As a result, a recent internal Dutch Ministry of Defence (MOD) evaluation report concludes, the International Security Assistance Force (ISAF) has demonstrated that the Netherlands Armed Forces, in cooperation with other departments, are capable of participating in high-intensity international operations, and if necessary do so as a lead nation. The high standard of performance in the Province of Uruzgan, in particular the so-called Dutch 3D Approach (Defense, Development, and Diplomacy), has been acknowledged, both nationally and internationally.[1]

It also notes, however, that the wealth of lessons need to be captured and institutionalized. It indicates that in the fields of training, force preparation, interdepartmental cooperation, and in particular timely adjustment of SOPs (Standard Operating Procedures) and doctrine, the four years of deployment suggest that work remains to be done. Other recent reports too note the need to improve the lessons learned cycle, highlighting that operational adaptation in Afghanistan failed to become institutionalized in the wider organization as a result of an ill-structured lessons learned process.[2]

That seems to echo a 1996 report by the National Audit Office, which revealed a similar structural weakness in the armed forces' ability to actually learn from operational experiences and adapt to the challenges of peacekeeping missions.[3] This resulted in the foundation of a department of evaluations at the

MOD that was given the twofold mission of informing Dutch politicians on operations as well as adapting the armed forces' procedures and processes to the requirements of future operations. The experiences of a decade of peacekeeping operations ultimately led to organizational adaptation of the Netherlands' Armed Forces; operational command was centralized under the Chief of the Defence Staff, and the troops were organized and received additional means in order to create a force capable of providing a sustainable contribution to international expeditionary missions.

The actual form these contributions took varied from small contingents of a few soldiers taking part in observation missions to enhanced battalions with robust means, as was the case in Bosnia, Ethiopia-Eritrea, and Iraq. Although the armed forces as a whole adapted to the organizational requirements of these new challenges, the centralized position of the department of evaluations caused a neglect of institutionalization of operational lessons, which were traditionally the domain of the specific operational services.[4] In 1996 for the first time a military doctrine emerged from these services (the army—the others followed later); the formal institutionalization of operational experiences, however, had never been a trait of the system. The Dutch mission in the Iraqi province of Al Muthanna (2003–5) revealed the consequences of this, as there were huge—sometimes contrasting—differences in the modus operandi of the different rotations.[5]

Thus, when the Dutch Armed Forces in 2006 were about to mount their most demanding venture since the Korean War, the deployment of a brigade-size task force to Afghanistan's Uruzgan Province, they constituted a small military (a total strength of fifty-one thousand) that had adapted to the organizational challenges of expeditionary missions. However, although a lot of experience on such operations was gained during the previous decade, operational lessons were noted but not further systematically documented, analyzed, incorporated in formal doctrines, and thus institutionalized. Adaptation in the Netherlands' Armed Forces —especially the army—was, apart from the larger organizational issues, a matter of individual units in specific operations.

This chapter explores how this small military went through a major experience, as the mission in Uruzgan was the largest Dutch military operation since the war of decolonization in the former Netherlands East Indies. Initially the mission in Uruzgan would last until 2008, but after fierce political debates it was prolonged to 2010. The debate over yet another—though much smaller—extension led to the fall of the Dutch cabinet in February 2010, and consequently it was decided that the mission would be terminated. Consequently, the specially commissioned Task Force Uruzgan (TFU) took over the lead from American forces in August 2006 and transferred command to its—again—American successors in August 2010.

Being responsible for all operations on the task force level required efforts that were new to the Dutch soldiers. The TFU clearly needed a larger operational framework and an organization fit to its tasks. Moreover, while the Dutch mission in Uruzgan was politically framed as a security and reconstruction undertaking (which will be discussed below), the reality on the ground can be best described as a counterinsurgency campaign.[6] This also required the Dutch soldiers to adapt to the challenges of population-centric counterinsurgency warfare. Therefore the Dutch military encountered two adaptation challenges in Uruzgan. First, it needed to conduct a campaign of unprecedented size and scope, and, second, the troops had to perform counterinsurgency tasks in addition to launching Stabilization and Reconstruction activities.

Studying these two major adaptation challenges encountered by a military that lacks a formalized lessons learned process and in which adaptation traditionally is an issue of individual units is difficult. In order to tackle this problem we will not focus on the kaleidoscope of units that contributed to the mission, but we will study the construct under which all these units operated: the TFU. This excludes specifically the substantial contribution made by transport and attack helicopter and F-16 units, which have encountered significant tactical and logistical challenges, resulting in interesting bottom-up innovations (such as the employment of iPads by pilots for assisting them in Close Air Support [CAS] missions and successful employment of the F-16 reconnaissance system to detect the placement of roadside Improvised Explosive Devices [IEDs]). A longitudinal analysis of four years of TFU campaigning allows us to identify the process through which the Dutch military tried to close perceived performance gaps and improve its performance. The main focus in this chapter, therefore, is on the operational aspects of the mission. We will explore changes in TFU strategy, organization, intelligence, operations, and training. Moreover, in the conclusion of this chapter the overall impact of the major TFU experience will be assessed by taking a look at recent MOD policy developments concerning the learning and evaluation process of the Netherlands Armed Forces. Before submerging in this all we will start this chapter with a discussion of the political and operational environment that shaped the Dutch mission in Uruzgan. Let us first start with the most elementary question" Why did the Netherlands' government deploy a task force to Uruzgan?

STRATEGIC MOTIVES

From the great variety of outlets such as parliamentary letters, scientific publications, and the media, one can discern a wide set of strategic motives why the Dutch deployed troops to Uruzgan. It betrays the features of the Netherlands' strategic culture, which abhors the notion of war, fosters the North Atlantic Treaty Alliance (NATO) and wants to be seen as a member of NATO's "A-team," is strongly Atlanticist, always justifies military missions in terms of humanitarian concerns, harps on international stability and the rule of law, invests in international institutions, and adopts a comprehensive 3D approach for the solution of security issues.[7]

One of the most prominent motives was to address the threats posed by terrorism. The terrorist attacks in New York, London, and Madrid seriously hit the world economy and thereby the Dutch, given its focus on services and logistics (for example, with the importance of Rotterdam harbor). The deployment to Uruzgan was therefore partly legitimized as the Dutch share of the burden.

This closely relates to the argument of NATO solidarity. The Dutch government highly valued its relationships with the United States, and NATO and believed—and still believes—this to be the most important fundament below its security policy. Despite Dutch acknowledgment of the American right to self-defense after 9/11, few Dutch politicians supported the mission Enduring Freedom. The Afghanistan case was, however, crucial for the cohesion and future of NATO. Some even boldly stated that *failure was not an option*. Pressure from the NATO headquarters, but also from Washington, Ottawa, and London to participate was therefore intense. And contributing to ISAF seemed like an appropriate way out for the Dutch government. Rather than a harsh antiterrorism mission, ISAF offered the perspective of stabilization and reconstruction.[8]

Within this perspective the next strategic motive was embedded—namely, to kick start the democratization of Afghanistan through stabilization and reconstruction. Without having a close connection with Afghanistan in a historical, economic, diplomatic, or emotional sense, the moral argument of the responsibility to protect was launched. It was our responsibility to assist the Afghan population. The Minister of Foreign Affairs, Verhagen, commented on this:

> Look at the history of Afghanistan: thirty years of war, the disastrous Taliban regime, horrendous violations of human rights and oppression of women. When you acknowledge the importance of human rights, then it is of great importance that the international community stays committed.[9]

The fact that it was a mission mandated by the United Nations (UN) and executed by NATO on the invitation of President Karzai also swayed the left side of the political spectrum. The mission met the requirement of an international legal mandate to contribute to the rule of law.[10] In the words of then Minister of Defence Henk Kamp, the various motives combined when he stated that

> the stabilization and reconstruction of Afghanistan, particularly in the south where the Taliban's roots lie, [are] of great importance to improving the international law and combating international terrorism which also threatens Europe.[11]

When the operation evolved the consequence that participation had on the reputation and influence of the Netherlands were openly introduced as a motive. It is difficult to establish direct causal relations, but the significant contribution to ISAF seems to have resulted in many benefits. The Netherlands participated in the G20 of economically powerful countries. Moreover, in March 2009 the Dutch were granted the organization of the important Afghanistan conference in The Hague. Former NATO Secretary General Jaap de Hoop Scheffer stated that the large contribution of the Netherlands made it easy for him to let Prime Minister Balkenende speak fifth or sixth at this and other important Afghanistan conferences with almost fifty countries participating.[12] In 2007, while discussing the extension of two years in Uruzgan, De Hoop Scheffer was also the one who put high pressure on the Dutch government:

> Don't do this to me. If the Netherlands will not extend their deployment it will make me, in my position, feel very awkward.[13]

Similarly, concern for diminished international standing was voiced in October 2009 by then Minister of Defence Middelkoop.[14]

The Dutch defense forces themselves constituted a final strategic motive in favor of the deployment in 2006. As the missions in the former Yugoslavia were unwinding, and the contribution to operations in Iraq had ended, Afghanistan offered a new area of operations. At a time of ongoing cutbacks this meant a new raison d'etre for the defense forces. Moreover, it offered an opportunity to operate in a relatively high intensity conflict area, a wish on the part of some top officials. With such a wide set of strategic motives, the TFU was sent to Uruzgan, a very complex province in Afghanistan's south, as we will see in the next section.

THE OPERATIONAL ENVIRONMENT: URUZGAN'S FRAGMENTED SOCIETY

Uruzgan's populace consists of an estimated 395,000 mainly rural inhabitants.[15] The vast majority of this population is ethnic Pashtun (91 percent), with Hazara (8 percent) and other ethnicities (1 percent) making up the minorities.[16] As the Dutch task force was deployed in the almost exclusively Pashtun-inhabited southern part of the province, we will focus on the Pashtun population. The Uruzgan Pashtun can be divided along three tribal confederations, the Zirak Durrani (57.5 percent), Panjpai Durrani (18.5 percent), and Ghilzai (9 percent), as well as some other tribes (6 percent).[17] Originally Uruzgan was dominated by the Ghilzai, but as Afghanistan historically was ruled by Durrani Pashtun, Durrani tribes were moved to Uruzgan at the end of the nineteenth century in order to end Ghilzai dominance.[18] This resulted in the Ghilzai being marginalized throughout the province, and today the district surrounding the provincial capital of Tirin Kot is the only area where Pashtun of both Ghilzai and Durrani descent are living. The Ghilzai-Durrani divide still forms a relevant source of tribal grievance in Tirin Kot District, as the former consider themselves victims of years of Durrani oppression.

Another societal cleavage resulting from this policy of Ghilzai marginalization is that between Zirak and Panjpai Durrani. The latter confederation is a construct created by the Zirak Durrani in order to assimilate Ghilzai tribes, and therefore Panjpai are often considered second-class Durrani. Especially in Uruzgan's western district of Deh Rawud this divide is "hot," as the Panjpai constitute the majority of the population, while governmental rule has traditionally rested with the Zirak. It therefore comes without surprise that the Taliban insurgency has found fertile soil in the Ghilzai and Panjpai Durrani areas surrounding Tirin Kot and Deh Rawud. It should also be noted that Mullah Omar, the overall Taliban leader, is a Ghilzai tribesman from Uruzgan.

To make things more complicated, the divides between different tribal confederations are important, but not as important as the affiliation with subtribes, which form the main solidarity groups and thereby define "patterns of loyalty, conflict and obligations of patronage."[19] When the communist government sought to establish control over Uruzgan by eliminating traditional tribal confederation leaders, local *mujahedeen* commanders rebelled against the communist governance and their Soviet allies. Since the demise of the communist regime, these commanders, who are organized along subtribal lines, have become the most influential power-holders in the province. As the different subtribe commanders each control specific areas and maintain their own supportive networks, Uruzgan's society has suffered from an increasing degree of fragmentation. In addition to traditional tribal grievances following the confederation divide, the process of fragmentation has led to conflicts within boundaries of confederations and subtribes. The result is a chaotic situation in which subtribe commanders fight against or alongside each other depending on what serves their interest best.

Former Uruzgan governor and Karzai trustee Jan Mohammed Khan might be considered the champion of these subtribe commanders.[20] Jan Mohammed became a prominent Popalzai (President Karzai's subtribe and part of the Zirak Durrani) militia commander during the communist era. After the fall of the communist regime he claimed his position as a Popalzai leader and succeeded in neutralizing the numerically superior subtribes of the Zirak Durrani by playing divide and rule, as well as in that he succeeded in subjecting the subtribes of the Ghilzai and Panjpai Durrani to suppressive measures.[21] He thereby effectively secured political leadership for the Popalzai, who account for only 10.5 percent of Uruzgan's population.[22] Jan Mohammed functioned as provincial governor until 1999, when he was imprisoned by the Taliban regime; he then resumed this position in 2003 with the use of the same methods that had brought him governmental power. Now, however, his position was stronger than ever before, as he enjoyed the support of President Karzai's Popalzai-dominated government, and US Special Forces fought alongside his militia against what Jan Mohammed labeled the Ghilzai and Panjpai Durrani "Taliban."[23] Thus when the Dutch were to deploy to Uruzgan they were confronted with a Popalzai provincial governor who had caused antigovernment resentment among a substantial part of the local population by exploiting traditional tribal grievances as well as subtribal conflicts in order to secure his rule. This complex environment of societal fragmentation and accompanying violent contention would be the TFU's arena for the next four years.

STRATEGY: POLITICS AND THE "DUTCH APPROACH"

The strategy of the TFU was only partly a reflection of this complex environment. One element shaping the ideas for the approach to be implemented was a keen political interest in differing from the US approach, which was considered too much focused on kinetic actions, too aggressive, too enemy-centric, too much informed by counterinsurgency (COIN) thinking, and too much associated with counterterrorism. Also in order to sway public support, the Netherlands wanted to de-emphasize the combat element, avoid search-and-destroy tactics and the use of air strikes, and put the focus on stabilization. As the Minister of Defence wrote to Parliament:

> Among large parts of the population there is no support for the behavior of coalition forces which is considered to be inappropriate. Their actions seem to impact the local situation negatively instead of positively. An operating style of ISAF, explicitly focused on winning the hearts and minds of the population is therefore necessary.[24]

In a similar vein the Ministry of Foreign Affairs told Parliament that "the international military presence over the past years has been directed at combating the Opposing Military Forces instead of improving the living conditions of the population." Hence the Dutch Approach would stress the Stabilization and Reconstruction (S&R) side of the mission.

This was entirely in line with the centrality of stability and a broad perspective on security. Stability projection was a key idea in a Defence White Paper of 2000 that put military missions in a larger scope of efforts to promote stability, such as development aid and economic policy. As the 2000 White Paper noted:

> The prevention and control of crises demands a broad, integrated approach. ... An integrated approach is necessary also for stabilization and reconstruction after cessation of fighting. Diplomatic, economic, humanitarian and, if needed, military instruments need to be deployed in an integrated fashion. ... Defense and development aid go hand in hand.[25]

In 2004 this approach was reinforced in a document formulating the Netherlands' "broad and integrated security policy," which became known as the 3D-approach.[26] This document also built on the Dutch experiences in the Balkans, Iraq (where Dutch units participated in SFIR—Stabilization Force in Iraq), and northern Afghanistan (where a Provincial Reconstruction Team, or PRT, had been deployed in 2003), and its earliest roots can be traced to an interdepartmental foreign policy review of 1995 which concluded that security and stability create essential preconditions for political, economic, and social development. The military was there to set the proper conditions.[27]

The Uruzgan strategy was a direct descendant of a white paper drafted in 2005 by the Dutch Ministries of Foreign Affairs, Defence, and Economic Affairs.[28] This policy paper outlined Dutch policy in future missions in failing states and identified three dimensions for an effective approach: (1) safety and stability, (2) governance, and (3) socioeconomic development. The letter that was sent by the Dutch cabinet to inform the Parliament on its intentions to deploy a mission to Uruzgan was based on this white paper.[29] It foresaw a deployment of two years to Uruzgan Province and stressed that positive developments in the area were expected, but that one could not realistically expect Uruzgan to be self-sufficient within such a short period of time.

It came to be labeled as the "Dutch Approach." As then Chief of Defence Staff General Dick Berlijn stated in 2006, "[It] is an operating style marked by knowledge of and respect for the local culture." It would not close its eyes to operational risks and would provide for sufficiently robust rules of engagement,[30] a hard lesson of the Srebrenica era. It acknowledged that the S&R mission might entail intense fighting for force protection, but it specifically was *not* to be confused with COIN, a term that was expressly avoided also in parliamentary debates, for it had too much of an offensive connotation that would undermine public support.[31] Reality within a few years would catch up with political wishful thinking.

THE TFU CAMPAIGN PLAN

In addition to these high-level policy guidelines, strategy in practice was informed by bottom-up planning. In preparation for the deployment the Dutch Ministries of Defence and Foreign Affairs commissioned The Liaison Office (TLO), an Afghan research-based nongovernmental organization (NGO), to carry out a civil assessment. Although the increased knowledge on the operational environment led to a replacement of Jan Mohammed as governor, the extensive analysis of the civil assessment and the policy framework did not lead to a clear strategy at the start of the operation. As a matter of fact the actual strategy was formulated in the field, during the first months of the deployment. Colonel Vleugels, commander of TFU 1, commented on this:

> We didn't have a campaign plan when we started, but later we got one from my higher headquarters that was close to ours, which is not surprising, as *they told us to do what we told them we would do.*[32]

The final strategy of TFU was laid down in the TFU Master Plan. In this document, the mission of the TFU was defined as:

> TFU assists the local government in building its capacity, authority and influence and prioritising and synchronising reconstruction and development programs with assisting the Afghan National Security Forces (ANSF), in order to set the conditions for a secure and stable Uruzgan Province.[33]

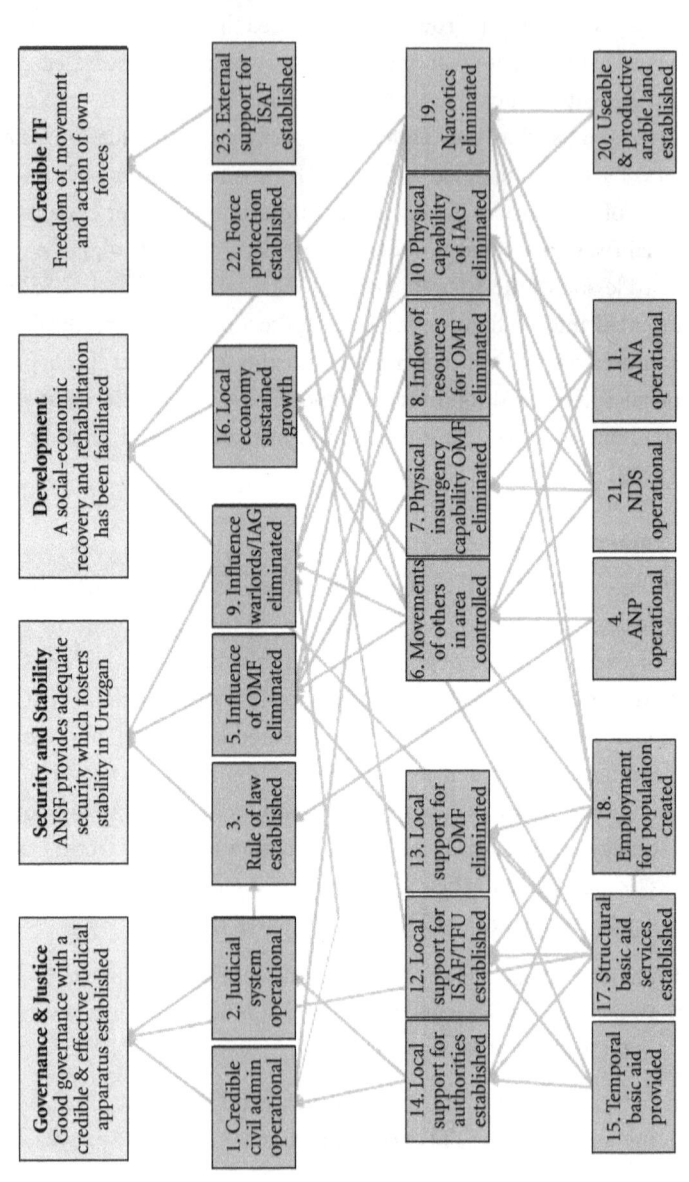

FIG. 1. The Effects of the Dutch Task Force Uruzgan[34]

The Master Plan functioned as a guideline for the planning and execution of the mission. Using an effects-based approach, the document focused on the overall effects that played a key role in the mission. The lines of operation (light grey blocks) and the desired effects (dark grey bocks) are illustrated in Figure 1.

The Afghan Development Zones (ADZs) concept played a crucial role in TFU's strategy. An ADZ is a geographical area in which improvements in security, governance, and development are delivered through an integrated approach by all relevant actors. Its aim is to rapidly meet the Afghan people's expectations in order to improve their confidence in the government of Afghanistan and the international community. The plan was to take a phased approach to the ADZ establishment. The first phase was to disrupt the opposing military forces (OMF) and to provide essential security. Having created a stronghold, subsequent operations would be conducted to secure and dominate the selected areas. These operations were aimed at convincing the local population of the legitimacy of the Afghan government and at promoting the development and reconstruction benefits.

The principle of the ADZ forced concentration of effort in an initial area, gradually expanding to other areas. For Uruzgan, the capital Tirin Kot and surroundings were identified as primary ADZ, and the town of Deh Rawud as a secondary ADZ. It was the objective to let the ANSF be the main responsible force for security within the ADZs. The TFU, together with the ANSF, would operate on the boundaries of the ADZ to establish security, while the Special Forces were to disrupt the OMF in the outer area.

ACKNOWLEDGING THE MILITARY REALITY: A COIN ORIENTED TFU PLAN

During the first two years, however, it soon became clear that the strategy and objectives as laid down in the Master Plan were too ambitious and not feasible with the available capacity. There was only limited geographical expansion, and most of the twenty-three defined effects were not met as planned. After heavy debate, in early 2008 the Dutch Parliament extended the deployment of the Dutch troops with an additional two years, until summer 2010. This led to the development of the second main strategy document, the so-called Focal Paper,[35] in which the strategy and objectives were updated and the time horizon extended until August 2010. In the Focal Paper the mission of TFU became as follows:

TFU, as part of the International Security and Assistance Force (ISAF), in cooperation with Afghan National Security Forces (ANSF) and in co-ordination with coalition forces is to conduct counter insurgency (COIN) operations resulting to the expansion of the Afghan Development Zones (Focal Areas) of Tarin Kowt, Chora and Dah Rawod in order to neutralise insurgency influence.

This adapted mission statement indicated an awareness that the scope in geographical area as well as nature of activities had changed and had to be changed. It reflected the tough experiences of the first two years. Already on February 3, 2006, Minister Kamp admitted that Taliban and Al Qaeda elements had to be eliminated in Uruzgan. By 2007 several daily journals ran articles highlighting the intense fighting that often went on. It also echoed previous comments of TFU commanders such as Colonel Theo Vleugels, commander of TFU-1, who in 2006 dispelled the notion that this was to be purely an S&R mission when he stated that "we'll do what is necessary and what is feasible." TFU-2 Commander Colonel Van Griensven in 2007 similarly had tried to circumvent the obvious naivety within the political debate when he noted that the dichotomy of reconstruction versus combat mission has no grounding in accepted doctrine. Also in 2007 Major-General van Loon explained that there was no tension between reconstruction and fighting because fighting might be required in order to establish security so as to allow reconstruction. While all military commanders acknowledged that activities for S&R generically dovetailed with those for COIN missions, they emphasized from the beginning that the actual operational environment in Uruzgan dictated operating more in the military dimension of the mission.[36] In 2009 the commander of Battle Group 8 explicitly conceptualized the comprehensive approach as the way to conduct the COIN mission he was asked to lead.[37] For the military it was and had been from the start a COIN mission.

Subsequently, while the TFU Master Plan tried to extend the ADZs to the outer regions of Uruzgan such as Khas Uruzgan, the Focal Paper clearly bounded this to three ADZs—namely, Tirin Kot, Chora, and Deh Rawud. And while the Master Plan referred to assisting the local government, the Focal Paper specifically focused on the conduct of COIN. The Focal Paper further stated that:

> In order to give the different actors, such as TFU, GoU [Government of Uruzgan] and NGOs/IOs [International Organizations], clarity on the effects and milestones to accomplish, the Focal Paper translates the three Endstates [safe and secure environment; socioeconomic development, governance] into seven detailed Lines of Effects (LoE): (1) security apparatus, (2) secure areas, (3) in-

frastructure, (4) basic living conditions, (5) health & education, (6) economic diversity and (7) governance support & capacity.[38]

All seven lines contain a series of desired effects, milestones, and objectives that must be achieved in order to reach the endstate. It is interesting to state that a total of ninety-one desired effects were formulated underlying the seven lines of operation.

When it became clear that the mandate most probably would not be extended for another period, the TFU began formulating a plan for transfer of its authorities. This plan, labeled the Uruzgan Campaign Plan (UCP), was delivered in mid-2009. The UCP

> is to provide a common ground for TFU and its Afghan and international partners within the province. In other words, the UCP is to facilitate, cooperation and to create unity of effort, which becomes even more important in a multinational context and with the increase of Afghan capacity and involvement in the mission.... The TFU campaign objective, within the context of the UCP, as part of ISAF, in partnership with ANSF and in coordination with GIRoA [Government of the Islamic Republic of Afghanistan], United Nations Assistance Mission Afghanistan (UNAMA) and the International Community, is to contribute to a reliable and effective government that can bring the government and the people closer together, and is able to provide a stable and secure environment and development progress in Uruzgan, in due course without ISAF support.[39]

The UCP also identified three major lines of operation, which were further subdivided into Reconstruction and Development (R&D) themes on which progress was desired. These R&D themes corresponded with the eight pillars of the Afghan National Development Strategy, with the exception that the pillar governance was split into governance and rule of law. The R&D themes were laid down in an *intellectual framework,* together with disablers hampering and enablers enhancing the progress on these R&D themes. Figure 2 illustrates this intellectual framework of the UCP.

Although the UCP clearly states that it provides a common ground, the document is by and large crafted by TFU personnel. After completion, the UCP became classified as NATO Secret, which made it very difficult if not impossible in cases to share it with other actors.

In sum, one observes a change in strategy driven by the adaptation of objectives and goals during the four year deployment of TFU. While the Master Plan

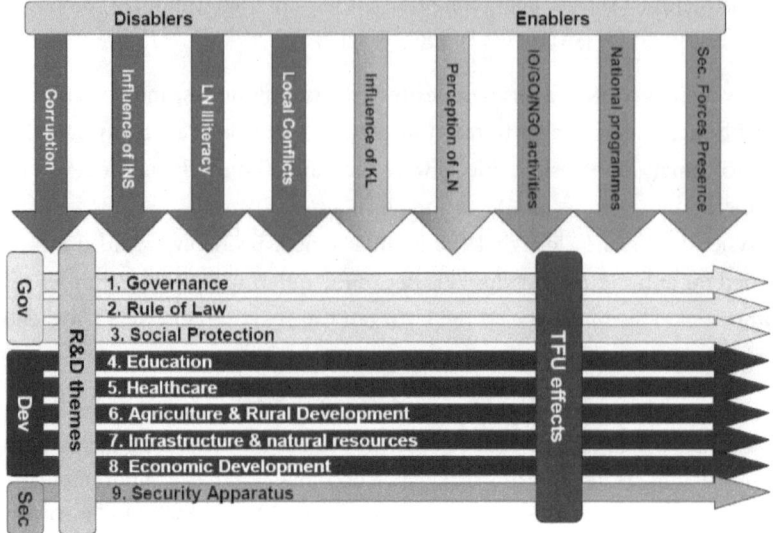

FIG. 2. Intellectual Framework of UCP[40]

was designed to foster security and development in the entire province within two years, the Focal Paper provided a much needed adaptation to the harsh reality of campaigning in Uruzgan. The Focal Paper not only acknowledged that the military was not merely providing security and development assistance, but actually involved in counterinsurgency, and also set some more realistic targets by focusing TFU efforts for the additional two years of the mission to the three main ADZs. Finally the UCP was formulated to arrange a smooth transfer of authority from the Dutch TFU to the local government and the ISAF successors. While the political climate initially dominated strategic thinking, and the need to develop the UCP, already early on bottom-up detailed operational planning informed actual military plans, and over the course of four years, the realities of the operational environment continued to shape military plans, necessitating a distinct deviation from initial guidelines and political expectations.

TFU'S COMPREHENSIVE ORGANIZATION

The strategy was expected to require a robust task force consisting of approximately one thousand troops.[41] In light of the security situation on the ground and the envisioned strategy, military planners urged seriously increasing this number. As a result permanent troops numbers were increased to fourteen hundred. This was still considered insufficient by military planners, but it was as far as Parliament would go. Until the redeployment of the TFU in August 2010, its size varied from fourteen hundred to a maximum of two thousand during peak periods such as a rotation.

The largest subordinate command of the TFU was the Battle Group (BG), which consisted of nearly six hundred combat troops. The BG was tasked to deploy in the ADZs, support the local authorities to increase their influence, support and monitor the deployment of the ANSF, provide force protection to other units of the Task Force, and prepare the activities of the Provincial Reconstruction Team (PRT), the second significant element of the TFU.[42] Aside from the BG and the PRT, the TFU had several other units under its command. These included an air detachment, a psychological element (PSE), a detachment of military police, an explosive ordnance disposal team (EOD), engineers, operational mentoring and liaison teams (OMLTs), Special Forces, and logistical capacity. In addition to the Dutch capabilities, several other countries, including Australia, Slovakia, Poland, and Singapore, contributed to the TFU.

While the BG emphasized security, the goal of the PRT was to promote good governance, and facilitate reconstruction. From 2006 until March 2009, the PRT was led by a senior military officer and consisted of approximately sixty troops. Its main capacity was made up of four so-called mission teams, each being responsible for a specific area of operation (that is, for different ADZs). In addition to military personnel, civilians from the Dutch Ministries of Foreign Affairs and Development Cooperation were embedded in the PRT. They were political advisers, development advisers, tribal advisers (later called cultural advisers), and counternarcotics experts.

Initially only two civilians were attached to the PRT. However, throughout the operation it was realized that their number was insufficient and gradually increased to thirteen. In March 2009, the command of the PRT was handed over to a civilian representative of the Ministry of Foreign Affairs, and from then onward the PRT had a civilian director rather than a military commander.[43] This was perceived as a significant step forward. It embodied the shift of focus of the TFU from security-driven operations to reconstruction and development. Having a civilian heading the PRT, who also acted as a co-commander of the TFU, greatly improved the status and influence of the civilians in the TFU. It was simply unique and unprecedented that a civilian commanded a military unit (that is, the PRT).

At the start of the operation in 2006, there were only six civilian international

NGOs in Uruzgan. Their number however gradually increased to thirty in 2009, and to almost fifty by 2010. The five most prominent organizations, HealthNet TPO, Cordaid, Save the Children, the Dutch Committee for Afghanistan (DCA), and ZOA,[44] united themselves in the Dutch Consortium for Uruzgan (DCU). Each of these organizations had a different focus, varying from education and rural development to health care and veterinary care.[45] DCU, just like most other international NGOs in Uruzgan, operated with very little international staff, and most of its activities were therefore implemented by local partners.

The involvement of UN agencies in Uruzgan was very limited. During the first years of the Dutch deployment there was hardly any UN representative working in Uruzgan. In time, that slowly changed after the Food and Agriculture Organization (FAO), for example, started operations in the province. After postponing the opening of a provincial office several times, UNAMA finally became active in Uruzgan from May 2009.[46]

In sum, after an initial increase in troop numbers we observe relatively few changes in the military capacity and size during four years of deployment. A development that stands out, however, is the change of the PRT leadership from a military to a civilian representative, who also acted as a TFU co-commander. This resembles the increased number and importance of actors on the civilian side, including both the NGOs and UN organizations as well as the civilian capacity that was embedded within the TFU.

INCORPORATING ETHNOGRAPHIC INTELLIGENCE

From the beginning the TFU incorporated ethnographic intelligence, a requirement for population centric S&R as well as COIN operations and entirely in line with policy guidelines.[47] This typically encompasses information on demographics, social and political organization, culture, religion, language, and so forth, which can be obtained through thorough ethnographical study of the local population. As already mentioned, the embassy hired an Afghan NGO to conduct ethnographic field research in May and June 2006 in order to enhance the Dutch understanding of the province.[48] This work culminated in the civil assessment of Uruzgan Province, which included a detailed context analysis of the local environment.[49]

A clear understanding of the political dynamics of Uruzgan province led to a successful lobby for the replacement of governor Jan Mohammed before the start of TFU operations.[50] When the Dutch first deployed to Uruzgan in March 2006, a new governor was installed. The insights that made possible this move were obtained mainly through the work of the Dutch embassy in Kabul. Early reports contained fine-grained information on the tribal divides, sources of conflict in the different districts, and individual power-holders. From this we might conclude that TFU was already properly informed about Uruzgan's societal landscape upon deployment in 2006. A report by the Dutch military intelligence service also points in this direction, as it concurs with the civil assessment and clarifies not only the divide between Ghilzai, Panjpai Durrani, and Zirak Durrani confederations but also contains information on subtribal conflicts and warnings concerning the still influential Popalzai network of former Governor Jan Mohammed.[51]

On the other hand, despite Colonel Vleugels (the first TFU commander) stressing the need for a population-centric approach, the intelligence section was still optimized for and focused on gaining and processing enemy-centric intelligence only.[52] Despite the fabled Dutch Approach, TFU daily business emphasized "kinetic" operations against the Taliban, a situation that would last until 2007.[53] The potential of the civil assessment and other reports on Uruzgan's societal context remained largely unexploited. The Dutch military needed time to adapt to the challenges of population-centric warfare. With the benefit of hindsight we can conclude that the Ministry of Foreign Affairs, which took the initiative for the civil assessment, could have accelerated this process by adding a sufficient number of civilian experts to the TFU staff and not limiting it to the two diplomats—a political and development adviser—who were initially dispatched to reinforce the soldiers in Uruzgan.[54]

Integration of an understanding of the local society into the military intelligence process and the design of operations improved at the end of 2006 when a civilian tribal adviser was appointed. Still, this adviser, with years of experience in Pakistan and Afghanistan, sometimes encountered stiff resistance when trying to change the military mindset.[55] The so-called Battle of Chora, the largest and most intensive combat action of the Netherlands Army since the Korean War, took place from June 15–19, 2007, and created awareness and an operational sense of urgency to adapt—that is, an "adaptive moment."

The battle involved a massive Taliban attack of an estimated strength of five to fifteen hundred fighters that, because of low-level initiative, was initially countered first by the small detachment of sixty Dutch soldiers in the town of Chora. Immediate reinforcement by additional TFU troops, including Australian soldiers, resulted in a battalion-size TFU fighting force that, along with dozens of air strikes, repelled the Taliban and inflicted approximately two to three hundred casualties upon them.⁵⁶ Not only did this make an important symbolic stand, demonstrating that the Dutch troops would indeed protect the local population if necessary (and in the process redeeming the Netherlands Army from the Srebrenica trauma) but, importantly, the TFU force was augmented by local Barakzai and Achakzai subtribal militias. The willingness of these groups to take up arms against the Taliban and support the TFU was the direct result of fruitful cooperation between the PRT and the tribal adviser, demonstrating the benefit of a proper understanding of the local population.⁵⁷

Meanwhile, civilian and military capacities were subjected to a process of further integration and, as already discussed, the civilian staff was gradually growing. In 2008 the tribal adviser was replaced by two cultural advisers, a university professor and a Dutch-Afghan national.⁵⁸ Interestingly, the TFU command was to be shared by a "duumvirate" of the senior civil representative and the military commander. Thus, integration of military planning with tribal and political analysis became the standard, which further enhanced the understanding of the operational environment.⁵⁹

The consequences of these measures were illustrated by Michel Rentenaar, who served as TFU "senior civil representative" from August 2009 until January 2010, when he stated that the emphasis of TFU operations had indeed shifted to "non-kinetic" operations.⁶⁰ The Dutch task force gradually learned to appreciate and use the population-centric intelligence that had already been available in 2006. Moreover, in addition to the civilian experts, military intelligence officers also became more proficient in the use of ethnographical data and methods.⁶¹ Population-centric intelligence thus became strongly incorporated in TFU's overall intelligence picture.

OPERATIONS

The tension between the imperatives (and political expectations) of the Dutch Approach and the immediate challenges of the tactical environment that made frequent kinetic engagements with Taliban elements unavoidable was only gradually diffused. Initially, as mentioned in the previous section on intelligence, the emphasis in TFU military operations out of habit and operational necessity was on kinetic operations against the Taliban. It maintained a COIN mindset. This situation gradually changed under the influence of the enhanced availability of population-centric intelligence. Eventually nonkinetic operations became the standard.

The work-up program for the TFU elements initially did not reflect the complex S&R-oriented mission, nor the complex operating environment of Uruzgan with a great variety of (civil) actors that would call upon many competencies and skills and much cultural knowledge. Adhering, to a large extent, to existing training programs, the work-up phase of the first deployments emphasized traditional military skills. With subsequent TFU rotation, vital insights were gradually incorporated. However, rather than an institutionalized improvement of the training programs, most adaptations were the result of informal bottom-up initiatives of the respective TFU, BG, and PRT and subordinate commanders who exchanged lessons learned in an informal manner by virtue of their personal acquaintance with each other.

Only from the fourth rotation onward was the training program for PRT personnel intensified and structured in a more formal way. Apart from military skills and drills, the redesigned training program paid more attention to conceptual issues, including a week with lectures on, for example, state building, governance, and development, and a two-week Civil-Military Cooperation (CIMIC) basic course at the Netherlands-based CIMIC Centre of Excellence. In cooperation with Dutch government officials a one-week role play named *Dutch Devotion* was developed to conclude the PRT training. A provincial governor, several mayors, and commanders of the police and fire brigades took part in this exercise and played the roles of Uruzgani key leaders. This enabled PRT personnel to practice their respective roles as if they were operating in Uruzgan.

The Battle Group training programs also initially focused primarily on infantry skills and drills. The program's center of gravity was platoon-level training, as platoons were considered the cornerstone of the organization. Already during the training programs platoons were complemented with engineers, explosive ordnance demolition experts, interpreters, and PRT personnel to practice as if they were working in Uruzgan. Issues that were heavily emphasized in the programs included dealing with ambushes (both with and without IEDs), convoy escorts, the handling of new weaponry and equipment, and improving shooting skills.[62]

Having completed their domain-specific training, the units jointly went through the final training modules. During these modules the Section Cultural Affairs and Information of the Netherlands Army provided a two-day Cultural Awareness Training.[63] On the first day the soldiers visited a mosque where they were informed about Islam. The second day consisted of role plays and lectures on Afghan norms and values. Just before actual deployment, all TFU units went through a joint exercise, labeled *Uruzgan Integration*. In this two-week exercise the units operated in an integrated fashion, practicing real-life challenges that were to be encountered in Uruzgan. These included counter-IED, house searching, information operations, joint planning, and riding convoys.

In line with the work-up program, and despite political rhetoric and the lexicon of stabilization and reconstruction, the first TFU to deploy was (correctly) assuming it would be conducting a counterinsurgency campaign in Uruzgan. Under influence of the Battle Group, counterinsurgency ideas (derived from, among other sources, a long list of classic COIN literature) were fed into the staff process, resulting in the adoption of the "oil-spot" concept and tactical experiments with the platoon house concept.[64]

An initial troop-to-task analysis, however, had already revealed that the TFU structurally lacked the means to maintain the sustained security presence that a counterinsurgency approach required in the ADZ's surrounding Tirin Kot and Deh Rawud (which were to be covered in the initial deployment). Moreover, TFU-1 was ordered to provide substantial support to operation Medusa, a large-scale military operation in Kandahar mounted by Regional Command South (RC-S). In addition, Taliban pressure on the densely populated Chora District forced the Dutch to include the Chora ADZ in their oil-spot prematurely. Consequently the TFU was overstretched from the onset, and this forced the subsequent commanders to disband the idea of permanent security presence. Instead, continuously patrolling and meticulously planned large-scale operations that concentrated the limited means and typically involved the RC-S reserve force would be the modus operandi for the TFU during its four-year deployment.

The execution of patrols remained relatively unchanged during the whole period of the mission in Uruzgan. Typically, an extended platoon (that is, infantry combined with PRT, PsyOps, and enablers such as EOD and forward air controllers) would embark on a multiday venture in which it would make use of either fixed patrol bases or temporarily watch-over locations for rest overnight. While the hard terrain severely restricted freedom of movement—and made patrols very susceptible to IEDs and ambushes, different commanders experimented with more flexible tactics such as mounting satellite patrols by foot from a mobile patrol base, night patrols, and so forth.[65] As these tactical experiments were initiatives of various individual commanders, they cannot be considered structural adaptations. What this does show, however, is that subcommanders were allowed reasonable freedom of action in accordance with the doctrinally propagated "mission-type order" or *auftragstaktik* principle.

In this phase of the mission hard tactical experience forced the TFU to seriously consider force protection. The popularized and often ill-understood focus on "hearts and minds" was supposed to indicate that the Dutch contingent would operate markedly differently from the warrior-style US tactics. In line with experience in the Balkans and Iraq, patrols would be conducted dismounted or in open patrol vehicles, and not in closed armored personnel carriers, which would prohibit making contact with the local population and either intimidate or convey fear and vulnerability. For similar reasons, and only seemingly a trivial detail, troops would wear transparent ski-goggles and not dark wraparound sunglasses, which were considered culturally insensitive and introducing a barrier between the soldier and Afghan civilians.

The threat of IEDs, however, and the frequent incidents with "TICs" (troops in contact) necessitated reverting to a more robust stance, and the acquisition of twenty-five mine-resistant Bushmaster armored vehicles (using a new Fast Track Procurement process created for so-called Urgent Operational Requirements) that would offer sufficient protection against IEDs and Rocket Propelled Grenades (RPGs).[66] In addition, F-16 fighters were equipped with a new reconnaissance pod that could detect road-side bombs. Moreover the air-ground cooperation between F-16s, Apache attack helicopters, and the Forward Air Controllers on the ground was enhanced through better training and the introduction of improved communications equipment, significantly facilitating force protection.

While patrols exposed the Dutch soldiers on the ground immediately to the reality of Uruzgan's fragmented society and tribal feuds, the primary mission they received from higher command was predominantly enemy-centric and typically concerned disruption of the Taliban, with nonkinetic tasks such as enabling PRT-actions and enhancement of the understanding of the local population only as supportive, secondary tasks.[67] This changed gradually with the growing availability of population-centric intelligence to the TFU-staff and its subordinate units. The resulting population-centric mindset is evidenced, for instance, in the following quotation:

> [T]he Taliban are less of a threat to the tottering structures of the Afghan state than feuding local tribes and predatory warlords. The Uruzgan insurgency is mixed up with a notably vicious tribal war between the Popolzai tribe and minority Ghilzai tribes. Jan Mohammad Khan, a Popolzai warlord and former

Uruzgan governor, marginalised the Ghilzais. This seems to have created lasting turmoil which is exploited by the Taliban.[68]

Instead of anti-Taliban patrolling, from 2007 onward patrols subsequently focused increasingly on enabling PRT development activities in order to gain influence over Uruzgan's various communities and liaise the local population to the provincial government. This became the dominant operational mindset within the TFU.[69] While hard to pinpoint a single causal factor for institutionalization of this new approach, the 2008 surge in civilian personnel at least offers the suggestion that the idea of primary population-centric patrols had indeed become accepted. These additional civilian experts were instrumental in providing the hard-needed expertise to efficiently conduct nonkinetic development activities.[70] Although patrols did not evolve dramatically in their appearance during the four-year deployment, the mindset and mission radically adapted to the specifics of the operational environment. Thus patrols became a strong tool for delivering nonkinetic effects.

A similar shift occurred in large-scale operations. Initially, operations were mounted to clear an area of Taliban presence, as there were no means available for follow-up. The consequences of such sweeping actions, such as Operation Perth in July 2006 and Operation Spin Ghar in December 2007, was that the insurgents immediately returned once an operation was finished. Moreover, the local population of such an area considered the Dutch incapable of providing protection and even held ISAF responsible for the insecurity they experienced upon the return of the insurgents.[71] These operations were of limited effect and certainly did not win influence over the population. During the first years of the mission there was only one large-scale operation that definitely won the local population's hearts and minds: the Battle of Chora. As they already maintained a permanent post in the district, the Dutch soldiers were able to exploit their victory through nonkinetic efforts such as cooption with various tribal authorities.[72] This combination of kinetic operations and a nonkinetic follow-up led to a spectacular rise in governmental control and provided the Dutch with a guideline for future operations.

At the end of 2008 sufficient Afghan security forces became available to provide the follow-up for large-scale sweeping operations. Operations were now planned with a clear idea of the ultimate effect: gaining influence over the population in a target area. Although this approach was pioneered during operation Bor Barakai in October 2008, it was not yet effectively practiced, as a follow-up patrol base was not constructed in the target valley but rather positioned at a hilltop on its boundary.[73] Operation Tura Ghar in January 2009 for the first time effectively executed the new approach, as not only a patrol base was established in the middle of the cleared valley but, in addition, establishing "below the radar" contact with key tribal authorities—of whom some were even met in Quetta, Pakistan—was already initiated in the planning phase.[74]

In sum, during the four years of operations in Uruzgan two adaptation processes can be observed. First, tactics were adjusted to the demands of the environment, an environment that dictated a more COIN- and kinetic-oriented approach, in contrast with the politically sanctioned strategy, a reality that took some time for the political level to come to grips with. Second, in particular from 2008 onward, the means available, the tactical learning that had taken place, including the awareness of the utility of deep knowledge of the societal structure, allowed and inspired an approach that combined both the tenets of COIN doctrine and the demands of the S&R mission. The idea of employing force as an enabler for nonkinetic activities in order to gain control over the local population gradually became firmly accepted. Where Dutch soldiers had first cynically joked that the TFU's so-called 3D-approach (Defense, Development, and Diplomacy) was an abbreviation for "Deter, Disturb, and Destroy," the Dutch adapted their patrols and large-scale operations to the requirements of Uruzgan's operational environment and ultimately conducted military operations that embodied all the components of the true 3D-approach.

LESSONS, AND LEARNING TO LEARN

The question that remains unanswered is how, ultimately, the small Dutch military has learned from this major experience. Although operational experiences traditionally remained unprocessed, efforts have been undertaken to institutionalize the operational experiences gained by TFU. This started as early as 2007, when the army's education and training center for operations issued an information bulletin to act as a guideline for counterinsurgency operations.[75] In February 2008 a concept information bulletin entitled "Observations on Operations in Afghanistan" was prepared by army officers who had served in the first two TFUs (including the two commanders, TFU-3 was at that time in the process of rotations), as well as by some officers who served with RC-S (including one former commander).[76] This document was carefully prepared and contained observations on new experiences from the technical, tactical, and operational level. The most important issue the thirty-three-page bulletin stressed was adaptation of the mindset from enemy-centric kinetic operations to nonkinetic population-centric counterinsurgency warfare. In accordance with the traditions of the army it was stressed that the publication should not be interpreted as a prescriptive doctrine; it rather was meant to transfer useful knowledge to future rotations and invited other officers to add their experiences in order to contribute to a new doctrine. Unfortunately, this meticulously prepared paper was never officially distributed within the armed forces, for no obvious reasons. Instead, a series of two articles were derived from it—as a result of the initiative of its authors—which were published only in mid-2009, in the military scientific magazine *Militaire Spectator*.[77] The new army doctrine, which was supposed to contain the insights mentioned in these articles, was published in November 2009, but it seems that the experiences were largely neglected; it is illustrative that the term "counterinsurgency" is scarcely mentioned in this document.[78]

Thus, the observations on operations in Afghanistan were never formally disseminated among the troops, and neither were they processed into the new doctrine. A study undertaken upon the initiative of former RC-S Commander Major-General De Kruijf also reveals this structural weakness of the Dutch military to institutionalize its lessons beyond the informal sphere of the technical and tactical level.[79] An analysis on the evolution of Information Operations and key leader engagement within the TFU illustrates this. While these nonkinetic activities rapidly became a standardized part of the tactical toolbox, the lack of structural embedding caused huge variations in the use of these instruments between the different TFU rotations.[80] Unsurprisingly, therefore, the first mentioned study urges rapid integration of recent operational experience up to the political and military strategic levels and mentions that the staff for evaluations will be strengthened.[81]

The lack of a formal, structured, and institutionalized lessons learned process has recently been acknowledged. A 2010 report lists several factors that have contributed to this state of affairs. First, not all personnel involved in the lessons learned process are actually sufficiently informed about recent developments and new operational insights. Evaluation and assessment are still often considered a secondary task. Second, and as a result, there is a somewhat skeptical attitude toward the products that those who are involved in the lessons learned process publish. It is insufficiently regarded as a tool that is of use for the tactical units, but instead regarded as a staff function where bottom-up ideas land in barren soil. Third, although key insights, results of practices, and ideas are often discussed informally in a variety of forums, these discussions do not result in follow-up analysis, doctrinal changes, or improvements in training and operating procedures, nor other deliberate decision-making. Dissemination of key insights occurred primarily through informal processes. Fourth, there is a disconnect between the lessons learned processes of the various services with each service having its own lessons learned process, dedicated personnel (or lack thereof), and culture. The process is decentralized, and there is insufficient cross-fertilization. The lack of formal lessons learned policy does not help in this respect.

Currently it seems that this call for institutionalization has landed in fertile soil. The army has called upon all former TFU commanders to identify crucial lessons that should be preserved for future operations.[82] In a similar fashion the Department of Evaluations of the Ministry of Defence has conducted an extensive analysis of the four-year Uruzgan campaign in order to collect lessons identified and best practices. The resulting report, which appeared in July 2011, is meant to be the first step to institutionalize these experiences and process them into lessons learned to be used in future operations.[83]

At the same time, upon request of the Ministries of Foreign Affairs and Defence, a committee of independent experts have carried out an extensive evaluation of the entire Dutch contribution to Uruzgan, thus including the efforts of Foreign Affairs.[84] This report, which appeared in late September 2011, focuses on the strategy of TFU, the operations, the results that were achieved, and the costs of the entire mission. The report concludes with a long list of lessons identified. A final step to improve and institutionalize the lessons learned processes will be the formation of service-specific Warfare Centers.

CONCLUSION: ADAPTATION IN URUZGAN . . . AND BEYOND?

What insights can we obtain from these analyses of various aspects of four years of campaigning in Uruzgan, which cost the Dutch taxpayer approximately 2 billion Euro? First, it is important to notice that the mission was politically framed in terms of reconstruction and development. Although the reality on the ground was that of a counterinsurgency campaign in a highly fragmented societal environment, the label of reconstruction and development linked the TFU mission to newly developed (though grounded in previous policy documents) Dutch policy on how to conduct such operations in fragile states. This conceptual framework was designed by the ministries of Defence, Foreign Affairs, and Economic Affairs and was based on experiences in previous missions.

Thus TFU strategy included providing security, fostering governance, and socioeconomic development from the onset. As such, a comprehensive approach was superficially entirely in line with the tenets of COIN doctrine, and seemingly there was no disconnect between political mandate and military strategy. However, whereas declared policy stressed the civilian side of the strategy, the military plan of necessity, and by inclination, initially emphasized the military aspects of the mission and considered it a COIN mission, a reality that only gradually was accepted at the political level. On the other hand, when the situation became more benign, the operation gradually obtained the character of the S&R mission, as originally envisioned in 2005.

During the entire campaign, strategy was formulated by the TFU staff at the grass roots level. This allowed for the initial strategy to be further developed with the reality of the operational environment as a guideline. The 2008 Focal Paper illustrates this as it adapts the initial strategy into a campaign plan for counterinsurgency in Uruzgan. In addition, the main driver for the change of strategy was the political decision from The Hague in 2008 to extend the mission, which in turn informed The Focal Paper, and the 2009 UCP was meant to provide a guideline for the transfer of authority in 2010. Thus, while TFU strategy certainly was adapted to meet the requirements of a plan for counterinsurgency in the specific operational environment of Uruzgan, domestic politics and government policy were the additional main drivers of this adaptation.

Second, a similar dually driven adaptation process can be observed concerning the organization. The TFU organization was initially tailored for a reconstruction and development mission with the BG and the PRT as its two main components. Uruzgan's contentious environment, however, forced an extension of BG assets, and thereby initially more emphasis was placed on the kinetic parts of the organization. Troop levels were increased up to a permanent contribution of fourteen hundred soldiers (a limit imposed by the political level in The Hague) and as many as six hundred temporarily deployed additional troops. Although the TFU strategy resulted from a comprehensive conceptual framework, involvement of civilian actors was minimal at first. The largest adaptation of the organizational aspect, therefore, was the influx of civilian personnel. The year 2008 especially saw a surge in civilian personnel and a radical change in the overall command structure as the senior civil representative became the commander of the PRT and the co-commander of the TFU. Moreover, the gradual increase of NGOs contributing to the mission and the establishment of the DCU further augmented the civilian capacity needed for the execution of a comprehensive campaign.

Another significant adaptation occurred in the field of intelligence. Despite the framework of reconstruction and development and the fact that Dutch soldiers rapidly recognized that they were fighting a counterinsurgency campaign, the focus in intelligence initially remained with the enemy. In Uruzgan, however, the Taliban was not the main reason for insecurity. The divide-and-rule politics of former Governor Jan Mohammed had caused violent contention, and the true problem was to realign marginalized societal fragments with the provincial government and prevent the Taliban insurgents from exploiting local feuds. To understand this required the population-centric intelligence typical for counterinsurgency campaigns. Under influence of civilian expertise a gradual shift occurred in the intelligence process. Especially the tribal adviser, and later the cultural advisers, provided useful insights on the local society and convinced military intelligence officers of the necessity of population-centric intelligence. Thus the emphasis in intelligence shifted from the Taliban to the local population.

The field of operations witnessed a similar shift under influence of the enhanced understanding of the operational environment provided by population-centric intelligence. At first kinetic activities were considered the primary tool to realize the TFU mission as patrols and large-scale operations were mounted to restore security by inflicting damage upon the Taliban. Although these actions remained much the same in physical appearance, the rationale underlying them changed radically. As the Dutch soldier came to realize that influence over the population in this environment had to be won by nonkinetic means, the idea of the use of force as an enabler for nonkinetic activities became firmly accepted within the TFU. In addition to a better understanding of the locale, the enhanced availability of ANSF was also of pivotal importance for this adaptation of military operations for counterinsurgency purposes.

Finally, the adaptation of the predeployment training program points at a remarkable feature of the small Dutch military. Despite the lack of a formal organization to institutionalize lessons learned, informal as well as professional personal contacts led to a quick adaptation of the training program. This made possible the use of the experiences of previous rotations in order to establish a tailor-made program that prepared the troops and civilians for the conduct of an integrated counterinsurgency campaign in Uruzgan's demanding operational environment.

To conclude this chapter, we can say that the Dutch military successfully improved its operational performance by adapting to the specific requirements of the four-year Uruzgan campaign. Although the modern Dutch military had never conducted a brigade-size operation, formulating a strategy for such a unit and fielding and sustaining a capable organization did create relatively few problems. Adapting to the reality of a comprehensive counterinsurgency campaign was more difficult and initially constrained by the political rhetoric and limited conceptualization of the mission, but was in the end successfully completed with the help of additional civilian support. The main drivers in the period were the political level in the Hague and the demanding operational environment in Uruzgan itself.

As of now the small Dutch military is facing another round of budget cuts. It has already been announced that tanks as a weapon system will be scrapped from the arsenal, and even the overstretched transport helicopter and F-16 fleet will suffer huge reductions. Although the Dutch military has taken its first steps to actually institutionalizing the major Uruzgan experience, the question whether or not sufficient funds will be available for this process might obstruct the completion of the learning cycle.

NOTES

1. Netherlands Ministry of Defence, *Lessons Identified ISAF, Final Report*, July 14, 2011, 60.

2. See, for instance, Netherlands Ministry of Defence, *Rapport Lessons Learned binnen Defensie* [*Report Lessons Learned within the Ministry of Defense*] (The Hague: Netherlands Ministry of Defence, February 17, 2010).

3. R. H. Sandee and P. W. W. Wijninga, "Lessen uit recente operaties," in E. R. Muller et al., eds. *Krijgsmacht, Studies over de organisatie en het optreden* ["Lessons from Recent Operations," in *Armed Forces: Studies on the Organisation and Operation*] (Alphen aan de Rijn: Kluwer, 2004), 719–20.

4. Rob de Wijk and Frans Osinga, "Innovation on a Shrinking Playing Field: Military Change in The Netherlands Armed Forces," in T. Terriff, F. Osinga, and T. Farrell, eds., *A Transformation Gap? American Innovations and European Military Change* (Stanford: Stanford University Press, 2010), 142.

5. Thijs Brocades Zaalberg and Arthur ten Cate, *Missie in Al Muthanna, De Nederlandse krijgsmacht in Irak 2003–2005* [*Mission in Al Muthanna: The Dutch Armed Forces in Iraq 2003–2005*] (Amsterdam: Boom, 2010), 297–302.

6. George Dimitriu and Beatrice de Graaf, "The Dutch COIN Approach: Three Years in Uruzgan, 2006–2009," *Small Wars and Insurgencies* 21, no. 3 (2010): 435.

7. For an excellent recent analysis of the strategic culture of the Netherlands, see

Rem Korteweg, "The Superpower, the Bridge-builder and the Hesitant Ally," Ph.D. diss. (Leiden: Leiden University Press, 2011), chs. 18, 20.

8. Christ Klep, *Uruzgan: Nederlandse militairen op missie, 2005–2010* [*Uruzgan: Dutch Military at Mission, 2005–2010*] (Amsterdam: Boom, 2011).

9. NRC-Handelsblad, September 19, 2007.

10. Korteweg, "The Superpower, the Bridge-builder and the Hesitant Ally," 288.

11. Letter to Parliament of the Dutch Minister of Defence, *Nederlandse Bijdrage aan ISAF in Zuid-Afghanistan* [*Dutch Contribution to ISAF in Southern Afghanistan*], The Hague, December 22, 2005, 3–4.

12. Klep, *Uruzgan.*

13. Jan van der Meulen, "Stemmen over Afghanistan en de risico's van het vak" ["Voices over Afghanistan and the Risks of the Job"], *Militaire Spectator* 178, no. 3 (2009): 135–45.

14. As quoted in "Nederland onder druk" ["The Netherlands under Tressure"], *NRC Handelsblad*, October 23, 2009.

15. Martine van Bijlert, "Unruly Commanders and Violent Power Struggles," in Antonio Giustozzi, ed., *Decoding the New Taliban: Insights from the Field* (New York: Columbia University Press, 2009), 155.

16. Susanne Schmeidl, *The Man Who Would Be King: The Challenges to Strengthening Governance in Uruzgan* (The Hague: Cingendael, 2010), 10.

17. The Liaison Office, *The Dutch Engagement in Uruzgan: 2006–2010* (Kabul: TLO, 2010), 3. The percentages are estimates in relation to the total population.

18. Schmeidl, *The Man Who Would Be King*, 10.

19. van Bijlert, "Unruly Commanders and Violent Power Struggles," 156–57; Martine van Bijlert, interviewed by Martijn Kitzen, Tarin Kowt, November 16, 2008.

20. Jan Mohammed was assassinated on July 17, 2011.

21. Antonio Giustozzi, *Koran, Kalashnikov, and Laptop: The Neo-Taliban Insurgency in Afghanistan* (New York: Columbia University Press, 2008), 56–58. See also van Bijlert, "Unruly Commanders and Violent Power Struggles," 156–58; and Bette Dam, *Expeditie Uruzgan, De weg van Hamid Karzai naar het paleis* [*Expedition Uruzgan: Hamid Karzai's Road to the Palace*] (Amsterdam: Uitgeverij de Arbeiderspers, 2009), 37–40.

22. Schmeidl, *The Man Who Would Be King*, 51.

23. van Bijlert, "Unruly Commanders and Violent Power Struggles," 158.

24. Dutch Minister of Defence, *Nederlandse Bijdrage aan ISAF in Zuid-Afghanistan* [*Dutch Contribution to ISAF in Southern Afghanistan*], The Hague, December 22, 2005, 10.

25. Netherlands Ministry of Defence, "Defensienota 2000" ["Defense White Paper 2000"], The Hague, 26.

26. Tweede Kamer, *Nederlandse Deelname aan de Vredesmissies* [*Dutch Participation in Peace Operations*], vergaderjaar 2003–4, 29521, nr. 1, April 8, 2004.

27. Tweede Kamer, *Herijking van het Buitenlands Beleid: Brief van de Ministers van*

Buitenlandse Zaken, van Economische Zaken, voor Ontwikkelingssamenwerking, van Defensie en van Financiën [*Alignment of Foreign Policy: Letter of the Ministers of Foreign Affairs, Economic Affairs, Defence and Finance*], 24337, nr. 2—1994–1995.

28. Netherlands Ministries of Foreign Affairs, Defence, and Economic Affairs, *Notitie: wederopbouw na gewapend conflict* [*White Paper on Post-conflict Reconstruction*] (The Hague: Koninklijke De Swart, 2005).

29. Dutch Minister of Defence, *Dutch Contribution to ISAF in Southern Afghanistan*.

30. General Dick Berlijn, speech at the departure of the F-16 detachment to Afghanistan, January 9, 2006.

31. Korteweg, "The Superpower, the Bridge-builder and the Hesitant Ally," 294.

32. Hans de Vreij (RNW), *Briljant citaat* [*Brilliant Quotation*], March 6, 2009, http://blogs.rnw.nl/vredeenveiligheid/2009/03/06/briljant-citaat, accessed May 24, 2011.

33. Task Force Uruzgan, *1 (NLD/AUS) Task Force URUZGAN MASTER PLAN* (Tarin Kowt: TFU, 2006); Ministries of Foreign Affairs and Defence, *Tussentijdse evaluatie ISAF 2008* [*Interim Evaluation ISAF 2008*] (The Hague: Ministries of Foreign Affairs and Defence, 2008).

34. Task Force Uruzgan, *1 (NLD/AUS) Task Force URUZGAN MASTER PLAN*; B. J. E. Smeenk, R. W. G. Gouweleeuw, and H. C. van der Have, "Effect gebaseerde aanpak in Uruzgan. Van het schaakbord naar een bord spaghetti" ["Effects-based Approach in Uruzgan: From Chess-board to Plate of Spaghetti"], *Militaire Spectator* 176, no. 12 (2007): 550–59.

35. Task Force Uruzgan, *Focal Paper: Foundations for the Future* (Tarin Kowt: TFU, 2008).

36. All quoted in G. R. Dimitriu and Beatrice de Graaf, "De Nederlandse COIN-aanpak: 3 jaar Uruzgan, 2006–2009," *Militaire Spectator* 178, no. 11 (2009): 616.

37. J. R. Schwillens, "Comprehensive Approach: de praktijk" ["Comprehensive Approach: The Practice"], *Militaire Spectator* 178, no. 11 (2009): 578–89.

38. The formulation is interesting, as it presumes a prescription of the Dutch ministries to the Government of Uruzgan that effects and milestones should be pursued.

39. Task Force Uruzgan, *Uruzgan Campaign Plan 2010* (Tarin Kowt: TFU, 2008); I. E. van Bemmel, A. R. Eikelboom, and P. G. F. Hoefsloot, "Comprehensive and Iterative Planning in Uruzgan," *Militaire Spectator* 179, no. 4 (2010): 196–209.

40. Task Force Uruzgan, *Uruzgan Campaign Plan 2010*; van Bemmel et al., "Comprehensive and Iterative Planning in Uruzgan." In the figure, LN means local national, and KL means key leader.

41. B. R. Bot, H. G. J. Kamp, and A. M. A. van Ardenne—Van der Hoeven, *Kamerbrief 22 December 2005: Nederlandse Bijdrage aan ISAF in Zuid-Afghanistan* [*Letter to Parliament 22 December 2005: The Dutch Contribution to ISAF in Southern Afghanistan*] (The Hague: Ministries of Foreign Affairs, Defence and Development Cooperation, 2005).

42. Piet van der Sar, "Kick the Enemy Where It Hurts Most," *Carré* 1 (2007): 10–17.

43. Kees Mathijssen and Peter Mollema, *De civiele organisatie in Task Force Uruzgan: Het Provincial Reconstruction Team (PRT) van de Task Force Uruzgan onder civiele Leiding* [*The Civil Organization in Task Force Uruzgan: The Provincial Reconstruction Team (PRT) of the Task Force Uruzgan under Civil Command*] (Tirin Kot: TFU, 2008).

44. Or "Zuidoost-Azië," meaning "Southeast Asia" in Dutch. The ZOA website points out that the increased geographical reach of ZOA beyond their original area of operations has meant this acronym has "lost some of its original meaning." See ZOA Profile: available at http://www.zoa-international.com/content/profile, accessed July 12, 2012.

45. ZOA Projects: available at http://www.zoa.nl/home/projecten/afghanistan/, accessed May 18, 2011.

46. Available at http://www.reliefweb.int/rw/rwb.nsf/db900SID/RMOI-7RVHZG?OpenDocument, accessed May 13, 2011.

47. David Kilcullen, "Intelligence," in Thomas Rid and Thomas Keaney, eds., *Understanding Counterinsurgency, Doctrines, Operations and Challenges* (London: Routledge, 2010), 142–43, 157.

48. Tweede Kamer, *Kamerstuk 2005–2006* [*Parliamentary Letter 2005–2006*], 27925, no. 221; see also Anonymous, "A Survey of Uruzgan Province," unpublished report (Kabul, 2006).

49. See Royal Netherlands Embassy in Kabul, *Civil Assessment* (Kabul, 2006); and Royal Netherlands Embassy in Kabul, *Context Analysis Uruzgan Province* (Kabul, 2006).

50. Kamer, *Kamerstuk 2005–2006, no. 213*; see also The Liaison Office, *The Dutch Engagement in Uruzgan*, 29.

51. Militaire Inlichtingen en Veiligheidsdients (MIVD), "SupIntrep Afghanistan, Stamverhoudingen in Uruzgan" ["SupIntrep Afghanistan, Tribal Relations in Uruzgan"], unpublished report (The Hague, 2006).

52. See Anonymous, "Leven in het Oude Testament" ["Living in the Old Testament"], in Wiebren Tabak, ed., *3D, de Nederlandse militaire inzet in Afghanistan* [*3D: The Dutch Military Engagement in Afghanistan*] (The Hague: Ministerie van Defensie, 2010), 15–16; and Smeenk et al., "Effect gebaseerde aanpak in Uruzgan," 555.

53. Hans Ariëns, "Dutch Approach klinkt nogal zelfvoldaan" ["Dutch Approach Sounds Kind of Smug"], *Internationale Samenwerking* 2010–11: available at http://www.ismagazine.nl/2010/02/17/dutch-approach-klinkt-nogal-zelfvoldaan/, accessed October 27, 2011.

54. Kamer, *Kamerstuk 2005–2006, no. 201*, 46.

55. Anonymous tribal adviser, interviewed by Martijn Kitzen, The Hague, March 1, 2010.

56. For detailed accounts of this important event, see Noël van Bemmel, "Infanteristen, commando's: iedereen vecht tegen Taliban" ["Infantry, Special Forces: Everybody Fights the Taliban"], *Volkskrant*, June 23, 2007; Nicolas Asfouri, "Over 100 Die in South-

ern Afghan Battle," *USA Today*, June 19, 2007; Paul Brill, "Ze schoten een magazijn op me leeg" ["They Shot a Round at Me"], *Volkskrant*, June 28, 2007; "Taliban hielden als beesten huis in Chora" ["Taliban Were Behaving like Animals in Chora"], *De Telegraaf*, June 22, 2007; Eric Vrijsen, "Uruzgan: het gevecht om Chora" ["Uruzgan, the Battle for Chora"], *Elsevier*, January 2007.

57. Commander PRT-3, Colonel Gino van der Voet, interviewed by Martijn Kitzen, The Hague, March 9, 2010, and Anonymous PRT-3 staff officer, interviewed by Martijn Kitzen, Wezep, September 21, 2009.

58. Ariëns, "Dutch Approach klinkt nogal zelfvoldaan."

59. Tweede Kamer, "Tussentijdse Evaluatie ISAF 2008" ["Interim Evaluation ISAF 2008"], annex of *Kamerstuk 2008–2009, 27925, no. 357*, 33–34.

60. Hans Ariëns, "Interview Michiel Rentenaar" ["Interview with Senior Civil Representative Michiel Rentenaar"], *Internationale Samenwerking*, 2010–3.

61. It must be mentioned that throughout the mission there was only one dedicated military human factors analyst in the TFU staff's vast intelligence section. Typically this would be an academically schooled junior officer. When awareness of the importance of population-centric intelligence grew, other intelligence officers also became involved with so-called white plate (population-centric) intelligence.

62. van der Sar, "Kick the Enemy Where It Hurts Most," 10–17.

63. Bas Ooink, "Cross Cultural Training: een nieuw model voor cultuurtrainingen" ["Cross Cultural Training: A New Model for Cultural Training"], *Militaire Spectator* 179, no. 3 (2010): 133–45.

64. Brigadier Theo Vleugels, interviewed by Martijn Kitzen, Utrecht, November 12, 2009; Colonel Piet van der Sar and Captain Ralph Coenen, interviewed by Martijn Kitzen, Weert, September 28, 2009.

65. Colonel Richard van Harskamp, "Observaties van Commandant Task Force Uruzgan IV" ["Observations of Commander Task Force Uruzgan IV"], *Infanterie* 14, no. 1 (2009): 7; Captain Ralph Coenen, interviewed by Martijn Kitzen, Tirin Kot, November 20, 2008; Martijn Kitzen, "Uruzgan Field Notes" (unpublished personal record, Tirin Kot, 2008), 45–48; Jerry Meyerle et al., *Counterinsurgency on the Ground in Afghanistan: How Different Units Adapted to Local Conditions* (Alexandria, Va.: CNAS, 2010), 145–50.

66. Korteweg, "The Superpower, the Bridge-builder and the Hesitant Ally," 295.

67. See, for instance, Lieutenant Colonel Andy van Dijk, "Personal Diary" (unpublished personal record, Tirin Kot, 2007); and Captain Gijs-Jan Schüssler, "Experiences Platoon Commander TF-7 in Uruzgan" (unpublished presentation for *13 Mechanized Brigade Counterinsurgency Seminar*, Breda, September 3, 2008).

68. Anonymous, "The Dutch Model," *Economist* (March 12, 2009).

69. Ariëns, "Dutch Approach klinkt nogal zelfvoldaan."

70. Peter Mollema and Cees Matthijssen, "Uruzgan: op de goede weg, civiel-militaire samenwerking in een complexe counter-insurgency operatie" ["Uruzgan: On the

Right Track, Civil-military Cooperation in a Complex Counterinsurgency Operation"], *Militaire Spectator* 178, nos. 7/8 (2009): 402.

71. Dimitriu and De Graaf, "The Dutch COIN Approach," 441.

72. Martijn Kitzen, "Close Encounters of the Tribal Kind: The Implementation of Co-option as a Tool for De-escalation of Conflict: The Case of the Netherlands in Afghanistan's Uruzgan Province," *Journal of Strategic Studies* 35, no. 5 (2012): 725–26.

73. Kitzen, "Uruzgan Field Notes," 106–7.

74. Kitzen, "Close Encounters of the Tribal Kind."

75. Education and Training Centre for Operations, "Informatiebulletin 07/2, Counter-Insurgency (COIN) en de Militaire Bijdrage" ["Information Bulletin 07/2, Counter Insurgency (COIN) and the Military Contribution"] (unpublished Army Information Bulletin, Amersfoort, 2008).

76. See Education and Training Centre for Operations, "Concept Informatie Bulletin 08/01, Observaties over Operaties in Afghanistan" ["Concept Information Bulletin 08/01, Observations on the Operations in Afghanistan"] (unpublished Army Study, Amersfoort, 2008).

77. See Lieutenant-Colonel P. B. Soldaat et al., "Observaties rond operaties in Afghanistan I and II" ["Observation on the Operations in Afghanistan"], *Militaire Spectator* 178, nos. 5/6 (2009): 252–66, 340–49.

78. See Education and Training Centre for Operations, *Land Doctrine Publicatie, Militaire Doctrine voor het Landoptreden* [*Land Doctrinal Publication, Military Doctrine for Land Based Operations*] (Amersfoort: OTCOpn, 2009).

79. Defence Staff, *Van Eredivisie naar Europees Voetbal* [*From Premier League to European Soccer*] (unpublished Nota, The Hague, 2010), 76.

80. See Lieutenant-Colonel Hans van Dalen, "Key Leader Engagement, Influence by Proxy" (unpublished thesis, The Hague, 2010), 139–47.

81. Defence Staff, *Van Eredivisie naar Europees Voetbal*, 76.

82. An initial confidential report was promulgated on November 15, 2010.

83. *Lessons Identified ISAF*, Internal Report, The Hague, July 14, 2011. This report incidentally confirms many of the observations made in this chapter. As the report is classified, no specific reference can be made.

84. A. de Ruijter, P. C. Feith, J. Gruiters, and M. L. M. Urlings, *Eindevaluatie Nederlandse Bijdrage aan ISAF, 2006–2010* ["Final Evaluation of the Dutch Contribution to ISAF, 2006–2010"] (The Hague: Ministries of Foreign Affairs and Defence, 2011).

8 Mission Command without a Mission: German Military Adaptation in Afghanistan

Thomas Rid and Martin Zapfe

In August 2008, a sergeant of the German Military Police (*Feldjäger*), manning an International Security Assistance Force (ISAF) checkpoint, mistook an approaching Afghan car for a vehicle-borne Improvised Explosive Device (IED). The soldier opened fire with the 7.62mm MG3 machine gun mounted on his *Dingo*-vehicle. Moments later, an Afghan woman and her two children were dead.[1] What happened after this tragic incident was emblematic of the German military experience in Afghanistan in its first seven years: the sergeant was subjugated to preliminary proceedings of a civilian German state attorney in his hometown, Frankfurt an der Oder. The proceedings on the suspicion of multiple involuntary homicides took several months, during which time the whole incident had been staged and replayed at a German training ground to give the civil attorneys a grasp of what had happened. At the end, the sergeant was cleared of any suspicion—according to German criminal law. The state attorney believed the soldier's statement that he acted in assumed defense of his comrades.

In the summer of 2010, after weeks of intensive and deliberate fighting, German infantry of the Quick Reaction Force 5 (QRF-5) defended a newly established outpost against numerous insurgents in the heavily contested Kunduz-Baghlan corridor. Supported by Afghan and US forces, the determined attackers were repulsed and, in a coordinated counterattack, finally routed, leaving several of them dead. Instead of being repatriated and investigated by civil attorneys, the QRF-commander became the twenty-second soldier, and the first staff officer, to receive the newly established cross of bravery from the minister of defense himself.

What has happened in the two years between the two incidents?

The two incidents illustrate the way the Bundeswehr operates in Afghanistan has undergone a significant change. This article therefore aims two answer two central questions. First: How did the German armed forces adapt to the operational challenges they encountered in Afghanistan? The Bundeswehr's contribution to NATO's (the North Atlantic Treaty Organization's) ISAF was the largest, the longest, and by far the most costly military operation that the Federal Republic of Germany has engaged in to date. The war has evolved significantly in the ten years since German special operations forces first set foot into Afghanistan, and so has the German military's approach to counterinsurgency (COIN)—even if American and British onlookers may get the impression that this adaptation started somewhat belatedly and then proceeded more slowly than in other allied countries. That impression would be correct: we find that the defining phase of adaptation in the German land forces began only in 2008 and 2009, seven years after the start of the war. That delay is longer than the entirety of World War II. Hence the second question: Why did the Bundeswehr, and especially the *Heer* (Army), change its ways that late and that slowly? In order to answer both questions, this article will explore the most significant drivers and, perhaps more important, the most significant constraints of the German Army's adaptation in Afghanistan.

The main argument is straightforward: German strategic culture produced inertia and stasis. This is not to neglect agency: political leaders in Berlin, as well as the nation's top military officers, hindered and delayed adjustment of tactical routines. Yet they did so while being collectively tied down by cultural constraints. These cultural constrains became visible as political limits and then translated into strategic, operational, and then tactical brakes. This is not to imply that bottom-up pressure was absent or that actual change did not happen. But in the face of what may be called strategic passivity—passivity of both civilian and military leaders at the top—such lessons remained confined to the level of companies and battalions. Innovation in German operations therefore ran the risk of remaining localized, short-lived, and highly dependent on individual entrepreneurial tactical leaders. The larger consequences are easy to spot: tactical improvisation may not receive the operational, strategic, and political backing it needs in order to rise to the level of *organizational* adaptation. In short: the German armed forces appear institutionally ill-equipped to absorb, retain, process, develop, and deploy the lessons it learned. If the Bundeswehr finds itself in a similar type of conflict a few years down the road, it therefore risks starting from scratch. But the frail lessons of Afghanistan demanded a heavy price paid with lost lives and limbs and, again, the personal trauma of war for many veterans. That price, we argue, has *politically* not been high enough to spur a reluctant elite into more unilateral action, either full withdrawal or full commitment; *as a consequence*, because of this reluctance not despite of it, Germany's experience with war in Afghanistan has been long enough and intense enough to start a process of *social and cultural* change of Germany's collective attitude vis-à-vis the use of force.

The argument is organized in three steps. First, the distinctive cultural background of German military adaptation will be analyzed and put into context. Second, the operational development of the Bundeswehr's mission in Northern Afghanistan will be portrayed, showing the stagnation before and the increased adaptation after the years 2008/9, respectively. Third, the strategic factors that hampered operational adaptation before these defining years will be presented and analyzed for their ongoing or diminished influence on Bundeswehr adaptation within and beyond the ISAF operation.

CULTURAL BACKGROUND

German "strategic culture"—"the sum total of ideas, conditioned emotional responses, and patterns of habitual behavior"[2] toward the use of military force in any circumstance—has distinctive consequences in real life. Following its total defeat in World War II, Germany has developed perhaps a unique and highly institutionalized cultural attitude toward the use of force.[3] When it comes to the state's use of armed force, even after the Cold War, prevailing public attitudes in Germany as well as the views of political elites in Berlin across party lines have been markedly different from those of their allies, most notably the United States.[4] The country's deeply rooted self-perception still is that German militarism brought unprecedented destruction and war to Europe—not once but twice during the bloody course of the twentieth century. Therefore the concept of a "civilian power" has a lot of currency in Germany.[5] That idea translates into "a culture of military restraint,"[6] as Foreign Minister Guido Westerwelle stated during the Libya crisis. Westerwelle's statement expresses a specific German attitude vis-à-vis military operations: any use of violence for political goals appears somewhat uncivilized, even under circumstances that should leave little room for doubt.[7] Germans apply the highest moral standards of humanitarian law to themselves first.[8] An extreme and therefore illustrative example is Muammar Gaddafi's expressed intention to slaughter the opposition in Benghazi, a rather strong casus belli for France, Britain, the United States, and many other NATO countries. Yet Germany refused to back the United Nations (UN) resolution authorizing the use of force. These attitudes may be changing slowly, and within Germany the government's position was highly controversial. But for the time being, and for analytical purposes, Germany's culture of military restraint has to be treated as a stable backdrop of military operations abroad.[9]

In concurrence with this restrained strategic culture, the German constitution equips this culture of military restraint with an extraordinarily brawny institutional arm: parliament. In 1949, the drafters of the Federal Republic's basic law did not foresee a renewed German Army. When the heated debate about German rearmament within NATO culminated in the decision to establish the Bundeswehr in 1955/56, parliament placed specific limits on the use of force. The new army was to be used for territorial defense only. The few exceptions then foreseen by lawmakers allowed for only unarmed humanitarian operations. This arrangement worked fine during the Cold War, when the Bundeswehr's sole purpose, standing up to armed Soviet aggression, never became a reality. After 1990, however, more and more expeditionary operations called for a new arrangement for civilian control of Germany's armed forces. Deploying the military involves parliamentary approval in one way or another in most Western democracies. Yet the parliament in Berlin, the Bundestag, has an exceptionally strong hand in military matters. So strong, in fact, that the modern German language, inspired by a landmark Constitutional Court ruling in 1994, has adopted a specific word for the relationship, the *Parlamentsheer*,[10] the parliamentary army. The Bundestag authorizes every armed deployment, specifying, inter alia, the exact mission, troop ceilings, and caveats.[11] Furthermore, it has become customary for legislators to grant an extension of a so-called mandate for only twelve months, unless the schedule was adapted to federal elections (the majority of Berlin's political elite was keen not to make Afghanistan a campaign issue). In addition, parliament reserved the right to end every operation for exceptional reasons at a time of its choosing.[12]

The German parliament may have plenty of political brawn in military affairs, but it doesn't seem to have a lot of brains. The German parliament, in sharp contrast to the US Congress, doesn't hold public hearings with generals and admirals discussing ongoing operations. Such hearings do take place, but officers meet lawmakers behind closed doors, without the public or journalists taking notice. When asked for the reasons for holding closed hearings, one of the Bundeswehr's top generals who regularly briefs parliamentary committees said candidly that this way the lawmakers' embarrassing lack of knowledge and expertise could easily be concealed (and he adds that occasionally the officers on the other side of the green table may benefit in the same way).

Current and former commanders with experience in Provincial Reconstruction Teams (PRTs) or on a Regional Command level agree: yes, German politicians have an honest interest in the operation in Afghanistan and the well-being of the German soldiers; but they mostly lack even the most basic military knowledge and insights into the conflict's local dynamics, let alone Afghanistan's mired regional politics. Such a differential of knowledge is to be expected in any protracted counterinsurgency campaign. Yet the gap is gaping exceptionally wide in Germany. One explanation for the legislators' predominant lack of notable expertise may be the federal government's predominant lack of notable interest in the Afghan campaign. Keeping Afghanistan on a low public profile, it should be noted, may be politically prudent: for many years, a common refrain among foreign policy experts has been that German politicians would "fail to explain to the public why we are in Afghanistan." But such an argument may be politically shortsighted. Arguably it was precisely the lack of public attention that has allowed many a government to *maintain* an ongoing commitment in Afghanistan against widespread but largely nascent popular doubt in the rationale of the operation. But this attitude has repercussions on operational performance. The result was that neither the government nor the Bundestag has been a driver of military adaptation. To the contrary: when Germany prepared to deploy elements of the *Panzergrenadierbataillon 212* and a mortar unit of the *Fallschirmjäger* as a QRF for Regional Command North (RC-N)—seven years into the conflict in mid-2008, the first German maneuver unit independent from the PRT—one social democratic member of parliament called up a member of the defense committee to plead: "Promise me that they won't shoot once they're down there!"[13]

This cultural and political backdrop had a number of more immediate consequences for the Bundeswehr's ability to learn and adapt. One consequence was that it muffled the public debate about war. Germany entered the war in Afghanistan in late 2001, when German *Kommando Spezialkräfte* (KSK) soldiers supporting Operation Enduring Freedom (OEF) fought against remnants of the Taliban and al-Qaida in the regions bordering Pakistan, little of which was well known publicly in Germany. Then, after the UN Security Council authorized ISAF, Berlin quickly became the third-largest troop contributor. At the time of writing, more than fifty-one hundred German soldiers were stationed in Afghanistan's Northern provinces as well as in neighboring Uzbekistan. By December 2011, Germany had lost fifty-three soldiers in battle.[14]

Yet, until 2011 politicians and opinion leaders shunned the idea that German soldiers were engaged in actual fighting, let alone fighting a war. Some insight into the nuances of the German language on post-1945 martial affairs may be useful. A common German word for combat is *Kampfeinsatz*, which may be literally translated as "battle engagement." The word used when a soldier is killed in action is *gefallen*, an expression with somewhat of a vintage aura, as no German soldiers had "fallen" in battle since 1945. Until 2009, politicians of the grand coalition then in charge in Berlin shunned both expressions, preferring instead to refer to the Bundeswehr's Afghanistan operation as a "stabilization mission" or "reconstruction." But slowly mounting fatalities increasingly put Germany's military engagement into the public spotlight. On June 23, 2009, for instance, three Bundeswehr soldiers were killed in a firefight with insurgents when their *Fuchs* armored personnel carrier overturned into a river. The funeral was held in early July in Bad Salzungen, a small town in Thuringia. Franz Josef Jung, then the minister of defense, attended the ceremony. "They were good soldiers and real patriots," the minister said, and added, "We are in a stabilization mission and, yes, it is true, we are in situations of combat, insofar as this is also a *Kampfeinsatz*."[15] Jung's rhetoric initiative was a significant step on the public debate on Afghanistan, and many newspapers gave his statement high public visibility. In October 2009, Jung again broke new ground in the German debate when he started talking about *gefallene Soldaten*.[16] But the minister drew a sharp line when it came to talking about war. *Krieg* is a heavy word in German. It still triggers painful memories of bombed-out cities and the noise of air-raid sirens. In Jung's view, calling the "stabilization mission" a war would play into the Taliban's hands. The Islamic militants, he argued, would like to spin the operation as a holy war. "They want us to talk about war," he said about the Taliban at the cemetery in Bad Salzungen, "because then they are combatants and can shoot at us with justification."[17] Therefore German politicians should see the operation as a *Kampfeinsatz*, a fighting mission, but not as war: *kein Krieg*.[18] The minister called the ongoing discussion if Germany was actually in a war "irresponsible." Not just the military situation would be decisive but also civilian reconstruction and gaining the trust of the people, he said, echoing a core tenet of the counterinsurgency theory that had then been widely discussed in the US defense establishment. When Guttenberg took over the defense ministry from Jung in 2009, he broke that taboo. At age thirty-eight, media-savvy, and energetic, the minister with star potential forged ahead in an interview with Germany's largest tabloid, *Bild*: "I do understand every soldier who says: 'This is war in Afghanistan, no matter if I get attacked, wounded or killed by foreign armed forces or Taliban terrorists.'" Guttenberg added, in a widely quoted line, "Without question, there are warlike conditions [*kriegsähnliche Zustände*] in parts of Afghanistan."[19] Even in April 2010, about half a year after that statement, it was still worthy of a headline in Germany's leading newspapers when the defense minister used the word "war" after three soldiers were killed in an ambush in Kunduz.[20]

Secondly, culture and politics stifle not only public but also internal debate. In military circles especially in the English-speaking world, German military writing is still highly influential. Carl von Clausewitz's writings are taught in civilian and military strategy courses the world over, and the names of reformers such as Gerhardt von Scharnhorst, Helmuth von Moltke, or Heinz Guderian are known beyond the narrow circle of professional strategists and military historians. It is perhaps a sign of the depth of the change in Germany's military culture that the defense establishment has no serious military journal today: there is no equivalent to *Military Review*, *Parameters*, *Armed Forces Journal*, the *Marine Corps Gazette*, or the *Army Times* in Germany. *Europäische Sicherheit & Technik*, a small journal with rather short articles, is seen largely as a glossy vehicle for defense industry ads, and *Y*, the house magazine of the Ministry of Defense (MoD), is virtually censored by the ministry's Public Affairs Department, carefully monitoring articles for statements that may be perceived negatively. Publishing a significantly critical article—for instance, a text like Nigel Aylwin-Foster's famous 2005 British critique of the US Army, "Changing the Army for Counterinsurgency Operations," in *Military Review*—is still unthinkable in Germany today.[21] And not just in a journal that would be funded by public money. The problem cuts deeper.

German officers, to make matters worse, seemingly generate little demand for intellectual exchange about tactical and operational issues. The dominant view among lieutenants, captains, majors, and lieutenant-colonels seems to be that any public statement, including in nonexisting professional journals, has the potential to get them into trouble and damage their career. In 2007, then-Federal President Horst Köhler, in a speech at the Command and General Staff College in Hamburg, invited the military leadership to participate in the public debate and speak out "to those responsible for foreign and security policy and to the public."[22] However, little has changed. The possibility of a constructive public debate that remains confined to professional, technical, and tactical issues—and doesn't spill over into the political—does not even register.²³ Some of these concerns may not be entirely unfounded: law—and policy-makers—are extremely sensitive when it comes to outspoken military leaders. Yet, as a result, it may be no exaggeration to say that the Bundeswehr has become a timid and anti-intellectual army.

Timid officers and a lack of internal debate perhaps help explain why the Bundeswehr still had no counterinsurgency doctrine in late 2012—ten years after German special operations forces entered Afghanistan, and six years after America's land forces published the now famous field manual FM 3–24, *Counterinsurgency*, in December 2006. Doctrine is essential to the development and adaptation of an army. While not an end in itself, it is of huge importance for the continuation as well as the widespread implementation of adapted tactics, techniques, and procedures. A small unit can adapt without doctrine—an army cannot. While ISAF, since 2009 closely coordinated by the US-led ISAF Joint Command, operates mainly on the basis of the US FM 3-24, Bundeswehr training and force generation needs a German variant. In short: a counterinsurgency doctrine would symbolize the official acceptance of "COIN"-templates by the German institutional army.

Some work has been done. Beginning in 2009, the Army Staff (FüH III 2) launched an initiative to formulate the "Conceptual Basic Thoughts on Counterinsurgency (*KGV Coin*)."²⁴ While the process began with a notable level of enthusiasm, it ultimately became mired in intra-army disputes and then in Berlin's interministerial process. The term "counterinsurgency" had to be replaced fairly quickly, in part as a result of objections by Guido Westerwelle, the foreign minister. The term was ultimately replaced with a three-line title, "Conceptual Basic Thoughts on the Military Contribution to the Establishment of Security and Public Order in Crisis Regions (KGV COIN)," while maintaining the expression "counterinsurgency"—in English, notably—throughout the actual document.²⁵ But even this concession to political correctness did not save the KGv; at the time of this writing, the only document relevant to counterinsurgency officially valid within the Bundeswehr is a preliminary handbook, restricted for use within the army and focused on the purely military tasks in a clear-hold-build operation ("Preliminary Basic Principles on the Contribution of Land Forces to the Establishment of Security and Public Order in Crisis Regions").²⁶ The document is cautious in tone and does not break new ground.

The idea of population-centric counterinsurgency nevertheless gained traction across the German armed forces. The "small"—that is, army-only—KGv was nipped in the bud by cautious process. But the Ministry of Defense decided to tackle the problem again on a higher level. In 2010 and 2011, the policy planning staff directed a series of four major workshops to lay the groundwork for a joint and interagency document on counterinsurgency. The process was led by Joachim Krause, a well-known professor of security studies at the University of Kiel, in cooperation with John Nagl at the Center for a New American Security (CNAS), a retired senior US Army officer and an expert in counterinsurgency who coauthored FM 3-24. The discussions in the framework of this process were of high quality but notable for their exclusive focus on American input into the German doctrinal process; official military or civilian representation from any other NATO ally except the United States is entirely absent. The ministry's goal was that this process would lead to a common understanding of counterinsurgency beyond the military realm, and beyond Afghanistan—if Berlin's interagency rivalries or changing political realities do not interfere again.

Another important influence on the German Army's adaptation in Afghanistan has been the legal context of military operations. An example may be illustrative: on Sunday, July 19, 2009, a Bundeswehr soldier killed a young Afghan civilian.[27] The victim was one of five passengers in a civilian pickup approaching with high speed a checkpoint manned by German soldiers. After firing a few warning shots, the German soldier stopped the car by firing into the motor block. The Afghan youth died on the spot, and the other four persons in the car were injured. In such and comparable situations, the public prosecutor of Potsdam is to open an investigation. The *Einsatzführungskommando*, the German military's highest operational headquarters, is located in Potsdam, southwest of Berlin in the state of Brandenburg. Therefore the public prosecutor in that region is responsible by default. Each time a soldier is involved in a "potentially criminally liable" situation, the Potsdam public prosecutor leaps into action. Germany's federal system then passes on the case and the proceedings to a public prosecutor in the respective soldier's home region. That happens about twenty times a year, according to the federation of the German Army, the *Bundeswehrverband*. It was not until mid-2010 that the Federal General Attorney became the single agency responsible for charges against German soldiers in Afghanistan. This was a direct result of the attorney's judging the Kunduz bombing of 2009 within the context of the Humanitarian Law of War instead of German criminal law, thus stipulating the existence of a noninternational armed conflict in Northern Afghanistan.[28] It took the German jurisdiction nine years to come to this conclusion. The judges, it should be noted, were still faster than the political leaders in Berlin.

For most of the German involvement in Afghanistan the result was "grotesque," in the words of one senior chancellery official. It was grotesque in a couple of ways. First, the procedure seemed odd because there didn't seem to be a legally binding requirement for the clunky prosecution arrangement. Secondly, soldiers often had to leave the area of operation as a result of doing the job the country was asking them to do. This had rather counterproductive side effects on morale. As one trooper put it, "[In] the field soldiers need courage, not fear from public prosecution."[29] Thirdly, the prosecutions themselves had an awkward feel. In all irregular operations the line between civilians and militants is difficult to draw, even for the soldiers in the field, where patrols and cups of tea with local elders provide the soldiers with significant intelligence about an area and their inhabitants. If the people present cannot properly tell combatants from noncombatants, what should a public prosecutor in rural Thuringia without much insight into either military affairs or Kunduz province do any better? For a long time it was this threat of legal procedures at home that tied down troops and their leaders. Perhaps more problematically, a general attitude of risk-aversion and caution seeped deep into the cultural fabric of the German Army, creating an attitude that starkly contrasts with the sometimes aggressive entrepreneurship that characterizes junior and mid-level officers who were socialized in counterinsurgency and counterterrorism operations of the stress-tested allied land forces.

OPERATIONAL LEVEL

Against the background of these constraining factors, in the years between 2001 and 2008/2009, adaptation of the Bundeswehr was piecemeal and largely confined to the tactical sphere. Germany was the first ISAF nation to leave the confines of Kabul and take over the role of lead nation for the new Regional Command North (RC-N) in 2003. Soon the Bundeswehr concentrated its attention and most of its efforts on two Provincial Reconstruction Teams, in Kunduz and Faizabad. The fact that stabilization was swiftly becoming the army's the main contribution carried a clear message: civil reconstruction, protected by passive military efforts, was to be the hallmark of Germany's Afghanistan contribution. And this civic approach—to a large degree politically motivated—dictated the overall thrust of Germany's strategy: shunning "war" at home meant making the military mission look less military abroad, and avoiding any military casualties.

Avoiding casualties was not only a result of tactical prudence or intentionally passive operational design, as it had been for the US Army during the dark days of 2007 in Iraq. The political and strategic level in Berlin explicitly handed down instructions to several German commanders to avoid casualties, especially fatalities, at all costs. These instructions resulted in operational and tactical micromanagement: PRT commanders regularly communicated directly with lieutenant-generals in Berlin, discussing what to do and what not to do, thus bypassing the German Regional Command as well as ISAF headquarters in Kabul. In mid-2006, Minister of Defense Franz-Josef Jung issued an often-cited order that all personnel leaving the fortified bases should ride in at least lightly armored vehicles, thereby constricting the freedom of action of tactical commanders.[30] More important, the ministerial guidance restricted the direct contact between German soldiers and the civilian population, in contradiction to some core tenets of modern counterinsurgency doctrine. Andrew Exum, a former ranger and advisor to McChrystal, had visited the Germans in Mazar-e-Sharif. In a short report to the ISAF commander, he described going on patrol in a German armored vehicle as "touring Afghanistan by submarine."[31] Exum:

> [T]he experience of traveling around Mazar-e-Sharif—a largely secure city in northern Afghanistan—in an armored German vehicle, whereby I could only observe Mazar and the Afghans themselves through a narrow two inch by four inch slit of bullet-proof glass, really bothered me. It was ... as if I was seeing Afghanistan through a periscope. And if this was how most German soldiers were seeing Afghanistan, I had no confidence that any of them really understood what was going on in northern Afghanistan at a time when the provinces under German responsibility were noticeably worsening.[32]

In the field German soldiers faced two practical problems that made things worse: too few protected vehicles and the tactical necessity to dismount in order to fight the enemy or contact the population. Jung's directive made both difficult, if not impossible. The guideline illustrates the then-prevailing disconnect between the tactical and the strategic level.

The situation even deteriorated. Ultimately fatal attacks on German troops did occur, as is inevitable for an army active in a hot war zone. So some German commanders decided ceasing all nonvital military activities, which in effect meant that virtually no one was allowed outside the wire. An added complication was the lack of airlift capacity. The Bundeswehr had only six medium-lift transport helicopters, with limited night-flying capabilities, which drastically limited the German tactical mobility and thus the pressure on the Taliban insurgency. Berlin's strategic passivity and caution created more than these immediate operational consequences; perhaps more important, the political and strategic environment created a freezing effect on the army's organizational culture: it structurally and culturally discouraged operational and tactical leaders from being proactive, dynamic, and self-confident—features that are essential to the concept of *Auftragstaktik* (mission command), that the German Army has traditionally held in high esteem since the days of Moltke the Elder.[33]

The German abhorrence of friendly casualties was, to an extent, mirrored by measures taken to avoid killing civilians and even enemy fighters. The German rules-of-engagement were extraordinarily strict—even more restrictive than the tactical directives issued by Stanley McChrystal, designed to limit Afghan frustrations and adverse political fallout from overly aggressive search-and-clear operations in Afghanistan's South. In Germany's area of operations, the pendulum swung too far: explicit as well as internalized restrictions even prevented soldiers from engaging known and identified Taliban elements without having been shot at. Fire discipline and discriminate shooting, essential in counterinsurgency campaigns, were overstated to a point that they posed a danger to the soldiers themselves. Only in mid-2009 were these rules-of-engagement changed so that German soldiers could "stop shouting and start shooting," as the *Times* put it.[34] The change was essential for the offensive operations that were to come. The long-imposed strict rules joined forces with a legal framework and an emboldened enemy. The result was a delicate situation: soldiers stood "with one foot in the grave, and with one in prison," as a saying went that circulated among German troops.

One illustration is fire support. For a long period, the German contingent had no organic fire support units. Their 120mm mortars, once deployed, were for the most part used for illumination. Every combat-related fire support operation had to be cleared in advance—not merely by the Regional Command but also by Berlin. And getting the green light from more politically minded officials several time zones away could be difficult and time consuming. German troops could therefore not rely on timely fire support, apart from allied assets such as fixed-wing aircraft stationed outside the area under Regional Command North's responsibility. Six German Reconnaissance Tornados were deployed from 2007 to 2011 in Afghanistan, each armed with two 30mm guns.

But German legislators explicitly prohibited the jets from assisting combat operations by either German or allied troops, "The reconnaissance planes are only used for reconnaissance. They will not be used for Close Air Support."[35] This suboptimal situation concerning the fire support for German operations was about to change no earlier than 2010, when the Ministry of Defense finally approved the deployment of 155mm howitzers.

For the first years of Germany's Afghanistan experience, adaptation in the Bundeswehr remained confined to the lowest tactical levels—squads or platoons coming to terms with enemy tactics, techniques, and procedures. Tactical leaders did analyze enemy ambushes and IED patterns, reacted to these changes, identified lessons, and the most exposed units implemented some of these improvements into their training within Afghanistan as well as in their respective garrisons back in Germany. But on any higher level, micromanagement combined with political caveats and risk aversion strictly limited adaptation. Bottom-up adaptation did exist, but it was held back by political and strategic considerations. Tactical adaptation without strategy was running the risk of being the noise before defeat, to paraphrase an adage often attributed to Sun Tzu.

But the Bundeswehr was to encounter something of a turning point in early 2009. The operational adaptation and tempo of German units increased significantly. Numerous factors contributed to this change, several of them beyond the direct influence of officers and commanders. First, ISAF became more centralized. The establishment of an ISAF joint command in 2009 led to tighter control of the various regional commands and thus narrowed the room for specific national arrangements. The US-led ISAF Joint Command, sometimes shortened to IJC, pressed for the implementation of an FM 3-24-inspired counterinsurgency campaign, based on the positive lessons from the "surge" in Iraq after 2007. On top of these changes, Sean Mulholland, a US Army brigadier general with extensive Special Forces experience, was appointed deputy to the German regional commander, thus further helping to inject an American counterinsurgency mindset as well as improved capabilities. Secondly, the influx of two US "surge" brigades needs to be highlighted. The arrival of one infantry and one combat aviation brigade into RC-North increased the operational tempo as well as the vital supporting capabilities and enablers: the number of air mobility assets jumped from six helicopters to around forty. The increased capabilities for tactical air transport, medical evacuation, and close air support made possible larger and sustained operations even within areas that previously had been off limits as enemy strongholds. Thirdly, enemy pressure grew more intense. The only option left was no option: minimizing the impact of sophisticated enemy attacks by hiding in the main bases, thus conceding the area of operations to the Taliban. More aggressive operations were seen as the only alternative. As a result, German military leaders above the lowest tactical level were forced to adapt. Enemy ambushes and attacks became more frequent and complex, combining IEDs, direct and indirect fire to inflict severe German losses.[36] One particularly deadly engagement occurred on April 2, 2010, which happened to be Good Friday that year. Some five kilometers outside Camp Kunduz, three German soldiers were killed in a sophisticated Taliban ambush. The event helped change the political climate back in Germany, and the Ministry of Defense agreed to deploy the Panzerhaubitze 2000, a heavy, self-propelled howitzer, for fire support. The decision had been delayed for fear of a public backslash in the face of heavy weapon systems being deployed for what was once a peacekeeping mission. Close Air Support swooping in from afar to support remote battles remains invisible to German journalists staying behind the wire in large, sprawling encampments with speed control, recycled trash, and a car-wash, like back in Germany; howitzers right in those camps are more visible. Berlin was comfortable deploying artillery only when it could justify the use of "big guns" in terms of protecting friendly forces, not destroying those of the enemy. From a military standpoint, the deployment of field guns was controversial: heavy artillery did not allow for enough flexibility in deploying increased firepower, and it probably didn't help the German Army to become more "population centric"—but the debate back in Germany would scarcely take note of such doctrinal nuances.

All in all, German land forces significantly improved their ability to learn and adapt on a tactical and operational level. The force structure of the German contingent was changed toward greater combat capabilities and enhanced civil-military cooperation. A new command structure improved the integration of ISAF's three lines of operations: security, good governance, and development. A major-general commanded what was now a de facto division-level headquarters, with a civilian of the equivalent rank of brigadier general serving as deputy commander for stabilization. This structure was still hampered by personal and organizational differences, as well as by conceptual ambivalence. But with enhanced civilian contributions the new structure improved the cooperation between the Bundeswehr and those German civilian ministries that contributed to the Afghanistan mission—mainly the Foreign Office and the Ministry for Economic Development.

For German flag officers, the main effort was no longer the PRTs, but maneuver forces. Two battalion-level task forces of combat infantry replaced the so-called Quick Reaction Force. These units were called *Ausbildungs- und Schutzbataillone*, training and protection battalions, in another nod to self-inflicted political correctness. The new task forces operated extensively and offensively in districts and provinces that had been tightly controlled by the insurgency. In concurrence with the commander of ISAF (COMISAF)'s guidance, the German units partnered with Afghan Forces on joint operations with the American contingent to establish combat outposts and separate the enemy from the population. German forces lived and operated "outside the wire," frequently using joint fire support assets to fight through ambushes or to engage attacking enemy forces. The Germans also adopted a method that is known as Operational Mentor Liaison Teams, shortened to OMLT and pronounced "omelette," like the egg dish. The OMLT approach, followed by NATO throughout Afghanistan, gained track: the teams became more specialized, infantry-heavy and, in part, gained the capability to coordinate ISAF fire support for the Afghan forces mentored. What began as a supposedly risk-free way to train and build up regular Afghan units turned into full-blown combat missions for some OMLTs.

The entry of US troops into the North had other operational effects. American special operations forces (SOF) continued ratcheting up a nascent kill-or-capture campaign to arrest or eliminate regional Taliban commanders. The raids, done mostly in darkness with night-vision equipment to limit unintended casualties, were tactically successful but strategically highly controversial. Raids began in 2008 and were increased in the following years. These attacks were effective in taking out the field-level commanders and degrading their capacity to operate effectively. As of 2011, more and more insurgent operations were being disrupted in their planning stages.[37] But the night raids also alienated the local population as well as Afghanistan's provincial and national political leadership.[38] German special operations forces participated in those missions. However, because of ongoing legal concerns, they were allowed to plan and execute only capture operations, thus limiting their participation.

In sum the effect of the increased American involvement in Germany's sector is astonishing. More and more active US forces could have enabled the German contingent to be even more passive. And indeed the experience of previous years before 2008/9 made such an outcome appear likely. So why the change? The explanation for the new German initiative provides important insights into the Bundeswehr's learning experience. More centralized ISAF control, more allied forces in the North, and more enemy pressure increased the bottom-up pressure that already existed to a critical level, as we have already pointed out. Platoon and squad leaders, the field grade officers in charge of companies or staff branches, and the junior staff officers were all veterans of previous tours in Afghanistan. These younger officers had mostly already learned their tactical lessons and did not need more time to further adapt and use the increased scope of action given by US enablers. Thus the Germans were ready to go. But readiness alone did not make the difference. The missing enabler thus far had been the reluctance of Berlin's political culture and strategic debate in Germany. But now Germany's strategic culture was catching up. As one senior German officer put it: the atmosphere had changed. All of a sudden, RC-North was not held back any more when they back-briefed Berlin on their upcoming plans. "Just do it!" officers were told instead, as one general recounted.[39] The reasons for this change are complex.

The ongoing change of culture on all levels is perhaps best illustrated by the remarkable renewed use of the cross of bravery, the *Ehrenkreuz für Tapferkeit*. The badge symbolically links the tactical changes with cultural change in Germany. The story of Jared Sembritzki best illustrates this point. The lieutenant-colonel was commanding a German Quick Reaction Force, then the main maneuver battalion in RC-North. The unit was made up of mountain infantry, *Gebirgsjäger*, from Bad Reichenhall in Bavaria and supplemented by mechanized infantry elements, *Panzergrenadiere*, with their armored infantry fighting vehicle, *Marder*. The force had fought numerous times since their arrival in Afghanistan in May 2010: first to clear the insurgents out of districts in the so-called Kunduz-Baghlan corridor that were formerly no-go areas for ISAF forces. Then, in September 2010, to defend a combat outpost near Shahabuddin, about seventy-seven kilometers southwest of Kunduz, against repeated Taliban attacks. The German soldiers, working closely with Afghans, were determined to hold the outpost and to use it to implement ISAF's strategy of protecting the population by living close to them.[40] In previous engagements the attacking insurgents had demonstrated tactical weaknesses, and the Germans had been able to counter the fighters fairly easily. But this time a sustained assault caught the Germans flat-footed. An estimated sixty attackers popped up within hand-grenade distance, having used the ever-present ditches to close in on ISAF forces. Supported by mortars, this rush threatened to break the small outpost's defenses. In this perilous situation, the German commander of the

QRF, Lieutenant-Colonel Jared Sembritzki, rallied his soldiers, organized the defense, and personally led the counterattack, which—in coordination with US SOF-elements, Afghan forces, and Close Air Support— successfully routed the enemy.

In September 2011, Thomas de Maizière awarded Sembritzki the new German cross of bravery. The lieutenant-colonel was the first staff officer ever to receive the badge. The minister's words described Sembritzki's courage and decisiveness "in the face of the enemy," *im Angesicht des Feindes*, as the citation reads.[41] Such language is remarkable. Much had happened since the days in which German defense ministers were unable to publicly pronounce the word *Kampfeinsatz*, let alone *kriegsänliche Zustände*, warlike conditions. But the change of culture, now obvious, had been in the making for a while. Indeed Jung, minister of defense from 2005 to 2009, did not only push a change of language but also initiated the new cross of bravery in August 2008, and Germany's president, Horst Köhler, officially endorsed it a few weeks later.

In all but color the *Ehrenkreuz* resembles the Iron Cross, the coveted German symbol of bravery awarded for the first time in 1813.[42] The cross is the only medal to be awarded for bravery on the field of battle. Before 2008, commanders had to improvise and use medals designed to be awarded for meritorious peacetime service. Thus the cross of bravery, awarded twenty-two times since its introduction in 2008, two of those posthumously, unmistakably symbolizes the escalating situation on the ground in Afghanistan as well as German society's efforts to come to terms with this development. Indeed the badge's introduction was accompanied by intense public debate. That debate also revealed a notable public unease with what many perceived as a renewed militarization of German foreign policy and, somewhat paradoxically, the Bundeswehr. In 2010, another medal exclusively for combat service was introduced: the *Einsatzmedaille Gefecht*, an equivalent to the US Combat Action Badge. Even historians with intimate knowledge of the Bundeswehr criticize this new medal as a distinct step toward an ethic of heroism, unhealthy warrior-devotion, and an "increasing militarization."[43] If this trend continues, it will only be a matter of time until the cross of bravery will be supplemented by new awards for repeated or higher bravery in combat—as is normal in any army. But this new "normalcy" of what is still referred to as a "fighting army" in the German press—the term itself contains a remaining trace of disbelief—at times appears to be at odds with the pacifist values and the emotional attitude vis-à-vis the use of military force that is common among Germany's political elite, especially the left-of-center but also among the more religious, and, perhaps to a somewhat lesser degree, in the country's general population. The cross of bravery itself epitomizes an ongoing negotiation of a new normalcy, a normalcy that seems to be oscillating between two poles: on the one hand military traditions that are standard in mature democracies, and on the other hand with what German commentators, borrowing from the American academic debate, have called the "post-heroic society," allegedly also a new standard in mature democracies.[44]

STRATEGIC LEVEL

Germany's ongoing cultural shift vis-à-vis the use of force on a social, political, and strategic level is complex and difficult to portray in the few brushstrokes available in a book chapter. No single event can express this change, but some events have been catalysts of change. The single most notable event was what became known as the Kunduz airstrike. To put it bluntly: until September 4, 2009, the Bundeswehr was on a peacekeeping and stabilization operation in Afghanistan. After September 4, 2009, at least in the eyes of the German public, the Bundeswehr was in a war.[45]

In the evening hours of September 3, two civilian trucks carrying gasoline were captured by insurgents near the city of Kunduz, home of one German Provincial Reconstruction Team. Gasoline is a valuable commodity in Afghanistan. After killing one of the drivers, the captors tried to cross the Kunduz River to bring the trucks and their booty to safety. Meanwhile ISAF had been alarmed. The German PRT-leadership did not want these trucks to fall into the hands of the insurgents. The officers feared that the gasoline trucks could be used as vehicle-borne IEDs against military targets, with potentially devastating effects. Fast action was required, but with the bulk of the PRT's available assets involved in another operation, a fast ground operation seemed unrealistic. The two trucks were stuck in a sandbank in the middle of the river in the early morning hours. At least one intelligence source had confirmed to the Germans that numerous insurgent fighters were present. So the German PRT-commander approved a bombing run by a pair of US Air Force F-15 jets. The targets were destroyed. But in the following days the gruesome truth became known: the bombs had eliminated not only a number of fighters. The hapless German colonel who authorized the airstrike, Georg Klein, had inadvertently caused the bombardment of numerous civilians from surrounding villages, potentially over one hundred.

The reasons for the presence of such a large number of civilians in the middle of the night remained unclear. What was not unclear was that Klein had ordered the deadliest German fire mission since World War II. Germany's public was in shock. The MoD, then still under Jung's leadership, tragically mismanaged the public relations side of the Kunduz airstrike, feeding information to a public hungry for news—and to a parliament desperately trying to get full disclosure of the strike's details—in piecemeal fashion. But what shaped public opinion in the months to come were not the errors the colonel in charge may have been responsible for—the airstrike was most probably a clear violation of McChrystal's orders concerning Close Air Support, and may have been justified vis-à-vis the pilots with false claims of emergency. What concerned the public was also not the ministry's cock-up of the affair. What counted was the simple fact that so many people had died, especially so many civilians, and that German forces were responsible for the tragedy.

"Kunduz"—Germany soon was on a first-name basis with the affair—had two consequences that may seem surprising. First, the Kunduz airstrike made one thing clear to the German public: what had begun as a peacekeeping mission to build girls' schools and protect a fragile peace had, over time and unbeknownst to most of them, developed into a serious war. Germany's leading politicians, as we argued above, had for a long time not acknowledged this simple but unpleasant truth. Colonel Klein's mistake made it impossible for political leaders to maintain double-speak on Afghanistan. Secondly: the complete leadership of the Ministry of Defense had to go as a result of the bombing and its subsequent mismanagement. The former minister of defense, Franz-Josef Jung, who in the meantime had moved on to another ministry, had become untenable as a cabinet minister and resigned. Soon thereafter, the new minister of defense, Karl-Theodor zu Guttenberg (who had come in after federal elections, not as a result of the airstrike), sacked the chief of defense, Wolfgang Schneiderhan, and the powerful deputy minister of defense, Peter Wichert. Angela Merkel, the chancellor, and zu Guttenberg himself barely survived the affair. The post-Kunduz shake-up, regardless of its justification, contributed to intraministerial processes of personal renewal and generational change at the highest levels.

This generational change requires a closer examination. The Kunduz affair not only helped reshuffle the ministry's political and senior civilian leadership. It helped bring about a generational change among the top-level generals as well. Generational change, as we have pointed out, had long been on its way inside the army. But it took a long time to trickle up to the senior officers, to those at the interface between the strategic and the political levels. During the first several years of Germany's involvement in Afghanistan, the civilian and military leadership in the Ministry of Defense remained, in essence, unchanged. This continuity does not include the ministers themselves, who changed several times. But it is important to note that Germany's political system across all ministries does not know the so-called revolving door that is common in the United States. When the government changes, or even when a new secretary takes over, a large number of senior positions will change—not so in Germany. When the minister or even the government changes, literally only a handful of people in the ministry may move office. Germany's governmental bureaucracy is therefore characterized by significantly more continuity across different administrations. The effect in the Ministry of Defense was delayed adaptation. Werner Freers, the first lieutenant-general with actual ISAF experience, was made the army chief of staff only in March 2010. Between March 2001 and March 2010, only two army chiefs of staff held office, and neither of them had personal leadership experience in Afghanistan.[46] The chief of defense, Wolfgang Schneiderhan, held office from June 2002 until November 2009, also without having served in ISAF command positions. This is not to imply that these generals actively stood against innovation and adaptation in the army. But these officers had their formative "out-of-area" experiences in the Balkans. And even that experience had become a somewhat distant one, after many years of slowly rising through the ranks of a conventional peacetime army. The quality that produced stellar military careers under the circumstances in the ministry during those years was avoiding risk, not taking risk. Therefore it is no farfetched speculation that these military leaders and those working under them were not as receptive and eager to implement the Afghan lessons as those who had served in Afghanistan's grueling heat and dust, losing their men to an enemy that proved more adaptive and determined than any previous one.

But even the Kunduz episode and the shake-up it produced left one larger problem largely untouched: the political confusion about the strategic objectives. Why was Germany engaged in Afghanistan in the first place? This question and the answer to it formed the political and emotional backdrop for whatever the Bundeswehr was doing in Afghanistan. And it was a question that proved more difficult to answer than meets the eye. Justifying the ongoing war has long vexed members of parliament, top politicians in the executive branch and their staffs, top generals, development officials, columnists, and of course the general public. The reason for the confusion was simple: there were multiple answers, at least seven, all of them relevant to a certain degree.

The first and foremost reason to be in Afghanistan was self-defense. The 9/11 event had been the key that triggered an American response in Afghanistan and European solidarity with a NATO partner. The West had to prevent the renewed establishment of a "safe haven" for al-Qaeda, the argument went. Peter Struck, minister of defense when Germany first sent troops to Southeast Asia, had coined the phrase that "Germany is also defended at the Hindu Kush."[47] Struck's strategic punch-line has dominated the debate across party lines ever since. In a widely anticipated speech to parliament after the Good Friday attacks in 2010, Merkel reiterated, "This famous sentence by Peter Struck hits the nail on the head for me. Until today nobody has expressed this more appropriately."[48] It is noteworthy that Islamic extremism progressively became more a problem at home in Germany as the Bundeswehr continued its operations in Afghanistan, with radicals quoting the German contribution to the war as a motivation to join the jihad. And more and more, the general population took notice of this unwelcome link.

A second reason was preventing a renewed Taliban takeover. This argument was related to countering Islamic extremism, but it had other dimensions. Initially, preventing a Taliban comeback was related to preventing an al-Qaeda safe haven. But to make the argument, the link between al-Qaeda and the Taliban needed to be clear and firm. As experts called the alliance between the two groups into question, and as the more local and patriotic motivations of the Afghan insurgency stepped to the fore, other reasons gained strength. Third, Germany had the responsibility to rebuild and stabilize the war-ravaged country, as many saw it. Routing the Taliban in 2001 by bombing Afghanistan with B-52s and smart missiles had also led to severe suffering among the civilian population. Therefore the international community had a collective responsibility to help the country back onto its feet. This argument appealed to the entire political spectrum in Germany, especially to the left, not just to defense-minded hawks. Fourth, the responsibility argument was flanked by those who highlighted values, often in contrast to the Taliban's "medieval" world view: with Western aid, modernity, education, and women's rights could be brought to a benighted Afghanistan. The proverbial girls' schools, often quoted by Jung as a sign of progress, symbolized this argument. These soft reasons explain why the left, especially the highly educated left, perhaps best represented by the Green Party, at times had the highest approval ratings for the war in Afghanistan.[49] A few other arguments did not receive the same level of public visibility, but were important for Berlin's political elite: a fifth reason to contribute in Afghanistan was alliance solidarity, and helping to prevent an embarrassing failure and potential damage to NATO. In Merkel's words: "[T]he international community went in together, it will leave together as well."⁵⁰ Then there were, sixth, sunk costs, especially for those Germans who suffered most there. In December 2011, the top British commander in Afghanistan, General James Bucknall, called on his country not to betray its "investment in blood" in Afghanistan. Top Bundeswehr officers rarely made such arguments in public, but the sentiment is there as well. Another argument sometimes brought forward in the United States but rarely mentioned in German discussions is Afghanistan's geopolitical significance.

The diverse coalition of reasons had a counterintuitive effect: the government was able to stick to what is best described as rather passive behavior for a decade, a "caveated" strategy that for a long time seemed designed to minimize damage, initially trying to "hide" from the war in the presumably calmer northern corner of Afghanistan. Germany kept a low profile in two ways: it kept a low profile in the sense that Berlin's political leaders avoided a fuller military commitment and a more aggressive high-stakes approach of the kind the United Kingdom chose to pursue. And on the other side German politicians also kept a low profile in the sense that they avoided a full withdrawal perhaps comparable to the Netherlands or Canada, though without the (in relative terms) significant losses that these two much smaller countries suffered in Afghanistan. The first scenario was politically not feasible; the second scenario also would have ended all military adaptation in Afghanistan.

Keeping Afghanistan on the lower ranks in the chancellery's political agenda therefore made possible the cultural change outlined in these pages—organizationally in the Bundeswehr and even in a broader sense at home in Germany. "German politicians need to explain Afghanistan to voters," had become a standard refrain at foreign policy events and political conventions in Berlin. But the hackneyed complaint always ignored the unpleasant political realities: support for the war in the general German population progressively withered away, and there was scarcely anything politicians could do about this. But the consensus of the political elite—minus the radical Left Party—was to support the campaign anyway. This was only possible because the German participation in the war never was narrowed down to one single reason, so every party and member of parliament was able to choose one or more arguments to support the campaign from the reasons listed above.

CONCLUSION

A decade of operational experience in Afghanistan changed the Bundeswehr. In the early 2010s it was hard to find a noncommissioned or commissioned officer up to the rank of brigadier general who has not served in Kabul, Mazar-e Sharif, Kunduz, or Feyzabad. In this sense Germany's armed forces more and more came to resemble those of other NATO allies. This was even more the case for combat and vital supporting units. That ongoing generational change is likely to leave a deep imprint in the Bundeswehr's cultural fabric, and it may define the army's personnel for the years to come, regardless of structure or doctrine.

Perhaps more important, changes in Germany's underlying national strategic culture have come to the fore. If there was an adaptive moment that generated broader momentum, it was the aftermath of the September 4, 2009, Kunduz airstrike. That tragic event, in manifold ways described above, opened a political window of opportunity that later translated, in counterintuitive ways, into more tactical flexibility to react to long-recognized operational challenges in the field. Germany's most powerful adaptation drivers were pressing operational challenges and, as a shaping factor, an evolving domestic strategic culture that ultimately removed the main obstacle to change: domestic politics.[51]

After 2014, the Bundeswehr will again have to deal with a new challenge: keeping the best and the brightest volunteers and career officers in active duty in an army that may again be confined to peacetime garrison life for a significant time. Significant challenges lie ahead: over the mid to long term, the institutional army has to work hard not to forget the lessons learned at such a painful price in Afghanistan. Without a thorough implementation of the adapted Tactics, Techniques and Procedures (TTPs) into official doctrine and, less important, force structure, the Bundeswehr could inadvertently shed its valuable experience and readapt to peacetime service faster than expected. The army's old ways could quickly come to the fore. This means training for conventional, regular scenarios, dictated by decades of Bundeswehr tradition. Adaptation in Afghanistan emerged bottom-up, and was later passively permitted on the political-strategic level. Easing these bottom-up pressures comes with an immediate risk of stalled adaptation. Afghanistan will, for the years to come, heavily influence the Bundeswehr. A substantial part of exercise scenarios up to the division level are designed around counterinsurgency scenarios. While Balkan-centric doctrine exercises have shaped the Bundeswehr of the 2000s, Afghanistan will do so for the 2010s. That creates an opportunity for continuation while avoiding operational single-mindedness. At the same time, the MoD's effort to finally create an interagency-drafted manual that integrates the lessons of Afghanistan while looking beyond the Hindu Kush may create an environment that stimulates further discussion within and outside the military. It remains to be seen how tighter-than-ever budget constraints will influence this process.

In Germany's future military operations, the Bundeswehr, and the Federal Republic as a whole, may not have seven years to engage in soul-searching on all levels and listening to the proverbial noise before defeat. This, after all, is part of becoming normal. It is also part of "normal" military operations to be able to adapt quickly and to learn faster. The irony is that the Bundeswehr's organizational adaptation may have a shorter lifespan and a more fragile existence than Germany's broader new cultural experience with using military force. Germans are meticulous also when dealing with their past, and Afghanistan will soon be part of that past. The country's larger cultural adaptation toward a new normalcy seems more durable, whatever the Ministry of Defense will do or not do.

NOTES

1. Ulrike Demmer, "Die endlose Sekunde," *Der Spiegel*, July 20, 2009, 38–40.
2. Alastair Ian Johnston, "Thinking about Strategic Culture," *International Security* 19, no. 4 (Spring 1995): 32–64.
3. For a historical treatment of Germany's strategic culture, see Thomas U. Berger, *Cultures of Antimilitarism: National Security in Germany and Japan* (Baltimore, MD: Johns Hopkins University Press, 1998); and John S. Duffield, *World Power Forsaken: Political Culture, International Institutions, and German Security Policy after Unification* (Stanford: Stanford University Press, 1998).
4. For a post–Cold War treatment of Germany's strategic culture, see Kerry Longhurst, *Germany and the Use of Force* (Manchester: Manchester University Press, 2004); Wilhelm Mirow, *Strategic Culture Matters: A Comparison of German and British Military Interventions since 1990* (Berlin: LIT Verlag, 2009).
5. Hanns W. Maull, "Deutschland als Zivilmacht," in Gunther Hellmann, Siegmar Schmidt, and Reinhard Wolf, eds., *Handwörterbuch zur deutschen Außenpolitik* (Opladen, 2006), 73–84.
6. Interview with Foreign Minister Guido Westerwelle, "Darf der Westen nach Bin Laden auch Gaddafi töten?" *Bild*, May 8, 2011: available at http://bitly.com/orPCSu.
7. For an especially eloquent critique, see Wolfgang Seibel, "Prinzipienlosigkeit als Prinzip," *Frankfurter Allgemeine Zeitung*, October 24, 2011, 7. For a longer view, see Klaus Naumann, *Einsatz ohne Ziel* (Hamburg: HIS Verlag: 2008).

8. For a conceptualization beyond Germany, see Theo Farrell, "World Culture and Military Power," *Security Studies* 14, no. 3 (2005): 448–88.

9. For a more instrumental understanding of German strategic culture, see Olivier Schmitt, "Strategic Users of Culture: German Decision-Making on the Use of Military Force in ESDP operations," *Contemporary Security Policy* 33, no. 1 (2012).

10. Ruling of the Constitutional Court concerning Bundeswehr participation in Operation Sharp Guard in 1994, BVerfGE 90, 286 (Marginal Number 322).

11. §3 II of the "Gesetz über die parlamentarische Beteiligung bei der Entscheidung über den Einsatz bewaffneter Streitkräfte im Ausland (Parlamentsbeteiligungsgesetz)," adopted in 2005.

12. Timo Noetzel and Martin Zapfe, "Der Einsatz im Fokus? Das Verteidigungsministerium und die Auslandseinsätze," in Robert Glawe, ed., *Eine neue deutsche Sicherheitsarchitektur—Impulse für die nationale Strategiedebatte* (Berlin, 2009), 187–94.

13. Christoph Schwennicke, "Verdruckste Krieger," *Der Spiegel*, June 30, 2008, 36–42.

14. See http://icasualties.org/OEF/index.aspx, accessed December 4, 2011.

15. Angela Gareis, "Afghanistan: Der unausgesprochene Krieg," *Westdeutsche Zeitung*, July 2, 2009.

16. "Jung: Ich verneige mich vor den gefallenen Soldaten," *Frankfurter Allgemeine Zeitung*, October 24, 2009.

17. Ibid.

18. "Kampfeinsatz, aber kein Krieg," *Frankfurter Allgemeine Zeitung*, July 2, 2009.

19. "Guttenberg hält Einsatz in Afghanistan für 'kriegsähnlich,'" *Zeit*, November 3, 2009.

20. Franz Solms-Laubach, "Warum spricht Guttenberg jetzt von Krieg?" *Bild*, April 5, 2010.

21. Nigel Aylwin-Foster, "Changing the Army for Counterinsurgency Operations," *Military Review* (November–December 2005): 2–15.

22. Speech delivered by Federal President Horst Köhler at the Command and General Staff College of the Bundeswehr (*Führungsakademie der Bundeswehr*), Hamburg, September 14, 2007.

23. Thomas Rid and Martin Zapfe, "Das Militär verdient Gehör," *Financial Times Deutschland*, April 13, 2010, 24.

24. For a closer look at the KGv, see Timo Noetzel and Martin Zapfe, "NATO and Counterinsurgency: The Case of Germany," in Christopher M. Schnaubelt, ed., *Counterinsurgency: The Challenge for NATO Strategy and Operations*, NATO Defense College Forum Paper Series 11 (Rome, 2009), 129–51.

25. "Konzeptionelle Grundvorstellungen zum militärischen Beitrag zur Herstellung von Sicherheit und staatlicher Ordnung in Krisengebieten (KGv COIN)," Bundesministerium der Verteidigung, unpublished draft manuscript, version 7, June 2010.

26. See Heeresamt, "Vorläufige Grundlagen für den Beitrag von Landstreitkräften zur Herstellung von Sicherheit und staatlicher Ordnung in Krisengebieten" (Köln, 2010).

27. "Bundeswehr tötet Jugendlichen in Afghanistan," *Hamburger Abendblatt*, July 21, 2009.

28. See Christian Schaller, "Rechtssicherheit im Auslandseinsatz," *Stiftung Wissenschaft und Politik*, Aktuell 67, December 2009, 2.

29. Demmer, "Die endlose Sekunde," 38–40.

30. "Vernetzung ziviler und militärischer Aktionen in Afghanistan beibehalten," *Welt am Sonntag*, Feburary 3, 2008.

31. Bob Woodward, *Obama's Wars* (New York: Simon and Schuster) 2010, 153.

32. Andrew Exum, "On Woodward's Book: A (Very Minor) Clarification," *Abu Muqawama*, September 27, 2010.

33. Werner Widder, "Auftragstaktik and Innere Führung: Trademarks of German Leadership," *Military Review* 82, no. 5 (September/October 2002): 3–7.

34. "New Rules Let Germans in Afghanistan Stop Shouting and Start Shooting," *Times*, July 29, 2009.

35. "Antrag der Bundesregierung" 16/6460 (2007), 3.

36. See Antonio Giustozzi and Christoph Reuter, "The Insurgents of the Afghan North, Afghan Analysts Network Thematic Report 04/2011": available at http://aan-afghanistan.com/uploads/AAN-2011-Northern-Insurgents.pdf.

37. Author discussion with senior US officer, Hamburg, September 1, 2011.

38. Rod Nordland, "For a Long-Term Afghan-American Accord, Night Raids Are a Sticking Point," *New York Times*, December 4, 2011, A14.

39. Author discussion with senior German officer, February 2, 2011.

40. "Schlacht um Shahabuddin," *Der Spiegel*, October 11, 2010.

41. "Ehrenkreuz der Bundeswehr für Tapferkeit verliehen," *Bundesministerium der Verteidigung*, September 6, 2011: available at http://bitly.com/qOupUR+.

42. Wikipedia has the images of the *Ehrenkreuz* (available at http://bitly.com/s1ZmRS+) and the *Eiserne Kreuz* (see http://bitly.com/ulsq6u+).

43. See Detlef Bald, "Ein Zeichen für zunehmende Militarisierung," *Tagesschau*, November 29, 2010: available at http://bitly.com/riDIg9+.

44. Herfried Münkler, "Die neuen Kriege," Berlin 2004.

45. See Timo Noetzel, "The German Politics of War: Kunduz and the War in Afghanistan," *International Affairs* 87, no. 2 (March 2011): 407.

46. March 2001 to March 2004, Lt.-Gen. Gerd Gudera; March 2004 to March 2010, Lt.-Gen. Hans-Otto Budde.

47. Dirk Eckert, "Die Sicherheit Deutschlands wird auch am Hindukusch verteidigt," *Telepolis*, December 13, 2002.

48. Hauke Friedrichs, "Merkel erklärt den Krieg," *Die Zeit*, April 22, 2010.

49. Thomas Rid and Timo Noetzel, "Germany's Options in Afghanistan," *Survival* 51, no. 5 (2009): 71–90, 77.

50. Friedrichs, "Merkel erklärt den Krieg."

51. As discussed by Theo Farrell in Chapter 1, this volume. In the German case study presented here, the role of new technologies can be neglected.

9 Canadian Forces in Afghanistan: Minority Government and Generational Change while under Fire

Stephen M. Saideman

Despite a history of peacekeeping efforts in the Balkans, Cyprus, and elsewhere, Afghanistan has been Canada's first encounter with significant combat since the Korean War. The mission has been controversial at home and dynamic in Afghanistan. We have seen significant variations in what the Canadians have been willing and able to do, moving from the days of so-called Can'tbats[1] to being among the most-forward leaning contingents in one of the most dangerous sectors and then back to its former status. Specifically, there have been four distinct phases in Canada's military effort in Afghanistan:

1. Deployed as a combat unit in Kandahar in 2002 as part of Operation Enduring Freedom (OEF).
2. Sent as a peacekeeping, stability operations effort in Kabul from 2004 to 2006, as part of the International Security Assistance Force (ISAF).
3. Redeployed to Kandahar from 2005 to 2010, engaged initially in intense conventional combat and then focusing more on counterinsurgency.
4. Moving out of the combat business and entirely into "Kabul-centric" training "behind the wire" on bases after July 2011.

To deal with these varying missions, the Canadians have reformed their command structures, adopted new weapons systems, and given increased discretion to the troops on the ground. While there were also efforts to adapt at the operational and tactical levels, these strategic adaptations are the focus of this chapter, as they were most surprising. The efforts have been very costly, producing one of the highest rates of casualties per capita among North Atlantic Treaty Organization (NATO) members (only the Danes have a higher rate) at a relatively novel time in Canadian political history—government by a series of minority parties. Indeed, we actually see (until 2011) minority governments placing Canadian troops in much more dangerous environments with fewer restrictions than the most recent majority government. Thus, any analysis of Canadian adaptation to this difficult environment must consider this puzzle: How do minority governments sustain flexible forces in harm's way? Delegating these decisions to the military is one way to deal with the challenges of minority government, but that leads to a second puzzle: Why have the leaders of the Canadian Forces (CF) changed the way they do business?

To understand Canada's efforts in Afghanistan, we have to address these dual puzzles: how minority governments fight a controversial war, and why the choices made by the Canadian Forces have changed over time. As a result, the combination of different shaping factors played a big role in how Canada adapted. The operational environment was so very difficult that politicians and military officers had to respond, but the directions of the responses were driven and constrained by domestic political dynamics and generational change within the military. Thus, in the Canadian case, distinct elements of the three categories of shaping factors specified in Chapter 1 influenced the patterns of adaptation at the operational level.

A CHALLENGING ENVIRONMENT

The focus of this chapter is on the Canadian experience in Kandahar from 2005 through 2011. Canada had joined the US effort in Operation Enduring Freedom in December 2001 with the deployment of its special operations unit, Joint Task Force 2, and then with a battalion into Kandahar the next month. After six months, the Canadian Forces rotated out and then were sent to Kabul to help ISAF transition to a NATO-led operation. The choice to move to Kandahar in 2005 has proven to be quite controversial in retrospect, as it proved to be more difficult than expected.

Deploying to Afghanistan repeatedly requires less explanation than the choice of Kandahar. Canada participated in every NATO mission before Afghanistan and in the Libyan effort more recently. Supporting this and other multilateral institutions fits squarely and consistently within Canadian national interests as a relatively smaller power neighboring the most powerful country in the world. So, the choice in 2005 was not Afghanistan or not, but where in Afghanistan to deploy as NATO moved out from Kabul to the rest of the country.[2] Going to Kandahar, despite the risks, made sense to both the civilians and the military. Prime Minister Paul Martin was trying to distinguish himself from his predecessor, Jean Chretien, by focusing Canadian efforts in a single place rather than having forty or so Canadian flags representing many small contingents across United Nations (UN) efforts around the world.[3] The Department of Foreign Affairs and International Trade pushed Kandahar as well. The military preferred Kandahar, as the Canadian Forces would be based at the major air base in the region with much support nearby, and because it would give the chance for Canada to have a visible and leading role. It is also the case that taking on a serious combat role would do much to demythologize the Canadian Forces as being good just for peacekeeping, something desired by much of the military. Others have argued that this effort was aimed at placating the United States. In all likelihood, there were a lot of reasons that motivated the decision to accept the Kandahar mission, with each actor varying about which ones mattered most.

Regardless of Canada's motives, it was quite clear that Kandahar, as the home of the Taliban, would be one of the most violent parts of Afghanistan. The geography has favored the Taliban, with the province adjacent to Pakistan and next to the poppy-rich province of Helmand. The terrain includes thick clay walls that separate compounds and form drying huts for the grape harvest, creating defensive positions that many have compared to World War I battlefields. The human "terrain" is also favorable to the insurgents, since the area is almost entirely Pashtun, with several of the key tribes having a deep history with the Taliban movement.

For the Canadians for much of this timeframe, the key problem has been of having too few soldiers in a very large and populous territory. In public briefings, Canadian commanders returning from the field would note how far the number of security forces (international and Afghan) were below those called for in standard counterinsurgency doctrine.[4] As a consequence, until 2010, there were never enough troops to hold the terrain that been cleared in the COIN sequence of clear, hold, build. This pattern was referred to as mowing the lawn or serving as a fire brigade,[5] both suggesting that little progress could be made as long as the Canadians were running all over the province (and beyond, to Helmand and Uruzgan),[6] without really controlling any territory for very long. Only in 2010 with American reinforcements were the Canadians able to implement counterinsurgency doctrine: living among the Afghans, spending more time in place.

The Taliban have clearly seen the city of Kandahar and the neighboring areas as a key focal point of their efforts, with the intent in 2006 of presenting the Canadians with the Afghan equivalent of Tet that turned into Operation Medusa.[7] Since then, some of the most dramatic operations have been aimed at key targets in Kandahar, including two large prison breaks.

CANADIAN ADAPTATION

The Canadian military has altered how it does business quite significantly. The traditional Canadian deployment, with a couple of notable exceptions (Kosovo), has been as part of a UN peacekeeping effort to separate combatants (Sinai, Cyprus), to support a mandate to manage a conflict (Croatia, Bosnia), or to keep the peace. The willingness to send special operations units in the fall of 2001 and then a battalion into Kandahar in early 2002 reflected a significant change in what Canada was doing, although this was somewhat temporary. The deployments to Kabul from 2004 to 2006 were closer to the old way of doing things akin to peacekeeping, but then the move to Kandahar redefined how Canadians and everyone else perceived the Canadian Forces. The combat ranged from very intense conventional warfare during Operation Medusa in 2006 to highly kinetic counterinsurgency until mid-July 2011.

This mission has required and been facilitated by a number of adaptations made by the Canadian Forces at all levels. In this section, I focus on three sets of adaptations—weapons systems, command structures, and discretion.

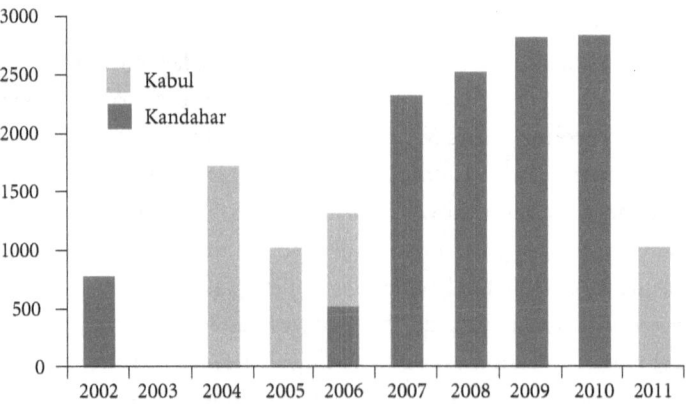

FIG. 1. Canadian Deployments[8]

Tanks and Helicopters

Perhaps the most visible adaptations made by the Canadians are the acquisitions of weapons systems that had been downgraded, ignored, and only recently developed. It is incredibly striking to hear that American Marines and perhaps the US Army have been inspired by the Canadian example of sending tanks to a counterinsurgency operation. The amazing thing from the Canadian perspective (other than Americans being impressed with Canadians) is that Canada had intended on doing away with main battle tanks. The Land Staff (the army) had planned to replace the old Leopard tanks with a mobile gun system. In 2003, Lieutenant-General Rick Hillier, Chief of the Land Staff at the time, asserted that tanks were "useless for soldiers in ... Kabul," and a "millstone that has hamstrung our thinking for years."[9]

This was overcome by events, as the cabinet decided to deploy the Leopards to Afghanistan with Operation Medusa underway. The former tank critic and then head of the Canadian Forces, General Hillier justified the turnaround in thinking by saying that the Taliban had "moved from a guerilla warfare–type style, a counter-insurgency, to some conventional techniques."[10] Tanks proved useful for clearing mines and penetrating the barriers created by drying huts mentioned above. Negotiations with potential sources of newer, air-conditioned Leopard 2s in October 2006 concluded in April of the following year with Canada buying one hundred tanks from the Netherlands and leasing twenty from Germany. The leased tanks would arrive quickly during fighting season, and the Dutch tanks would arrive in the fall. This decision was announced after another set of attacks that caused a spike in Canadian casualties.[11]

The decision to buy and lease tanks to be deployed to Kandahar was controversial both because it reversed the previous policies and because it was not subject to debate within parliament. Because the opposition was in disarray, they could not effectively challenge these decisions. The rushed process was criticized by Auditor General Sheila Fraser, as a number of the tanks sent to Afghanistan were not ready for operations there, nor were there enough spare parts.[12] As a result, the troops stripped some of the tanks to repair others.

Because Canada is one of very few contingents to deploy tanks in Afghanistan and because Canada was in the process of doing away with its tanks, the decision to deploy them and then go through much effort to lease a new generation is evidence of significant adaptation. Sending tanks was not a standard procedure but required special effort, including cabinet involvement but not parliamentary input. The deployment was clearly a reaction to the intensity of combat, the difficulty of the terrain, and especially the vulnerability of lighter armored vehicles to improved explosive devices (IEDs). Bureaucratic politics may also have played a role, as this situation gave the armor branch of the Canadian Forces an opportunity to push for keeping tanks in the inventory and in a visible role in Afghanistan. Still, given Hillier's previous desire to eliminate tanks, it seems clear that the nature of the operations drove this decision more than the interests of the tankers in the Canadian Forces.

Similarly, there developed an urgency for Canada to deploy more helicopters to Kandahar, as convoys on the ground were getting hit hard by both IEDs and suicide bombers. The problem was that Canada had underinvested in helicopters during the bad budget days of the 1990s, so the Canadian Forces were ill equipped to move troops and equipment around the province by air. Indeed, they were one of the few sizable contingents to have no helicopters of their own. The Manley Panel insisted on more helicopters as part of the conditions for extending the mission beyond 2009: "[To] improve the safety and operational effectiveness of the Canadian Forces in Kandahar, the Government should secure for them, no later than February 2009, new medium lift helicopters."[13] Finding deployable helicopters quickly became quite a challenge, with Canada buying six used Chinooks already flying in Afghanistan from the United States and leasing six commercial helicopters. The CF also deployed eight Griffon utility helicopters to escort the Chinooks.

The two reinforcement decisions of tanks and helicopters share much in common. The focus was on sending equipment that would reduce Canadian casualties, especially those posed by the IED threat. In both cases, past decisions that had limited Canadian capabilities were reversed rather quickly. To expedite matters, in each case there was an innovative mix of purchasing and leasing. While operational requirements enunciated from the field spurred these acquisitions, bureaucratic actors back in Canada clearly had significant stakes involved. The Armor Branch of the Canadian Land Staff clearly favored getting back in the tank business,[14] and the Air Staff got a jump start in the process of finally getting medium-lift helicopters after more than a decade of struggles with previous governments. Both programs ultimately faced criticism from the Auditor General, as the rush to get the systems to Afghanistan violated procedures in a variety of ways.[15]

Again, operational factors—the IED threat combined with a relative scarcity of Canadian troops and assets—were the most significant factors shaping the timing of and the precise nature of the acquisitions. The politicians and military officers back in Ottawa favored specific purchases as they focused on getting what was necessary to keep the mission going—minimizing casualties and maintaining enough support from the opposition.

Command Structures

Canada not only changed the equipment the soldiers were using but also significantly altered how they were being commanded both in Ottawa and in the operational theater. Until 2006, the Deputy Chief of the Defence Staff had a variety of duties, including overseeing all operations abroad. The second in command of the Canadian Forces, thus, had a great deal of responsibility with serving as the operational commander as just one of many roles. Moreover, this office existed within the building housing the Department of National Defence, including the civilians under the Minister of National Defence.

One of the most significant innovations during the Afghanistan war was an institutional one back in Ottawa, with the creation of Canadian Expeditionary Forces Command (CEFCOM), in addition to Canada Command, standing up in early 2006. The Commander of CEFCOM was a three-leaf officer,[16] and acts as commander of all operations outside North America and reported directly to the head of the Canadian Forces—the Chief of the Defence Staff. CEFCOM was similar to institutions elsewhere, including US combatant commands, Australia's Joint Operations Command, and Britain's Permanent Joint Headquarters, where the commander has operational control over units from all (or most) of the services.[17]

CEFCOM was based in an office park on the outskirts of Ottawa. While there were other defense-related offices in the same building, it was physically quite separate from the civilians in the Department of National Defence that General Rick Hillier criticized quite frequently for micromanagement in his book.[18] This, of course, was no accident, as it gave the military guys a freer hand in running operations without having to check back for approval.

In interviews with former Prime Minister Paul Martin and former Minister of Defence Bill Graham, they made it clear that the new command was Hillier's initiative. They thought it was a good idea, but felt that how the military is organized is up to the military.[19] To be clear, Hillier took office as Chief of Defense Staff (CDS) with this kind of transformation of command in mind, rather than reacting to events on the ground. And he had a free hand to do so.

Discretion

The most significant change in how the Canadian Forces fought in Afghanistan involved the limits placed upon the commanders in the field. The CF went from having the reputation of being among the less useful and reliable allies to being among the most, precisely because the military commanders in Ottawa reduced the restrictions on the forces in the field.

Canada was one of the first non-American units to southern Afghanistan in early 2002, but the rules placed on the commanders significantly limited what they could do. Lieutenant-Colonel Pat Stogran told then-Lieutenant-General Rick Hillier, "We'll never be able to fight this way."[20] Stogran was facing the same rules as a bomber pilot: any mission that might risk collateral damage needed to be approved ahead of time. This essentially meant a phone call home anytime the battle group was to leave the base, since collateral damage is always a possibility when hundreds of soldiers move out. Stogran was frustrated by the restrictions imposed by Ottawa, fearing that he might have to face a choice of violating his orders or watching atrocities be committed in front of him.[21] The result, for Canadian operations, was more time on guard duty at the Kandahar Airfield in 2002 than Stogran (and others) would have preferred.[22]

After pulling out of Kandahar in the summer of 2002, the Canadians were sent to Kabul as part of the transitioning of ISAF to becoming a NATO operation. Major General Andrew Leslie went into Kabul as Deputy Commander of ISAF and as the Canadian contingent commander in 2003.[23] Leslie had to ask Ottawa for permission for operations where there was a significant chance of collateral damage, or the potential for lethal force, significant casualties, or strategic failure.[24] Given the kind of operations at the time, the limits were not tested much.

Brigadier General Lacroix led the NATO effort in Kabul during the first half of 2004. Lacroix received as his official national guidance: "*NDHQ* [*National Defence Headquarters*] *authority is required*, prior to committing CF personnel to *any operations*, wherein there is a reasonable belief that CF units or personnel may be *exposed to a higher degree of risk*."[25] Responses to field requests could take up to twenty-four hours or more.[26] On a few occasions, Lacroix had to face the galling situation of needing to find an alternative to the Canadian contingent while waiting for deliberations in Ottawa to conclude.[27]

When Lieutenant-General Rick Hillier became overall ISAF Commander, he was very frustrated. The leaders of Canada's armed forces gave Hillier the authority to act as a NATO Commander but little influence over Canadian forces in Afghanistan. Instead, a Canadian colonel was the commander of the nation's contingent, who was operating under relatively strict caveats. "I didn't turn to Canada as my go-to nation when I wanted a job done, because of the complex and cumbersome system in Ottawa and bureaucratic approach to operations."[28]

The move back to Kandahar in 2006 occurred nearly simultaneously with a cultural revolution in the Canadian Forces. Instead of micromanaging from the other side of the globe, the new generation of military commanders in Ottawa delegated more to the field, led by a new CDS, none other than Rick Hillier. The "operational authority matrix" was greatly simplified, reducing Ottawa's role in operations.

Colonel Steve Noonan had significant room to operate "wide arcs of fire," as he called it. Noonan found himself facing a new command philosophy. Noonan was allowed to act first if necessary and then explain his actions later.[29] His successors had similarly wide degrees of discretion. Brigadier General David Fraser said, "Everything I did over there was notification, not approval.... If I had to go outside the boundaries of the CDS intent, then I would have to get approval. I never got to a boundary."[30] In the official *Letter of Intent* given to Fraser by the CDS, Fraser was told:

> Within the bounds of the Strategic Targeting Directive, you have *full freedom to authorize and conduct operations as you see fit*. In the interest of national situational awareness, *whenever possible* you are to inform me [CEFCOM] in advance of the concept of operations for any planned operations, particularly those likely to involve significant contact with the enemy.[31]

Given that Fraser was leading at the time of Canada's most intense combat since the Korean War (Operation Medusa in late summer of 2006), this is quite striking.

This pattern of increased discretion and delegation continued until the end of the Kandahar mission in 2011. Brigadier Tim Grant replaced Fraser, and found that he "was empowered to make 99% of the ops-related decisions in theatre."[32] Grant could and did send Canadian troops out of Kandahar Province to the other parts of Regional Command South to assist the British in Helmand. At no point did Grant have to reject a NATO request, although he did engage in some discussions with his NATO commanders to "achieve the desired effect."[33] Interviews with subsequent commanders, including Brigadier Guy Laroche and Major General Jon Vance, indicate that the increased discretion continued under the new CDS, General Walter Natynczyk.[34]

Only with the new "Kabul-centric" training mission in July 2011 do we see commanders facing narrower rules and limited discretion. Behind the wire has clearly meant that the new mission would have little combat capability and no authority, but, unlike the earlier restrictions, these limitations came straight from the Prime Minister's Office, as we will see below.

The pattern of these adaptations, with the civilians playing a very heavy role in shaping the acquisition of weapons and with the military officers largely deciding how the troops would be deployed, is very much in the tradition of Canadian civil-military relations. The politicians tend to be deeply involved in the purchases of weapons systems, but usually provide few instructions and little oversight about how the forces are used.

AN UNLIKELY RECIPE: MINORITY GOVERNMENT AND MILITARY CHANGE

Canada experienced two significant changes in how it could operate in the world, and both were out of character: a series of governments that fell short of majority support in the parliament; and an assertive military. These two developments greatly shaped not just where the Canadians were deployed but how they fought. In this section, I first consider how minority government actually provided significant freedom for Canadian Forces, as long as the Prime Minister was interested in manipulating the competing parties. Then I address why the military changed how it used this freedom, moving from being among the least useful to serving as among the most reliable allies in southern Afghanistan.

Minority Government as Opportunity and Constraint

One might expect a government in a parliamentary system that falls short of a majority to be quite weak. That actually depends on whether the government needs to get votes through the parliament and whether the opposition can get its act together. While there has been some debate, the Canadian constitution does not require votes in parliament for foreign deployments. The realities of minority government have meant that Prime Minister Harper had to seek support from other parties to continue the mission. This would seem to empower the opposition, but the opposing parties could not agree to much but a time limit of 2011 for combat forces in Kandahar. Thus, divisions among the three opposition parties gave the Prime Minister a surprising degree of room to maneuver. Ultimately, we saw Canada stay longer and had a far freer hand in the conduct of military operations than we might expect from a minority government fighting an increasingly unpopular war.

A quick review of the political chronology and the players leads to a discussion of the Manley Panel, which served as conduit between the challenges in the field and the politics back home. When the war started, in 2001, Jean Chretien led a Liberal majority government (elected in 2000). Paul Martin succeeded Chretien as Liberal Party leader and Prime Minister and remained in those positions despite having only a plurality of seats after the 2004 election. Stephen Harper became Prime Minister in 2006 when the Conservatives gained the highest number of seats but fell short of majority then and again in 2008. Only in May of 2011 did Harper's party gain a majority in seats.

In terms of policy positions, the Liberals had long been a supporter of peacekeeping, but Martin sought to change Canada's traditional strategy from deploying many small missions to many different efforts to a more focused effort whereby Canada could make a difference and be more visible.[35] Harper's Conservatives supported the war effort when in opposition. The two other parties operating at the federal level, the Bloc Quebecois and the New Democratic Party, tended to be pacifist and opposed to Canada's participation in the Afghanistan missions. The Liberal minority government did not hold votes to provide mandates for the mission, but doing so would not have been problematic, given Conservative support. However, once the Conservatives were in power, the Liberals became a less reliable partner, hesitating at first and then beginning to oppose the mission. Harper rushed a vote through the parliament in 2006 to extend the mission from 2007 to 2009. This happened so soon after the Liberals left office that they could not really oppose the policy that they had initiated. But it was a different story in 2007–8, when Harper had to figure out how to get enough Liberals to side with the government to extend the mission until 2011.

Harper's strategy was quite clear: to divide the Liberals to get enough support to renew the mandate. He appointed a senior Liberal politician, John Manley, who had previously served as Foreign Minister and Deputy Prime Minister, to head a commission, the Manley Panel, that would examine the mission and determine what Canada's next steps should be. The panelists visited Afghanistan, held hearings, and solicited reports. The report and the process proved to be an opportunity for a variety of actors to win key policy battles.[36]

The panelists clearly heard from the field about the growing threat of improvised explosive devices and the requests for ways to minimize the casualties produced by these weapons. As a result, the report indicated that the Canadians in Kandahar needed helicopters, unmanned aerial vehicles, and reinforcements. The helicopters would allow for less risky movements by air, especially to distant outposts, rather than sending convoys over dangerous roads. Unmanned Aerial Vehicles (UAVs) were also seen as essential in the IED fight, as they could observe the roads that would be used so that the ground forces would not be surprised. Being spread thinly over the ground was also seen as increasing Canadian vulnerability, so that led to the demand for at least one thousand allied troops.[37]

These requests from the ground essentially became conditions for renewal of the mandate, leading to the purchasing and leasing programs discussed above, as well as the deployment of an American battalion (freed up by the deployment of a French unit to Kapisa). Thus the Manley Panel process connected the battlefield to the domestic political contest by making the casualty-reducing recommendations a necessary condition for passage of an extension.

The Liberals did get something out of the report as well—a commitment for more development and reconstruction efforts. One of their reasons for opposing a mission that they had started was that Harper was viewed as not as serious about the civilian side of the effort. With the report calling for more resources and better integration of the military and civilian efforts via a "whole of government" approach, the Liberals could vote for the extension without appearing to be hypocritical.

This new mandate was more specific than the previous one.[38] Yet it did not include any caveats. An idea was floated to limit the Canadians from engaging in offensive operations, but the opposition parties could not cooperate enough to make that happen. Only some Liberals were interested in such a restriction, while the other two opposition parties were interested only in voting against the mission. Ultimately, enough Liberals voted with the Conservatives to extend the mandate until 2011. That deadline of 2011 became the most important part of the mandate, as it ended the combat mission.

It is important to note that Harper apparently had two distinct goals when launching the Manley Panel—to get the mission extended and to gain more control over the Canadian Forces. Setting up better coordination between civilians and commanders, including a senior cabinet post held by David Mulroney, was viewed as a way to give the Prime Minister a better grasp of military operations.[39] However, the military resisted, as will be illustrated further below.

Before moving on, we ought to consider public opinion. Were the changes in

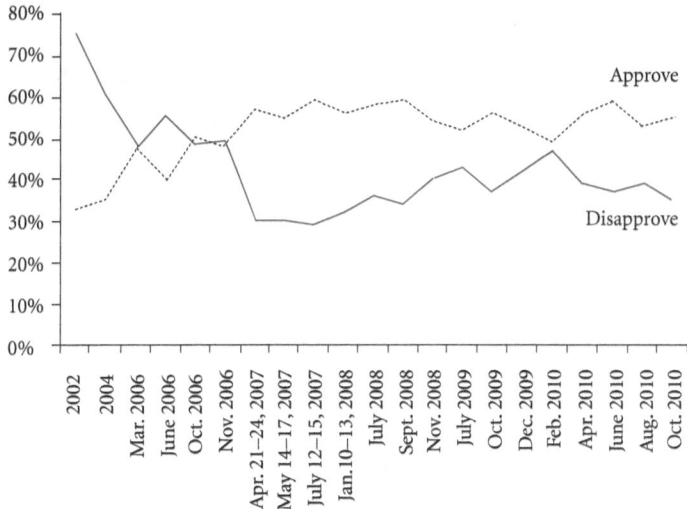

FIG. 2. Casualties and Public Opinion[40]

how Canada fights related to the standing of the mission back home? In a word: no. While the decline in support for the mission might explain why Canada left Kandahar in 2011, such trends do not correlate with Canadian behavior on the ground. Figure 2 illustrates this quite clearly. While caveats might have ensured popularity with high restrictions and public support in the early years, it is quite clear that declining popularity did not lead to increased restrictions after the mission declined in popularity.

Generational Change: From Somalia to Afghanistan

The Canadian Forces have undergone a cultural revolution in a very short time span. After a "decade of darkness,"[41] where the highest-ranking officers were more concerned with avoiding failure, new generation changed to a philosophy that quickly became common sense to the next generation—mission command.[42]

In 1993, during Canada's participation in the UN mission in Somalia, an event occurred that would have lasting effects. Specifically, members of the Canadian Airborne Regiment beat and killed a Somali that they had captured while he had been trespassing on their base. The initial efforts to cover up the event led to a series of investigations that ultimate led to the disbanding of the entire regiment, the firing of two different Chiefs of the Defence Staff, and one Minister of Defence losing his position.[43]

This sequence of events helped to foster a senior officer corps that was risk averse and likely to impose restrictions upon troops on the ground. In Bosnia, Canadian forces faced significant restrictions on how they could operate, even after NATO replaced the United Nations in late 1995.[44] Indeed, this is where the nickname of Can'tbat was first applied. It insinuated that the Canadian Forces would always say that they "can't" when asked by NATO commanders to do something. US Commander General Montgomery Meigs told Canadian General Rick Hillier: "[T]he Canadian troops were the most proficient under my command in the NATO mission. It was a pity that Canada would not let them do anything."[45]

These restrictions, imposed by the Canadian commanders back in Ottawa, frustrated not only the allies but also a generation of colonels who would move up the ranks. As these new leaders were rising, they were still hemmed in by the more risk averse generals and admirals who had come before them. General Ray Henault and Vice Admiral Maddison were the CDS and DCDS (Deputy Chief of Defence Staff) from 2001 to 2005, and they clearly bore the scars of serving at high levels during the post-Somalia years.[46] To be fair, Afghanistan in 2002 was Canada's first real combat mission in about fifty years, so Henault was especially cautious during the first deployment under Stogran, discussed above. "Stogran is probably correct about how tight it was," said Henault.[47] Maddison indicated the importance of the Somalia experience: "We had to be absolutely clear we would not lose control of people executing missions."[48] This concern led to an elaborate targeting process to address the problem of collateral damage, often involving the DCDS and sometimes the CDS, essentially requiring significant phone calls home for permission to operate when risks were higher. Conversations with senior civilians who served in the Department of National Defense (DND) at the time support the view that Henault and Maddison were quite risk averse.

This all changed with the appointment of Rick Hillier as CDS. Hillier's memoir is full of frustration with how things used to work in the Canadian Forces.[49] "I had some frustrations in Bosnia when I had no role to play [serving as the NATO regional commander], with the Canadian contingent commander, a Colonel, who had complete authority. Same as in Afghanistan, as a three star, I literally had no ability to influence the Canadian contingent."[50] Once in a position to do something about it, Hillier did exactly that. He reformed the organizational structure, creating CEFCOM, to move operational command of missions overseas to a greatest distance not just from the civilians in the

Department of National Defence but also from himself. While phone calls and email made this change symbolic, it had a real effect, giving the new institution significant independence in planning operations.

Hillier had an abundance of army officers whom he could appoint to positions in Ottawa and in Afghanistan who had similarly frustrating experiences in Bosnia, Croatia, Kosovo, and Afghanistan. They all found the tight constraints of previous caveats, particularly in the Balkans, enormously frustrating. In Afghanistan, as mentioned above, Hillier, as Commander of ISAF, had to ask permission from a colonel to use the Canadian Forces and was often refused. As a result, they quickly bought into the mission command philosophy he articulated.[51]

When asked about the adoption of mission command, Hillier had a hard time articulating an alternative philosophy.[52] He saw delegating to the folks in the field as common sense, and that anything less would be a tremendous waste of the resources and time spent training Canadian officers. Hillier asserted that "having flexibility to change is a sign of a mature organization and of the command team involved."[53] He attributed the changes as resulting "partly from my frustration but also of the command group, who have been around this numerous times ... over the past fifteen years."[54]

It is likely that Hillier's reforms, even in the face of pressures to cut budgets—that is, transformation—are likely to endure until a new shared experience produces a new culture shock. His successor, General Walt Natynczyk, and his likely successors have the shared experiences of tight caveats in Bosnia and much more operational freedom in Afghanistan. They certainly have reported much less frustration and much more satisfaction with the way things are, compared with the way it used to be.[55]

Toward the end of Canada's time in Afghanistan, changes in circumstances and a change in command led to much news. With the Americans surging into the province of Kandahar and with the Regional Command Headquarters taking a more directive role, Brigadier General Jonathan Vance had new opportunities and constraints. Canadians were no longer largely responsible for security in the province, but were now to be concentrated in certain sectors. Vance saw this as a chance to dedicate resources to a single inkspot to prove that both Canada and the government of Afghanistan can deliver. The model Deh-e-Bagh village project in the Dand District became a focal point for the Canadian military effort in 2010.[56] It was an effort to show how ISAF could be successful in building confidence by deploying forces to a particular area for a longer period of time, rather than the old way of running around the province to put out whatever fire might be burning. Vance was also motivated by the need for the Canadians to make a big difference at least in one spot, as time was running out on the mission with the 2011 deadline.[57] The CF not only remained in place to hold the area but also focused development resources to build projects to gain favor, such as solar-powered street lights.[58] This ran into a certain amount of trouble back home, as the other branches of government had set up targets and processes focused on the entire province. More schools in Dand did not necessarily mean that Canada was meeting its stated target of fifty schools, since the added ones in Dand were not on the lists generated by the Afghan Task Force back in Ottawa.[59]

In the interagency battle that ensued between the military on one side and the Department of Foreign Affairs and International Trade on the other, the higher ups in the Canadian Forces supported the initiative of the ground commander. Adapting to changing circumstances and more specific directives, Vance developed a new plan, more faithful to counterinsurgency doctrine and to the NATO effort even though it was contrary to much of the civilian planning. It is not clear that Vance could have had such freedom to pay heed to his multinational chain of command and develop an operation that conflicted to that degree with the civilian masters before 2006. Vance had more discretion as he was working under a new generation of commanders who welcomed initiative rather than risk avoidance and valued adhering to the NATO campaign plan.

Kabul-centric Training: When a Prime Minister Seizes Control

One of the paradoxes of the Canadian experience has been the latitude the Canadian Forces have largely had to determine how they operated in Afghanistan while a minority government ruled back in Ottawa. Above, I show that the opposing parties could not cooperate enough to impose restrictions, especially when Prime Minister Harper used a senior Liberal, John Manley, as a wedge to divide the Liberals. Yet the issue of Afghanistan challenged the Harper government through a series of elections, as the mission probably did cost the Conservatives votes in Quebec. So then, why did Harper not impose restrictions upon the Canadian Forces? Given his reputation as a control freak,[60] this is quite a mystery. In this last section, the connections between the two previous pieces of minority government and generational change are tied together.

In Canada, the formal Commander-in-Chief of the Canadian Forces is the

Governor General, not the Prime Minister. This has tended to create the perception that Canada is a near perfect model of Huntington's civil-military relations, where the civilians say where the military goes but the military says how it operates. The Prime Minister appoints the Chief of the Defence Staff and can fire him, but that can be very costly. Rick Hillier created a healthy amount of heartburn in the Prime Minister's Office, as he was quite public in his opinions, even when they ran against the Minister of National Defence. It is no accident that Hillier was eventually replaced by a softer-speaking Walt Natynczyk. Still, the Prime Minister did not appear willing or able to impose restrictions on the ongoing mission in Kandahar. Perhaps he sensed that this would make him more responsible for any subsequent casualties. Instead, Harper tried to control the messaging of the mission, with the various agencies having their talking points vetted by the Prime Minister's Office.

So Harper could not or would not try to alter a mission while it was underway. When given the opportunity, however, to craft a new mission, Harper grabbed it. The new training mission that began in the summer of 2011 is focused on Kabul and other places that are less dangerous and in billets that are entirely behind the wire. That is, these new units are heavily caveated—they will not serve as mentors in the field, nor will they engage in any combat operations. "There is zero offensive combat action involved in the mandate here whatsoever."[61] Harper gets to satisfy his alliance partners without incurring significant domestic political costs. Learning that the civilians determine the mission, Harper has crafted a mission that significantly constrains how the military can operate. Indeed, the initial proposals shaping the deployment were made entirely within the Prime Minister's Office with the military being let into the process only after all of the decisions were made.[62] Thus Harper has adapted to the difficulties of operating a modern military in a complex environment: by developing a new mission that limits the military's discretion.

CONCLUSION

Canadian Forces faced intense combat, more challenging and more costly than at any time since Korea, in Kandahar. Undermanned in difficult terrain and focused on the IED threat, the Canadians adapted by deploying tanks and helicopters. Newly empowered ground commanders used these new assets in cooperation with NATO forces with little interference from back home. Prime Minister Harper had to innovate to keep the mission going, even if only for a few years, via the Manley Panel process. This effort had only a modest impact on giving him more control of the military, but Harper innovated by developing a new mission that came with restrictive caveats that the military could not alter.

Unifying all of this is how context combined with pre-existing preferences to drive the direction of adaptation. Frustrated by higher than expected casualties, even more problematic in a time of minority government, the civilians were willing to provide new equipment and significant leeway. This opened up opportunities for the military to demand and receive equipment that it had long sought (especially helicopters), and they used the equipment as they saw fit, increasingly following NATO directives rather than instructions from home. This caused some tensions with the civilians overseeing the governance and development efforts, but the military chain of command supported their commander in the field.

The combination of minority government and generational change facilitated adaptation. The second dynamic is less surprising than the former. We would expect that individuals who were put into difficult positions by fearful superiors would become more willing to take risks and to delegate more freely when the opportunities arose. The surprising element here is that Prime Minister Harper had more latitude because his opponents could not cooperate with each other. They did impose the 2011 deadline, but they could not agree enough to force changes on how the mission was to be conducted. The politics of the Manley process also meant that the opposition could not stop the expensive and somewhat reckless acquisition of new equipment.

What does this case suggest about how military adaptation works at the highest levels? First, national capitals work slowly in peacetime, but the urgency of operational requirements sped processes up a great deal. While Auditors-General may have problems with this, the reality is that politicians and policymakers will skip steps when the alternative is to be held responsible for casualties in the field. The debates about helicopters in the 1990s, or the F-35 more recently, illustrate how extended political decision-making can be. By contrast, decisions on tanks, helicopters, and UAVs were made quickly—because opposition would be seen as endangering the troops.

Second, while bureaucrats seeking more resources and politicians seeking to grandstand can try to take advantage of these opportunities, the moments of adaptation are created by events on the ground. Canada had been in Afghanistan on and off for several years before tanks or helicopters became priorities. Operation Medusa put tanks on the agenda. The feedback that the Manley

Panel received from the troops in Afghanistan put helicopters and UAVs on the national to-do list. Adaptation inherently involves reacting to challenges and events. The most salient would be those on the ground where the stakes are the highest—not just for the grunts on the battlefield but for the politicians back home.

Third, in a complex, multidimensional effort such as counterinsurgency in Afghanistan, all three sets of factors highlighted in Chapter 1—operational challenges, domestic politics, and civil-military relations—will shape adaptation. Not every factor matters equally or all the time. In the Canadian case, the operational challenges were so severe in terms of the mismatch between capabilities and the physical and human terrain that adaptation at the strategic level was quite likely. In terms of domestic politics, public opinion mattered only indirectly, but the four-cornered combat that had been Canadian politics in the 2000s deeply shaped the mission, particularly its length but also the addition of new capabilities and the granting of increased discretion. The third category of civil-military relations also mattered, as the existing rules and norms empowered the Chief of the Defence Staff so much so that a new one could bring about a cultural revolution.

To be sure, Canada could adapt only so far. It had limited land forces to deploy, so it could never adequately cover its assigned area of responsibility without substantial assistance. There would be no significant increase in the size of the forces and only limited added capabilities. Not only was the size of the mission limited, but the patience of Canadian politics was finite. While the Conservatives could count on the opposition to disagree about most things, ending the mission in Kandahar did produce consensus among the other parties. Indeed, the Prime Minister also lost his patience with the mission, so that the end in 2011 became a long foregone conclusion rather than a date for a new political fight.

What is certainly true is that the Canadians faced their greatest military challenge since World War II. While the debates will continue about the merits of the mission, there is no doubt that the understaffed Canadians learned how to fight this new kind of war and developed new ways of working both in the field and at home.

NOTES

Acknowledgments: David Auerswald is my coauthor on the larger project from which this paper is drawn, and has influenced much of the thinking here. The Social Sciences and Humanities Research Council has funded much of this project, as has the Canada Research Chair program, including the very helpful research assistance of Bronwen De-Sena, Chris Chimm, Alexia Jablonski, Jenyfer Maisonneuve, Mark Mattner, Ora Szekely, and Lauren Van Den Berg. The Security and Defence Forum of Canada's Department of National Defence was instrumental in arranging the first sets of interviews and many other important contacts (as well as my trip to Afghanistan in 2007). Errors are those of the author. *The views expressed here are those of the author and not the Canadian Department of National Defence.*

1. Hillier also noted this dismissive nickname during a speech to the Conference of Defense Associations Institute, February 22, 2008.

2. The alternatives were in RC-West, which would have been safer but either working under the Italians in Herat or isolated in Chaghcharan, where the Lithuanians have served rather quietly.

3. Interview with former Prime Minister Paul Martin, Montreal, Canada, March 29, 2007.

4. Brigadier General Denis Thompson was most explicit about the numbers problem when he briefed students at McGill University, March 24, 2009.

5. I have heard these phrases used often in interviews with Canadian officers and civilians.

6. See Lee Windsor, David A. Charters, and Brent Wilson, *Kandahar Tour: The Turning Point in Canada's Afghan Mission* (Mississauga, Ontario: John Wiley and Sons Canada, 2008).

7. Operation Medusa, in the fall of 2006, was not the first combat the Canadians faced in Kandahar, but it was the first extended offensive operation with a significant escalation in the pace and severity of casualties. See Bernd Horn, *No Lack of Courage: Operation Medusa, Afghanistan* (Toronto: Dundurn, 2010).

8. The post–July 2011 training mission is "Kabul-centric," with most of the trainers in Kabut but some in Herat and Mazeer-e-Sharif. The numbers used to build this slide come from the NATO placemats (available at http://www.isaf.nato.int/isaf-placemat-archives.html, accessed October 24, 2011), slides on presentations given by Canadian officers that I have seen over the years, and newspaper accounts of the new training mission.

9. "The Return of the Leopard." *Ottawa Citizen*, July 8, 2006: available at http://www.canada.com/ottawacitizen/news/observer/story.html?id=95b4c9e5-de13-4425-bc87-218c1031583c, accessed September 20, 2011.

10. "Canada Sends More Troops, Tanks to Afghanistan," *The Age*, September 17, 2006: available at http://www.theage.com.au/news/world/canada-sends-more-troops-tanks-to-afghanistan/2006/09/16/1158334734082.html, accessed September 20, 2011.

11. "Ottawa Announces Tanks to Battle the Afghan Heat, and Enemy," *Canada.com*: available at http://www.canada.com/topics/news/world/story.html?id=bee62caa-0f88-4cc9-a5a1-5947dfe73b5d, accessed September 20, 2011.

12. "New Tanks Hobbled by Glitches and Lack of Spares," *Canadian Press*: available at http://www.ctvbc.ctv.ca/servlet/ArticleNews/print/CTVNews/20091103/AG_tanks_0 91102/20091103/?hub=BritishColumbiaHome&subhub=PrintStory, accessed September 20, 2011.

13. Independent Panel on Canada's Future Role in Afghanistan: available at http://dsp-psd.pwgsc.gc.ca/collection_2008/dfait-maeci/FR5-20-1-2008E.pdf, accessed October 28, 2011, 35.

14. One officer with command experience in Kandahar indicated skepticism about the real need for tanks, suggesting bureaucratic politics.

15. Office of the Auditor General of Canada, "2010 Fall Report: Chapter 6, Acquisition of Military Helicopters": available at http://www.oag-bvg.gc.ca/internet/English/parl_oag_201010_06_e_34289.html#hd4b, accessed September 20, 2011.

16. Thus far, each of the three CEFCOM commanders has been from the land staff (army) and served previously in Afghanistan or in the chain of command of the Afghanistan operations.

17. Special operations forces for some countries operate under a different chain of command much of the time.

18. Rick Hillier, *A Soldier First: Bullets, Bureaucrats, and the Politics of War* (Toronto: Harpercollins, 2010).

19. The issue of command structures became more political in the summer of 2011, when Lieutenant-General Andrew Leslie criticized the new commands as causing duplication and essentially a waste of personnel.

20. Hillier, *A Soldier First*, 248.

21. Interview with Colonel (ret.) Pat Stogran, April 25, 2007, Ottawa, Ontario. See also his article: Patrick Stogran, "Fledgling Swans Take Flight: The Third Battalion, Princess Patricia's Canadian Light Infantry in Afghanistan," *Canadian Army Journal* 7, nos. 3–4 (2004): 14–21, for his blunt but career-threatening views on his experience.

22. Hillier, *A Soldier First*, 248.

23. Interview with Lieutenant General Andrew Leslie, Ottawa, Ontario, March 8, 2007.

24. Strategic failure refers to the possibility of a tactical effort potentially undermining the NATO mission and/or the Afghan government. Interview with LTG Leslie.

25. DCDS Intent Task Force Kabul, December 19, 2003, A0241084, 6, acquired via Access to Information request. Italics added.

26. Interview with Brigadier General Jocelyn Lacroix, Kingston, Ontario, February 2, 2007. Major General Peter Devlin ranked Canada at this time as being in the middle of three tiers of countries in terms of their flexibility to respond to events and their domestically induced constraints. Interview, May 15, 2009.

27. For more on restrictions Canada faced while in Kabul, see Sean M. Maloney, *Confronting the Chaos: A Rogue Military Historian Returns to Afghanistan* (Annapolis, MD: Naval Institute Press, 2009).

28. Hillier, *A Soldier First*, 257–58.
29. Interview with Colonel Steve Noonan, Ottawa, Ontario, March 26, 2007.
30. Interview with Brigadier General David Fraser, Edmonton, Alberta, January 29, 2007.
31. Commander's Directive to Commander, Task Force Afghanistan, Rotation 2, (3350-165/A37) A0232107, acquired via Access to Information, 14. Italics added.
32. Interview with MG Tim Grant, Ottawa, Ontario, February 7, 2008.
33. Ibid. Grant did point out that allies had not only caveats but also their own agendas, of which one had to be conscious.
34. Given Natynczyk's comments when we interviewed him, when he was the Vice-Chief of the Defence Staff, we did not expect significantly decreased discretion. Interviews with a series of officers after Natynczyk became CDS, along with the letter of intent that BG Jon Vance received in December 2009, bear this out.
35. Interview with Paul Martin, Montreal, Canada, March 29, 2007.
36. The Independent Panel (2008) toured Afghanistan in the late fall of 2007. I was part of a group of academics that happened to be visiting Afghanistan in December of the same year, so we ended up receiving many of the same briefings as the panel.
37. This was also seen as a way to mitigate the perception that the Canadians were fighting alone in Kandahar, which was never true. Despite the reinforcements, Canadians still saw Kandahar as a Canadian problem.
38. For the first mandate, see House of Commons of Canada, 39th Parliament, 1nd Session, Journals no. 25, May 17, 2006; for the second, see House of Commons of Canada, 39th Parliament, 2nd Session, Journals no. 66, March 13, 2008.
39. Interview with senior civilian official, May 2011.
40. Angus-Reid Public Affairs Surveys, 2002–10: available at http://www.angus-reid.com/issue/afghanistan, accessed September 13, 2011.
41. Hillier, *A Soldier First*.
42. For more on mission command, see Eitan Shamir, *Transforming Command: The Pursuit of Mission Command in the U.S., British, and Israeli Armies* (Stanford: Stanford University Press, 2011).
43. For the Somali affair, see David Jay Bercuson, *Significant Incident: Canada's Army, the Airborne, and the Murder in Somalia* (Toronto: M&S, 1996): and Sherene Razack, *Dark Threats and White Knights: The Somalia Affair, Peacekeeping, and the New Imperialism* (Toronto: University of Toronto Press, 2004).
44. These restrictions were reported in interviews with a number of officers who served in command capacities in both Bosnia and Afghanistan.
45. Hillier, *A Soldier First*, 158.
46. Indeed, I had been interviewing Canadian commanders for quite some time before the word "Somalia" was brought up—repeatedly by Maddison. Interview with Vice Admiral (ret.) Greg Maddison, Montreal, June 19, 2007.

47. Interview with General (ret.) Ray Henault, Ottawa, November 4, 2010.
48. Interview with Maddison.
49. Hillier, *A Soldier First*.
50. Interview with General Rick Hillier, Ottawa, March 11, 2008.
51. I met with nearly every Canadian commander in Afghanistan, and most expressed a lot of frustration at the restrictions they faced in earlier deployments.
52. The interview was conducted while he was still the CDS, so he was apparently reluctant to criticize his predecessors. Obviously, this became less of a constraint when he wrote his memoir (2010).
53. Interview with Hillier.
54. Ibid.
55. I interviewed General Natynczyk when he was the Vice-Chief of the Defence Staff, both commanders of CEFCOM, and nearly all of the officers who commanded in Afghanistan. To a man, they report being quite pleased with the discretion they had after 2005, contrasting sharply with those who served before 2005.
56. This paragraph is based on interviews with Major General Jonathan Vance on June 22, 2011; Lieutenant-General Marc Lessard, Commander of CEFCOM, on August 2, 2011; Elissa Golberg, the former representative of Canada in Kandahar and then Director of the Stabilization Task Force in Ottawa on June 7, 2011; and a few anonymous military and diplomatic officials.
57. Interview with Vance.
58. "Canada's Afghan 'Model Village' Praised by U.S. General," CBC News, June 25, 2009: available at http://www.cbc.ca/news/world/story/2009/06/25/mcchrystal-afghanistan-canadian-village025.html, accessed October 25, 2011.
59. Some of the other consequences of the Manley Report were quarterly reports and signature projects. The resulting civilian planning was focused very much on the province of Kandahar, as the website suggests: available at http://www.afghanistan.gc.ca/canada-afghanistan/index.aspx, accessed September 12, 2011. Interview with Vance.
60. For instance, see the CBC backgrounder on Harper: available at http://www.cbc.ca/news/background/harper_stephen/, accessed September 15, 2011. An early book on his time as Prime Minister is Lawrence Martin, *Harperland: The Politics of Control* (Toronto: Viking Canada, 2010). See also Murray Brewster, *The Savage War: The Untold Battles of Afghanistan* (Mississauga, ON: Wiley, 2011).
61. James Cudmore, "Canada's Afghan Training Mission Details Revealed," CBC News, May 17, 2011, http://www.cbc.ca/news/politics/story/2011/05/17/pol-afghan-training-mission-details.html, accessed February 18, 2013.
62. This was clear from a number of interviews with people who would have otherwise been involved in the design of the deployment.

10 Military Adaptation by the Taliban, 2002–2011

Antonio Giustozzi

INTRODUCTION: ORIGINS AND EVOLUTION OF THE TALIBAN INSURGENCY

The start of the Neo-Taliban insurgency in 2002–3 was slow and fraught with difficulties. The Taliban had been demoralized and dispersed by their quick defeat in Operation Enduring Freedom (OEF), and the help of foreign jihadist organizations, mostly Arab or Pakistani, was crucial in the relaunching of the insurgency. Apart from cross-border raids mostly launched by their Pakistani allies, the Taliban focused their effort on reorganizing and remobilizing their former members and associates, whether they had taken refuge in Pakistan or had returned to their villages in Afghanistan. A key factor in enabling the Taliban to gradually enjoy some success in their remobilization effort was the harassment and discrimination to which many former Taliban were subjected by the winners in the 2001 war, who by 2002 had turned into chiefs of police and governors and were therefore in a position to target their former rivals. As a result, the insurgency in its early days was quite fragmented, with some pockets emerging inside Afghanistan, either autonomously or as a result of the work of Taliban political agents, and cross-border raids mainly meant to convey the image of a resurgent Taliban and to distract the attention of the Americans and of the Afghan government from the political work of laying the ground for the insurgency inside Afghanistan.[1]

This fragmented origin of the Taliban, the lack of a capable organizational infrastructure, and the predominant cultural and political patterns of southern and eastern Afghanistan meant that from 2002 onward, as the insurgency gradually expanded, it tended to organize along a patrimonial pattern, with charismatic leaders gradually developing a kind of integration of the different fragments through personal networking, either with each other or through a central hub, the political leadership of the insurgency. A combination of military requirements and of political pressure ended up shaping the insurgency into a relatively coherent whole: each charismatic leader developed his own network of commanders, personally loyal to him, usually recruited along tribal lines; each commander in turn recruited his subcommanders among his personal connections. The networks all come together at the top, sometimes having in common just the loyalty to Mullah Omar as the leader of the movement and the legitimate ruler of Afghanistan. Some networks were closer to each other than others, while some were actively hostile to other networks, a hostility fed by personal and tribal rivalries. In sum, in these early years the insurgency could be described as a network of networks and could relate to each other only through the central hub. About twenty main networks could be counted in 2011, with the number fluctuating as a result of the fate of the charismatic leaders and the emergence from time to time of new rivalries, as well as the expansion of the insurgency.[2]

While the network-based organization granted the insurgency great resilience, in particular making it able to absorb major punishment without lasting damage, it also came with significant shortcomings. It made the implementation of a coherent strategy by the leadership difficult (although not quite impossible, as in the very early stage, before the network of networks developed); tactically it hindered cooperation, particularly where networks overlapped, which was very frequently. The degree of control that the leadership could exercise over the network was limited and achieved mainly through the co-optation of the network leaders into the leadership itself. Decision-making had to be done mostly by consensus.[3]

This character of the insurgency has sometimes led analysts to describe the insurgency as fragmented and the Neo-Taliban insurgency as essentially an opportunistic umbrella label, covering in fact a variety of local insurgencies, driven by different motivations. Undoubtedly, some of the networks were more autonomous and assertive than others; the Haqqani network, originally based in southeastern Afghanistan, being a case in point (but there are other examples as well). Sometimes, the fastest growing and better supported networks, such as the Haqqani's, tried to renegotiate the terms of the relationship with the leadership, gathered in the so-called Leadership Council; such efforts peaked in 2008, but, as we shall see, petered out afterward. Nevertheless, the evidence suggests that the insurgents managed to operate with national aims in mind and were not focused on a few specific localities. All the network leaders were old Taliban; the participation of local communities to the insurgency is a fact, but it was subsumed into the Taliban, who on the whole succeeded in exploiting the communities to a greater extent that the communities managed to exploit the Taliban. The different networks were brought to contribute to a common strategy, even if they might have interpreted that strategy and their role within it in their own ways. Therefore, the fragmentation of the insurgency should not be exaggerated; the "network of networks" model seems a better model to describe the insurgency.[4]

Until 2005 it cannot be said that the Taliban were under serious military pressure. Their strategy of deception combined with the distraction of their main adversary (the US) to ensure that US military units continued to focus their efforts on the cross-border raids of the jihadist allies of the Taliban in the southeast and east, neglecting the inroads the Taliban were making in the south. In 2005 NATO (the North Atlantic Treaty Organization) was noticing some deterioration in the south, a fact that contributed to the decision to deploy larger numbers of troops in that region the following year. In 2006 the Taliban felt sufficiently strong to move down from the mountains and enter the heavily populated areas of central Helmand and those surrounding Kandahar city. This move rang the alarm bell in Washington and in NATO headquarters, soon leading to efforts to contain the Taliban and improve the efficacy of counterinsurgency. The increased presence of Western troops on the ground caught the Taliban tactically unprepared: the fighting was still being conducted in an amateurish way, with guerrilla and asymmetric tactics being rarely used. The introduction in 2007 of Special Forces raids against Taliban commanders, largely in the south, further increased the pressure on the amateurish Taliban. We can identify this as the point of departure of Taliban military adaptation.[5]

The motivation of the insurgents has been the topic of much debate; ISAF (International Security Assistance Force) and the Afghan government in particular have been trying to cast them as a bunch of mercenaries, motivated by the desire to loot or fighting for money. While this might be useful propaganda, it makes for poor analysis. The first evidence of the fact that the Taliban are mostly not motivated by greed is their willingness over the years to take heavy casualties and keep fighting—not the behavior of mercenaries. The Taliban casualty ratio is difficult to calculate because the exact size of the insurgency is not known, nor are their casualties known, but based on anecdotal evidence it can be estimated in the range of 20 to 30 percent of their fighting force each year. Furthermore, evidence of looting and of the misappropriation of property has been very limited, and when it has occurred it has caused a reaction from the leadership: a wave of looting in 2008 in provinces such as Wardak, Logar, and elsewhere, where the ranks had been expanded very fast in that year, was punished with executions and other punishments in 2009. There is evidence, in various provinces, of the Taliban's not only banning the theft of property from civilians not linked to the government but also actively intervening to prevent it. The picture is somewhat confused by the presence of thousands of bandit gangs everywhere in Afghanistan, but particularly where the insurgents operate. Often such gangs cast themselves as Taliban in order to discourage police of local communities from resisting them. The Taliban have also often encouraged the activities of gangs in areas where they were just trying to get established, possibly increasing the confusion. Another piece of evidence is the interrogation of prisoners by ISAF, which has seemingly indicated that economic gains do not figure prominently among their motivations, despite the fact that it would be quite convenient for them to claim that in order to secure a faster release. The prisoners usually claim to be fighting against a corrupt government and against foreign presence in the country, and appear to have a strong sense of belonging to the Taliban as a movement.[6]

While it is clear that in the south the Taliban are supported by narcotics traders and many Taliban themselves grow poppies, this does not necessarily indicate that they are motivated by economic gain.[7] In fact, the Taliban's casualty rate in Helmand has been probably the highest in the country; joining the Taliban in Helmand is almost a death sentence. No prospect of economic gain can stand against such a high chance of getting killed in the space of a few years.

STRATEGIC LEVEL ADAPTATION

Below we shall show changes that occurred at the tactical and organizational levels, as well as at the political one. Proving that adaptation occurred within the Taliban's strategy is difficult, because information about the character of their strategy is so fragmentary. Changes can be identified, but it is difficult to determine if they represented adaptation or whether they had been planned from the start. In terms of the main directions of the Taliban's effort, the effort to expand in the north of the country might have been a decision taken in 2006 to relieve pressure in the south, although it might also have been meant to happen from the start once conditions were ripe. The growth in the allocation of resources to eastern Afghanistan in late 2010 might similarly be a strategic decision, meant again to relieve the south of an ever-growing pressure and at the same time to exploit perceived weaknesses of the enemy in the east. Finally, a debate over the possibility of negotiation with the enemy occurred in 2010–11, but even in that case it is not clear whether this can be described as a new development, as feelers about the possibility of talks were already sent as early as 2003.[8]

Changes to Strategy

Although details are scant, we know that strategy debates have been taking place within the Taliban leadership, reportedly with the assistance of insurgency experts from Pakistani military intelligence (Directorate for Inter-services Intelligence, or ISI). In addition, from 2006 onward there has been growing evidence of an ability by the Taliban to implement a degree of strategic prioritization. Although as described above the networks were an impediment in the formulation of a coherent strategy, the leadership worked with them (the two in fact coincided to a large extent, as most network leaders were incorporated into the Leadership Council) to obtain their consensus; in addition the leadership had a number of tools at its disposal to help and incentivize the development of the insurgency in specific areas, according to its strategy. Such tools included the dispatch of political agents and military advisers or specialists to regions where they were most needed. As a result, seasoned southern fighters and non-Afghan jihadists started appearing in areas where the insurgency was simply trying to assert itself, such as Wardak in 2006 and northern Afghanistan in 2008. Therefore it can be argued that even in 2005–8, if not earlier (see previous reference to the "strategy of deception"), the Taliban did have a strategy and a degree of ability to implement it, even if sometimes they faced serious difficulties in getting the various networks on board when important decisions had to be made. In the section about organizational developments we shall discuss how from 2008 onward the Taliban as an organization have tried to reshape the insurgency in such a way as to increase central control over the fighting units, hence expanding the strategic options available to fight their enemies.[9]

Recruitment

Throughout the period taken into consideration here, the Taliban have consistently subordinated tactics to strategic considerations. A change in attitude that has been possible to detect is a shift in the focus of recruitment away from the clergy (the original social base of the movement) toward other constituencies: former jihadist commanders active in the 1980s and 1990s and, more recently, the disenfranchised youth. The rationale for this change seems to be that the Taliban wanted to accelerate the expansion of the insurgency and boost its ranks. A number of former jihadists from other political groups did join the Taliban, including some from the Taliban's erstwhile rival, Jamiat-i Islami.[10] The recruitment of the nonclerical youth appears to have intensified in 2011, when the Taliban were reported to systematically target high school students for recruitment at least in a few provinces around Kabul.[11] Another change (which might in fact be the result of delayed implementation of a plan conceived at the beginning of the insurgency) has been the rapidly growing investment in governance structures by the Taliban from 2006 onward: shadow governors and judges.[12]

Throughout 2002–11 the Taliban have also constantly sought to expand their ranks. From 2010 onward, however, it appears that the effort intensified, perhaps in direct response to the announced US surge (see above on the targeting of new social categories for recruitment). Although measuring the size of the Taliban insurgency is a hard task, particularly as from early 2010 ISAF and the intelligence agencies have refused to release new estimates, the ever-growing number of insurgent-initiated attacks suggests that the insurgency has kept expanding to date.

In addition to the data about violence, other indications of a more determined effort to expand ranks and accelerate the growth of the insurgency consist of a seemingly determined effort to intensify cooperation with the Iranian Pasdaran during 2010, when at least three meetings took place between Iranians and Taliban leaders, and of an escalating Pakistani support for the insurgency.[13]

Another important shift has been the growing investment in the professional training of Taliban recruits. This has taken two shapes. As far as common fighters are concerned, at two main sites in Pakistan by 2011 each recruit was sent through a training process that made him proficient in the use of the weapon assigned to him (rifle, machine gun, rocket launcher, and so forth). This is in clear contrast to the practice of the early days, when every fighter was supposed to learn how to use his weapon on his own and little or no actual training was provided. The result was improved marksmanship and fewer accidents caused by accidental discharge. Reportedly, Mullah Omar issued a decree banning the deployment of untrained fighters in 2010.[14]

With regard to the cadres of the insurgency, evidence emerged in 2010–11 that years earlier the Taliban had started inducting teenagers in training and indoctrination courses lasting years; by 2010 such courses were outputting the first classes of graduates, used to lead tactical units and to man special units in charge of the most daring attacks, almost invariably resulting in the death of the attackers (many of whom were equipped as suicide bombers).[15]

Concern for Image

Gradually from 2007 onward the Taliban have also modified their political stance, presumably with the aim of facilitating the cooperation of the population with their fighters. After an aggressive campaign against state schools in 2006, perhaps motivated by the fact that these were soft targets that were abundantly available throughout the country, the Taliban gradually softened their attitude from 2007 onward. This culminated in a decision to allow the boys' schools to reopen in 2011, on condition that the government adopt the Taliban curriculum.[16] In 2007 Mullah Omar also issued a decree that allowed his commanders to ignore his social edicts dating back to the 1990s, whose strictures were alienating the population of the areas recently come under Taliban control. Finally in 2010 the Taliban started adopting a softer approach toward NGOs (nongovernmental organizations) and development aid in general: under pressure from the communities they started allowing NGOs, and even government agencies, to carry out projects, although always collecting "revolutionary taxes" on the projects.[17]

In this regard it should also be mentioned that the Taliban have over time shown a growing concern for their image vis-à-vis the civilian population, even in their internal documents; the three versions of the Code of Conduct issued thus far have gradually tightened the rules of engagement, ordering the fighters to avoid whenever possible involving civilians in the fighting. The rules also prescribe trying suspect spies and collaborators before executing them. The leadership struggled to impose the tightened rule on its field commanders, and cases of commanders disregarding the rules were reported through 2011, but the process is nonetheless significant of certain concerns at the top of the movement. It is also true that at least in some areas the Taliban have been trying to devise Improvised Explosive Devices (IEDs) that can be armed and disarmed as needed, in order to avoid unnecessary civilian casualties.[18]

Adjusting Target Lists

Another likely strategic choice in response to the growing challenge posed by the adversary (ISAF) has been the intensification of the targeted assassina-

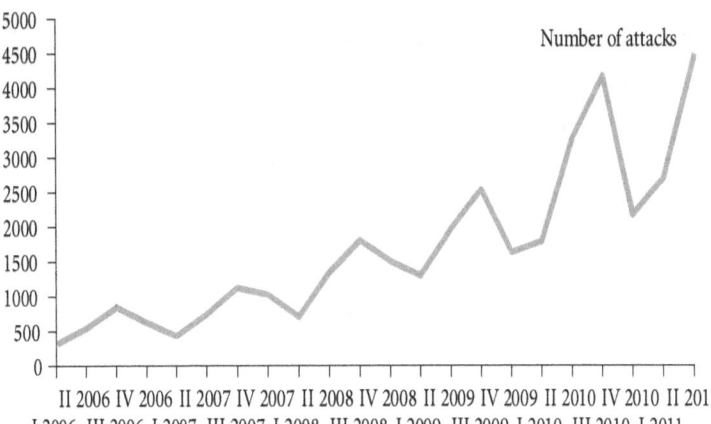

FIG. 1. Number of Insurgent Attacks[20]

tion campaign that the Taliban have been carrying out since 2006. In 2010, assassinations increased by 50 percent and in 2011 by another 25 percent on the basis of preliminary data, despite the ever-greater scarcity of targets such as government officials in the areas most affected by the insurgency as a result of the outflow to Kabul of those threatened.[19] Sources within the Taliban hint that the intensification, in fact lowering the targeting bar to more junior government officials, but also targeting for the first time people very close to President Karzai (such as his step-brother Ahmad Wali), was decided in response to the intensification of ISAF's own targeting of Taliban commanders for capture or killing.[21]

ISAF sources claimed in 2011 that a 20 percent decrease in attacks against their units was recorded in the first half of that year, compared with the same period of 2010;[22] Figure 1 and even JIEDDO data (Joint Improvised Explosive Device Defense Organization) on IED attacks (see below) seem to contradict that, but certainly ISAF casualties (at least Killed in Action) were lower in the first eight months 2011 than they had been in 2010 (416 vs. 490). It is also true that the number of attacks on ISAF units did not continue growing in 2011 at the same pace as in the previous year, in part for operational reasons (no major ISAF operation in 2011, nor any major attempt to expand the territory held) and in part because of the growing role of the Afghan security forces, whose casualties instead continued to rise.[23]

OPERATIONAL LEVEL ADAPTATION

Changes to Force Preparation (Training and Doctrine)

In addition to distinct improvement in training from 2010, already from about 2005 foreign fighters were, instead of being deployed in autonomous units, increasingly integrated in Taliban-led units as advisers and specialists, adding a mentoring dimension to the task of training. In particular, IED specialists in southern Afghanistan were often of Punjabi origin, even as late as 2011. Locals were convinced that these were Pakistani Army specialists, filling the gaps created by the ever-expanding Taliban use of IEDs while more Afghans were being trained for that task. The utilization of a few trainers from Iraq in the training camps in Pakistan was also reported in 2009. The combination of advising and mentoring with basic predeployment training certainly had an impact on the proficiency of the Taliban in the field, as recognized by ISAF officers as well. Marksmanship improved, as did the ability to carry out complex ambushes, deploy IEDs effectively, and so forth.[24]

It is difficult to comment on the effectiveness of the integration of advisers and specialists with Taliban units; whatever information has come out suggests that the fighting ability was enhanced, allowing the Taliban to beat the Afghan Army in the few occasions when they could meet it without ISAF being present.[25]

By 2008 the Taliban's military technique manual was already in its fourth edition, suggesting continuous effort to refine tactics and operations.[26] The development and implementation of standardized techniques and tactics did not proceed without meeting resistance from some commanders, who did not believe that they needed to be taught how to fight or objected to asymmetric tactics in principle, sometimes because of their cowardly character and sometimes because of their indiscriminate character. Some commanders went so far as defecting (or withdrawing) over these issues.

The Use of IEDs

The evolution of the Taliban toward asymmetric tactics (that is, mostly IEDs, although not exclusively) appears to have started in Helmand and more in general in the south, where military pressure from ISAF went up first from 2005 onward. Indeed, the debate on asymmetric tactics started in 2006, coinciding with the Taliban's engaging ISAF for the first time in significant numbers. The initial debate featured two aspects: IEDs and suicide bombing.

Compared with suicide bombing, reverting to the use of IEDs was relatively noncontroversial and must have appeared a suitable response because they can be managed without a complex command and control structure, which ISAF's pressure made it difficult to maintain. Moreover, the transfer of knowledge and know-how can happen horizontally with just some facilitation from the leadership. Still, it necessitated the creation of a category of specialists handling the manufacturing and deployment of IEDs, de facto separated from the other fighting units even if connected to them. IED users in Afghanistan were organized in teams with a clear hierarchical structure (in contrast to Iraq), with a leader and trained emplacers under his command. Other members of the cell included a finance specialist, a bomb-maker, and the "triggerer." The typical workshop features two or three workers.[27] At the tactical level, the system seems fairly disciplined, with a cell leader taking all the decisions about where and when to deploy an IED, and the members executing. When an IED specialist is killed, a decline in IED activities is usually observed for some time, indicating a scarcity of specialists. They tend to be replaced within weeks by people sent from the leadership, again indicating a degree of central management of the IED effort.[28]

The Taliban were quite ruthless in adapting their IED deployment methods to ISAF's pressure. The deployment of growing numbers of drones to monitor the roads led them to employ children as emplacers, in the knowledge that ISAF's rules of engagement forbade firing at them. Similarly, the technology of IEDs evolved to match countermeasures: low- or no-metal-content IEDs were developed and deployed, while larger IEDs were utilized to inflict damage on armored vehicles.[29] Initially, the Afghan insurgents had been relying on military explosive, usually taken from unexploded devices and stocks of artillery rounds; as the IED production expanded they had to find alternative sources of explosive and found them in ammonium nitrate. While most IEDs until 2007 were made with military explosive, by 2008, according to the Afghan Ministry of Interior, it accounted for just 38 percent of all IEDs, and that percentage shrank further to 20 percent in 2009.[30]

The reliance on fertilizer for the production of IEDs (as opposed to military explosive) hampered the development of very large IEDs, because it takes about 20 kilograms of fertilizer to obtain an explosive power comparable to that of 1.5 kilograms of military explosive. About 250 to 750 kilograms of explosive derived from fertilizer were needed to destroy the new armored vehicles deployed by the Americans and other ISAF contingents from 2008 onward.[31] That of course reduced the tactical situations in which high destructive power IEDs could be deployed.

The Taliban developed their own "cottage industry" of IEDs to deploy them in the thousands around Afghanistan, but mostly in the south, where military pressure on them was higher (again confirming the character of adaptation of IEDs). Some ISAF officers even argued that by 2010 the Taliban had gone beyond the cottage industry stage and had developed a veritable industry of IED-making. The original workshops, producing IEDs in batches of ten and then deploying them, were being replaced by larger outfits, at least in Helmand, able to produce an IED every fifteen to twenty minutes.[32] The Taliban say that they have an IED development center in Pakistan, where new techniques are developed and tested, and ISAF officers confirm that the Taliban do experiment on new IED concepts, although not always successfully.[33] ISAF officers also believe that the Taliban cannot have developed the IED techniques alone; the help of the Iraqi insurgents is openly acknowledged by the Taliban, but ISAF sources believe that Iranian and Pakistani help was also at hand.[34]

Interestingly, unlike Shiite insurgents in Iraq, Afghan insurgents never developed or adopted shaped charge technologies. This might be due to a lack of sufficient technical skills, as high-precision manufacturing is needed for shaped charges. There might, however, be other reasons as well for the lack of shaped charges in Afghanistan: although potentially very effective against armored vehicles, shaped charges are in fact very difficult to deploy tactically: they are effective only at a range of two meters or so; if detonated too far or too close, they do not work.[35]

The development of IED employment did not proceed evenly across Afghanistan; indeed, a disparity has been observed in IED techniques and in the tactical sophistication of their deployment even from district to district within Helmand. This may be due to two reasons:

- Lack or slowness of transfer of tactical experience;
- Impact of deployment of newly trained specialists and lack of training updates for deployed specialists.

The massive and rapid shift toward IED use was delayed by the difficulty of training sufficient numbers of specialists fast enough to replace losses (which were heavy, as ISAF targeting was relentless) and expand the ranks. As of 2011, a veritable IED industry had appeared only in Helmand, Kandahar, and Khost. In 2011 it was still in a relatively early stage of spreading to other provinces. Even in Kandahar Province, in 2011 districts such as Panjwai and Zhari had few IED specialists, and they were often Punjabis, suggesting shortages.[36]

On the whole, however, the Taliban have demonstrated themselves quite apt

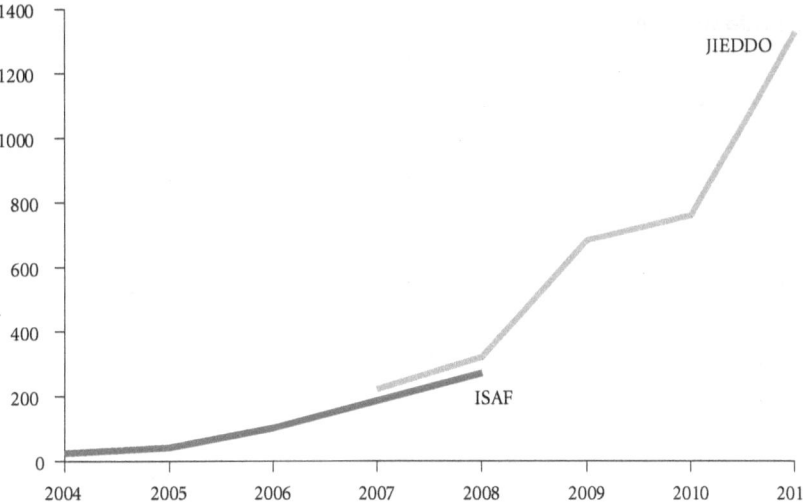

FIG. 2. IEDs Detonated or Neutralized, Monthly Rates[3]

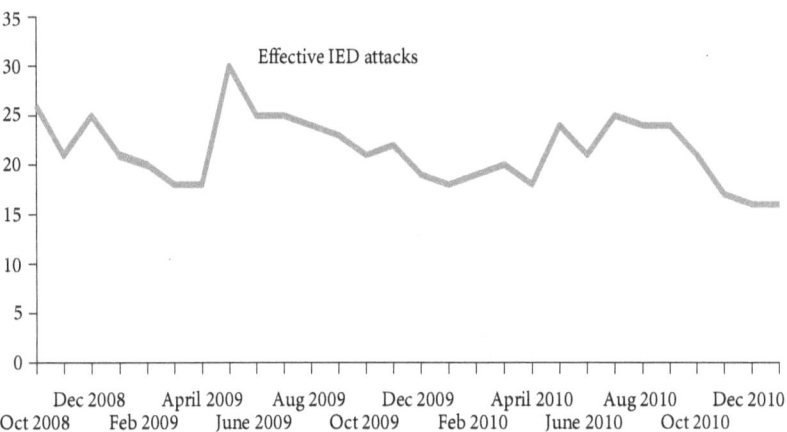

FIG. 3. Effective IED Attacks, Percentage Rates out of All IED Attacks[38]

competitors in the arms race with Western armies, which were investing billions of dollars in the development of countermeasures to IEDs. Figure 3 suggests that Western armies (and in particular the Americans) were slowly gaining an edge over the Taliban's IEDs, whose rate of effectiveness declined a little amid many fluctuations; however, at the same time the Taliban were deploying many more IEDs (Figure 2), so that the actual number of successful attacks increased steadily.

Suicide Bombing

Suicide bombing, a tactic inspired by the Iraqi jihadists and Al Qaeda and subject to the active lobbying of one of the main network leaders, Mullah Dadullah, proved more controversial within the Taliban's ranks. Dadullah argued in favor of the Iraqi model of indiscriminate terror in order to destabilize the cities and hit at the base of support of the new regime, but his position was in the minority within the leadership. Eventually a compromise was reached whereby suicide bombers would be deployed as tactical weapons, the idea being that one volunteer could inflict disproportionate damage on the enemy at little cost to the Taliban. In reality the first few generations of suicide bombers turned out to be quite ineffective because of casual recruiting and lack of training. Only gradually did the Taliban manage to create a cadre of specialist suicide bombers and deploy them effectively for raids against key enemy installations, increasingly integrating them into special teams who supported the bombers with light arms fire. More and more the deployment of the suicide bombers and of the supporting teams took place within a framework of careful planning, helped by a more and more widespread network of informers, infiltrated well into the government apparatus.[39]

Although commentators have viewed the special teams and the suicide bombers mainly as a symbolic effort, there are indications that it was always meant to achieve tactical results. In several attacks the teams succeeded in creating panic within the security forces of the Kabul regime and disrupting the chain of command by targeting the command centers; they might have been experiments that could lead to a later integration into wider operations meant to capture a garrison or a town.

Planning and Coordination

In the area of command and control significant technical but also organizational adjustments have been observed. Coordination and intelligence sharing have been facilitated by the creation of VHF (Very High Frequency) radio-relay networks that use "hundreds of small antennas linked to big solar panels," mostly relying on equipment that can be bought off the shelf in Pakistan or "stolen from NATO trucks and assembled in the field."[40] While the VHF network facilitated communication and coordination, the development of an increasingly capable intelligence apparatus helped the Taliban in a range of other tactical situations, often allowing them to avoid meeting the enemy in conditions of tactical inferiority, or to mount ambushes.

The most significant aspect of Taliban operational adaptation is to be found, however, not in technology but in the development, from 2008 onward, of a centralized system of command and control. Until then instances of Taliban centralized planning of large operations existed, although such instances were rare. The best known example is the operation leading to the battle of Pashmul. The Taliban reportedly conceived the Pashmul operation in 2005, when it became known to them that Canadian troops would have taken over responsibility for Kandahar province.[41] The plan, conceived by Mullah Dadullah, was to use Pashmul as the springboard for stepping up Taliban operations in the province. It does not appear to have been in any sense a detailed operational plan: its grand lines featured intensifying activity in order to inflict as many casualties as possible on the Canadians and then seize control of a section of Kandahar city, including a symbolic building. The aim would have been to demoralize Canadian public opinion, cause panic among Canadian politicians, and force a Canadian withdrawal, with large-scale political repercussions. Although it appears that the plan faced opposition within the Taliban leadership, it was eventually approved. Dadullah was placed in charge of it.

During the summer of 2006 an unprecedented buildup of Taliban presence in the area of Pashmul started taking place; the area was chosen because of its vineyards and ditches, which offered cover to the Taliban. Bomb production workshops were established and fortifications built; it might be that the plan actually involved luring or provoking the Canadians into attacking Pashmul, where terrain is more difficult than in the rest of Kandahar, getting them to fight on a ground of the Taliban's choice. The confidence that the Canadians could not take Pashmul derived from the belief that even the Soviet Army had not been able to take the area in the 1980s. As the buildup reached into the several hundreds and perhaps over a thousand Taliban, attacks in and around Kandahar intensified.[42]

The Taliban's operational planning in Pashmul was sophisticated, yet flawed in a number of ways. The myth of the Soviet Army's being unable to take Pashmul, for example, misled them into thinking that Pashmul was unconquerable. In reality, there is no indication that the Soviet Army ever tried anything more than battalion-size probing operations there. It is not clear whether the big operation planned near Kandahar in the summer of 2006, culminating in the battle of Pashmul, was actually meant to have a kind of "Tet offensive" impact in demonstrating the Taliban's emergence as a major military force, or whether that was simply achieved by chance and the operation was decided on the basis of an unrealistic assessment of their own capabilities vis-à-vis the Canadian troops. In any case, the Taliban effectively exploited well the situation and prioritized the return of their cadres in the area of Pashmul after their tactical defeat, symbolizing their resilience and challenging ISAF's statement of a crushing victory. Significantly, the Taliban never tried to repeat Pashmul-style operations afterward; this could be read as a sign of adaptation to an environment dominated by ISAF's conventional superiority, even if Pashmul had been a one-off for which Dadullah had to lobby energetically.

From 2008 onward a deliberate drive toward and capacity for more centralized planning and coordination has become evident. The decision to invest in the creation of such a system appears to have been dictated by the belief that without more efficient coordination and integration it would not have been possible to meet the challenge posed by ISAF and Enduring Freedom. In this regard it worth pointing out that ISAF sources themselves considered how the Taliban could have been more effective if they had been able to transfer knowledge and experiences more effectively across units; the isolation of some groups of insurgents, often among the most effective, hampered tactical development: "[If] they were a pan-Afghan cohesive force . . . , they would pose a much greater threat to NATO."[43]

Sources within the Taliban hint that a major role in advocating the innovation was played by Pakistani Army advisers, and in fact the model adopted resembles the structure developed by some of the *mujahedeen* parties in the 1980s when fighting the Soviet Army (chiefly by Hizb-i Islami), again under Pakistani mentorship. A Military Commission was established in Peshawar to handle the day-to-day coordination of the fighting units. The commission wanted to advocate the right to make appointments and to distribute supplies, and appears to have enjoyed Pakistani support in this regard. This proved difficult in the face of the dominant presence of the patrimonial networks described at the beginning of this chapter. It was most successful where the networks were weaker—that is, in eastern Afghanistan. The commission also had the ability, thanks to its growing control over supplies, to dispatch its own commanders to the strongholds of the more reticent networks as a way to break their monopoly and put them under pressure. Some of the networks, such as Haqqani's, could be reined in using this type of tactics.[44]

One of the main points of contention between the Military Commission and the networks has been the desire of the commission to rotate military commanders regularly (typically every six months). As it was conceived, this decision would have had devastating effects on the ability of the network leaders to maintain a following, as the patrimonial character of the networks would have been blown apart. In the old system, the commanders "owned" their fighters and the network leaders "owned" their commanders. The networks unsurprisingly resisted bitterly, and friction at times boiled over—a number of network figures were arrested by the Pakistani services to "discipline" them. Eventually a compromise was worked out, whereby the network leaders were allowed to influence the patterns of rotation of the commanders. By 2011 rotation seemed an established practice even in southern Afghanistan, where the networks were strongest. The intent was manifold:

- Make the command structure more difficult to track down by ISAF and the government forces (local commanders were widely known to the population and vulnerable to informers reporting their identity and whereabouts);
- Make the military structure more responsive to the center and less to local communities, network leaders, and local interests;
- Prevent defections (leaders finding it increasingly difficult to take their men with them when defecting);
- Strengthen the esprit de corps of the Taliban military machine as opposed to local identities and loyalties.

Although rotations were mainly within the commanders' region of provenance (hence a southerner being appointed in another district in the south), in some cases northerners have been dispatched to lead the fight in the deep Pashtun south.[45]

It is too early to say whether the networks are doomed and will gradually dry up as resources are increasingly channeled through the Military Commission. Commanders loyal to the commission and independent of the networks are reportedly better supplied and funded, a strategy presumably aimed at gradually attracting commanders away from the old patrimonial system. One negative side effect of the attempted centralization might have been to send commanders reluctant to be disciplined scurrying around for resources. There are also indications that, aside from the direct negotiations between the Iranian Pasdaran and the Taliban leadership, the Pasdaran have had considerable success with their offers of patronage targeted at individual commanders, perhaps exactly because of the shortage of funding and supplies for the noncompliers.[46]

The Military Commission might have found it understandably difficult to expand its control over the areas more remote from the Pakistani border, where communications are difficult (the Taliban rely on couriers for the most important mail) and, more important perhaps, where in many cases military pressure from ISAF and government forces was not as high. It has not been yet possible to test the strength of the Military Commission in a sufficient number of provinces to establish a clear pattern of its strength or weakness; however, a number of other signs of military adaptation, such as the shift toward asymmetric tactics, have spread unevenly even in the south (see discussion of Kandahar compared with Helmand, above). Here implementing rotation might at times have been problematic. Moreover, the allied jihadists, whom the Military Commission also tried to bring under its control, have been even more reluctant to accept the integration into the new command and control structure than the networks, perhaps because they maintained wholly separate funding channels. One case in point is the Islamic Movement of Uzbekistan, which is in principle part of the command and control structure, but continues to operate de facto autonomously in northern Afghanistan and sometimes even clashes with Taliban commanders.[47]

Difficulties aside, the drawback of the new command and control structure and of the associated rotation of commanders is that it weakens the relationship of the field commanders with the civilian population and tends to militarize the Taliban, Interviews in southern Afghanistan seem to indicate that the increased military efficiency of the Taliban has come at the cost of a loss of popularity among the villagers.[48]

The Taliban have also tried to adapt by improving relations with other jihadist groups operating in Afghanistan. We have already discussed attempts to integrate the IMU and the Pakistani jihadist organization into the command and control structure. While this has not been possible with the second largest (after the Taliban) Afghan jihadist group, Hizb-i Islami, efforts have been clearly in place to patch up relations with them, particularly after a wave of clashes occurred between the two groups in various locations in 2010. At least along the Pakistani border, in eastern Afghanistan, the two groups were seen cooperating closely in operations against ISAF and Afghan government forces in late 2010 and early 2011, although it appears that in the Kabul regions problems between the two continued.[49]

CONCLUSION

The Taliban has invested a lot of energy and has taken considerable political risk in striving for military adaptation in the face of rising enemy pressure. From the beginning, the Taliban have been facing an immensely superior enemy in terms of technology and tactical proficiency and have always been in conditions of numerical inferiority. In a sense, their adaptation started as early as 2002, when they had to convert from a semiregular force used to fight conventional battles into a guerrilla force, something that they had never been except as individuals or small groups in the 1980s, under different labels. Since then the Taliban have shown adaptability and resilience. Particularly when we consider the cultural and social context in which the Taliban operate (the most conservative sections of the Afghan population), it is clear that the Taliban's task was a difficult and contradictory one. The Taliban claim to be fighting to defend Afghanistan's old way of life, its religion (as they interpret it), and its customs. However, in their efforts to keep fighting, the Taliban had in a way to modernize themselves and adopt some of the features of centralization that they had once abhorred. Resistance to centralization had been a prime mover of clerical resistance to the state since the 1920s, and particularly in the 1970s and 1980s. This "modernization" of the Taliban alienated some of their constituencies in the villages and created a potential weakness, which however the Afghan government did not manage to exploit because of its lack of committed cadres and its weak governance apparatus.

The development of an almost industrial effort to inundate parts of Afghanistan with IEDs is one of the most noteworthy aspects of this modernization; the Taliban of 2011 were quite different from the original model of Pashtun warriors on which their operations were initially based. The construction of elaborated tunnel systems, at least in Helmand, is another striking development much at odds with the original ethos of the recruits—a Pashtun warrior shuns manual work. Pakistani help and advice certainly played an important role in getting the Taliban where they were in 2011, although telling the role of the advisers apart from that of the leaders is of course very difficult.

Adaptation therefore has enabled the Taliban to keep threat at a consistently high level from 2006 onward. Such adaptation took different shapes: strategic, tactical, and organizational. In all cases it was a painful process, leading to friction and divisions. As of September 2011, the Taliban had been able to get over all the splits created by the process. The transition from a patrimonial, decentralized insurgency toward a centralized, bureaucratically managed one was still an ongoing process at the time of writing, and its final outcome could not be known. The very fact that the Taliban managed to embark on this ambitious transformation and achieved at least a degree of success is in itself remarkable.

NOTES

1. See A. Giustozzi, *Koran, Kalashnikov and Laptop* (New York: Columbia University Press, 2008).
2. See A. Giustozzi, *Negotiating with the Taliban: Issues and Prospects* (New York: Century Foundation, June 2010), for a discussion of the Taliban network system.
3. On friction among the Taliban, see, for example, M. van Bijlert, "Unruly Commanders and Violent Power Struggles," in A. Giustozzi, ed., *Decoding the New Taliban* (New York: Columbia University Press, 2009).
4. See Giustozzi, *Koran, Kalashnikov and Laptop*; and Giustozzi, *Decoding the New Taliban* for a discussion of the Taliban's "fragmentation."
5. On ISAF's deployment to southern Afghanistan, see the other chapters of this book. For a chronology of the Taliban insurgency up to 2007, see Giustozzi, *Koran, Kalashnikov and Laptop*; and Giustozzi, *Negotiating with the Taliban*.
6. This issue is also discussed in Giustozzi, *Koran, Kalashnikov and Laptop*. See Giustozzi, *Decoding the New Taliban*, for different views on the Taliban's relationship with financial gain.
7. Interviews with Taliban commanders in Musa Qala, Helmand, July–August 2011; G. Peters, "The Taliban and the Opium Trade," in Giustozzi, *Decoding the New Taliban*.
8. See Antonio Giustozzi and Christoph Reuter, "The Insurgents of the Afghan North" (Kabul/Berlin: Afghan Analyst Network, May 2011); and Giustozzi, *Koran, Kalashnikov and Laptop*.
9. See Giustozzi, *Negotiating with the Taliban*.
10. On the recruitment of jihadi commanders, see A. Giustozzi, "The Taliban's Marches," in Giustozzi, *Decoding the New Taliban*; and A. Giustozzi, "The Taliban beyond the Pashtuns" (Waterloo: CIGI, 2010).
11. Interviews with high school teachers, Kabul, Ghazni, and Wardak, 2011.
12. See Giustozzi, *Negotiating with the Taliban*.
13. Interviews with Taliban commanders, Kabul and Helmand, 2010; personal communication with ISAF officers and diplomats, 2010–11.
14. Interviews with Taliban commanders, Helmand, summer 2011.
15. Personal communication with ISAF officers and diplomats, April 2011.
16. On schools, see A. Giustozzi and Claudio Franco, "The Battle for the Schools: The Taliban and State Education" (Berlin: AAN, 2011).
17. Giustozzi, *Negotiating with the Taliban*.

18. Kate Clark, "The Layha" (Berlin: AAN, 2011); Thoman Johnson and Matthew C. DuPée, "Analyzing the New Taliban Code of Conduct (Layeha): An Assessment of Changing Perspectives and Strategies of the Afghan Taliban," *Central Asian Survey* 31, no. 1 (2012); Sean Rayment, *The Bomb Hunters* (Harper and Collins, 2011), 38, 60.

19. Targeted assassinations per year were as follows: 271 (2008); 225 (2009); 462 (2010); 190 (first half of 2011). Source: UNAMA.

20. Source: ANSO.

21. Interviews with Taliban and former Taliban, London and Kabul, February and September 2011.

22. Tom Van den Broek "IED Planting, Seizures Rise in Afghanistan," *USA Today*, September 6, 2011.

23. Sources within the respective ministries place losses of the police and of the army in the summer months of 2011 at the highest level ever: Joshua Partlow, "Afghan Police Casualties Soar," *Washington Post*, August 31, 2011; "109 Afghan Army Personnel Killed in One Month: Official," *Xinhua*, July 27, 2011.

24. Interviews with Taliban commanders and local elders, Kandahar and Helmand, June–August 2011; interviews with Taliban cadres in Pakistan, 2009; personal communications with ISAF officers, 2010–11.

25. On one example in the summer of 2010, see "Afghan Army Offensive Goes 'Disastrously Wrong,'" BBC News, August 13, 2010; and Ahmad Waheed and Bill Roggio, "Taliban, Al Qaeda Forces Repel Afghan Army Assault in Eastern Afghanistan," *Long War Journal*, August 13, 2010: available at http://www.longwarjournal.org/archives/2010/08/taliban_al_qaeda_for.php.

26. Imtiaz Ali, "Preparing the Mujahidin: The Taliban's Military Field Manual," *CTC Sentinel* 1, no. 10 (September 2008).

27. Interviews with Taliban commanders, Helmand, Pakistan, and Kandahar, 2008–11; Lieutenant-General Michael Oates of JIEDDO, cited in "Bigger, Badder IEDs in Afghanistan": available at http://defensetech.org/2010/03/16/bigger-badder-ieds-in-afghanistan/; Thomas Harding, "Children of Five Used by Taliban to Lay Bombs," *Daily Telegraph*, June 30, 2010; Walter Pincus, "Military Moves High-Tech Tools to Afghanistan," *Washington Post*, June 23, 2009.

28. Personal communication with ISAF officers, 2010.

29. Rayment, *The Bomb Hunters*, xix, 59–60, 144; Ben Gilbert, "Afghanistan's 'Hurt Locker': Facing Off with IEDs," *Minnpost*, February 10, 2010.

30. "Most Bombs in Afghan War Made from Banned Fertiliser," *AFP*, July 12, 2010.

31. Rayment, *The Bomb Hunters*, 128.

32. Ibid., 7, 61.

33. Ibid., 74, 253; interviews with Taliban commanders, Helmand, 2011.

34. Rayment, *The Bomb Hunters*, 85.

35. Personal communication with military officers, specialists in IEDs, 2011.

36. Interviews with Taliban commanders in Panjwai and Zhari, summer 2011.

37. Source: JIEDDO, ISAF. Note: data for 2011 is preliminary and partial.

38. Source: ISAF.

39. Simon Robinson, "The World's Worst Suicide Bombers?," *Time*, July 28, 2007; Sami Yousufzai and Ron Moreau, "The Fallout from the AWK Murder," *Daily Beast*, July 14, 2011; Dion Nissenbaum, "Taliban Attacks in Afghanistan Show Growing Sophistication," *McClatchy Newspapers*, July 15, 2010.

40. Roy Gutman, "We've Met the Enemy in Afghanistan, and He's Changed," *McClatchy Newspapers*, March 14, 2010.

41. On the Pashmul Operation, see C. Carl Forsberg, *The Taliban's Campaign for Kandahar* (Washington, DC: Institute for the Study of War, 2009), 25–27.

42. See Chris Wattie, *Contact Charlie* (Toronto: Key Porter, 2008).

43. Rayment, *The Bomb Hunters*, 219.

44. Interviews with Taliban leaders, Kabul and Pakistan, summer and autumn 2011.

45. See note, 40 above.

46. Interviews with Taliban leaders and commanders, Ghazni, autumn 2010.

47. Interviews with Taliban leaders, 2011; Giustozzi and Reuter, cit.

48. Interviews with elders and former Taliban commander, Helmand, summer 2011.

49. "Tensions Rise Between Hizb-i-Islami and the Taliban in Afghanistan," *CTC Sentinel*, 1 August 2010; personal communication with diplomats in Kabul, 2011.

11 Shoulder-to-Shoulder Fighting Different Wars: NATO Advisors and Military Adaptation in the Afghan National Army, 2001–2011

Adam Grissom

The Afghan National Army (ANA) is the centerpiece of the North Atlantic Treaty Organization's (NATO's) military strategy in Afghanistan. While combat operations by coalition forces attract most of the attention from media and policy-makers, the alliance's campaign plan recognizes that the West's ultimate objective in Afghanistan can only be accomplished by Afghan security forces operating on a long-term, steady-state basis to deny sanctuary to transnational terrorist organizations. Consequently, the main effort for the International Security Assistance Force (ISAF) is, despite appearances, to develop the capabilities of Afghan forces rather than fight insurgents directly.[1]

The alliance is devoting substantial resources to this mission. A three-star headquarters based in Kabul, NATO Training Mission—Afghanistan (NTM-A), oversees several thousand Western advisors embedded *shohna ba shohna* (Dari for "shoulder to shoulder" and NTM-A's credo) with ANA units, dozens of training centers and schoolhouses, and an annual assistance budget of more than $10 billion.[2] The ANA has been expanded massively since 2003, from a few hundred ragged former militiamen to more than 170,000 professionals and 120 combat battalions equipped with Western weapons and body armor, tens of thousands of vehicles, and stationed at hundreds of new and refurbished bases and training facilities throughout Afghanistan.[3] The results are undeniably impressive. As of late 2011, NTM-A can fairly point to the new model ANA as one of the largest and best equipped armies in the region.

Yet there remains a puzzle. Despite massive assistance from NATO, the operational effectiveness of the ANA remains dismal. Indeed, as of late 2011, after years of training and advice, and tens of billions in Western investment, not a single ANA unit is capable of operating independently of NATO support.[4] More critically, from the perspective of long-term Western interests, the ANA is still widely considered to be outmatched by the Taliban.[5] How can this be?

This chapter explores the puzzle of poor ANA operational effectiveness by examining how the institution has adapted during the last decade. It seeks to explain why the ANA has adapted as it has in response to a particular set of external stimuli, including Afghan domestic political pressures, battlefield dynamics, and external advice. NATO's role in this story is important but ultimately exogenous, and its own adaptation in the advisory role is covered elsewhere in the volume. Here we focus on the ANA as an agent of adaptation.

MILITARY ADAPTATION IN THE ANA

The Afghan National Army has experienced three major phases of adaptation since 2001. The first phase corresponded to the early years of the interim Afghan government, from 2001 to 2003, when the ANA was viewed primarily as a symbol of the new political order. The second phase, from 2004 to 2006, was motivated by a growing conviction that residual instability in the Afghan hinterlands required an ANA with more than merely symbolic capability. The third phase, from 2007 to 2010, was driven by the rapidly escalating challenge posed by the Taliban. It is the third phase, when the West committed to building a large and capable Afghan Army, that is the primary focus of this chapter.

The Lost Years: 2001–3

The Western project to rebuild the Afghan Army began with the Bonn Agreement of December 2001, which called for "the reintegration of the mujahedin into the new Afghan armed and security forces" with the help of the NATO International Security Assistance Force.[6] The former militiamen were to be retrained to create an army that would be, "nationally respected, professional, ethnically balanced, democratically accountable, organized, trained, and equipped to meet the security needs of the country, and increasingly funded by Government of Afghanistan revenue."[7]

After Bonn a debate occurred in Kabul and Western capitals regarding the appropriate size of the new army. The newly appointed Afghan Minister of Defense, Northern Alliance strongman Mohammad Qasim Fahim, advocated for an army of 250,000 that would incorporate the major warlord militias into government on a wholesale basis.[8] Western donor countries balked at the cost of supporting such a large army as well as the political implications of directly integrating the formerly warring factions into government service. The United States and Britain countered with a proposal for a relatively small, highly trained and multiethnic army free of links to strongmen.[9] Given the apparently benign security situation in the country, the primary purpose of this army would be to serve as a symbol of the new Afghanistan and, eventually, replace ISAF contingents providing security in Kabul. It would be complemented by a new national police force that might provide work for many former militiamen as well as a robust Disarmament, Demobilization, and Reintegration (DDR) program to convince as many as possible to lay down their weapons.[10]

The Anglo-American concept was adopted at a security donors conference in April 2002 sponsored by the G8 (group of eight largest economic powers).[11] Responsibility for developing the Afghan security architecture was distributed among several major powers under a "lead nation" concept. The United States accepted responsibility for individual and collective training for the army and for overseeing the development of the intelligence service, called the National Directorate for Security (NDS). Britain took the lead in counternarcotics efforts. Germany agreed to train Afghan National Police (ANP) officers. Japan agreed to manage the Disarmament, Demobilization, and Reintegration of militias, and Italy would lead the development of the judiciary.[12]

By this time Britain had already begun training the first unit of the new national army, which became known as the 1st Battalion Afghan National Guard.[13] American Special Forces began training a second battalion in May and a third battalion at the end of the year.[14] During this period the basic format of an Afghan army battalion was established by consensus among the US, Britain, and the nascent Afghan Ministry of Defense. Numbering approximately six hundred personnel, the new battalions (called *kandaks* in Dari) featured a small headquarters element and four maneuver companies (or *toli*) of three platoons each. The battalions were light units lacking motor transport and heavy weapons. Initial training lasted approximately ten weeks and covered only the most rudimentary basics of soldiering and small unit tactics.[15]

Morale and retention proved to be enormous challenges for these early units. More than half of the personnel deserted their assigned units within a year of the Afghan National Guard's establishment, and by late 2002 the British-trained 1st Battalion was down to just two hundred soldiers.[16] The few understrength national guard battalions were dwarfed by key warlord militias, including Fahim's own Jama't-e Islam, which boasted twenty thousand full-time fighters during this period.[17] Tellingly, US special operations forces continued to cooperate closely with militias (which they labeled Afghan Military Forces) as they hunted Taliban and al Qaeda remnants in eastern Afghanistan.[18]

Even these meager accomplishments were wiped completely clean in September 2003, however, when American officials became convinced that the Afghan Ministry of Defense had fallen under the de facto control of Fahim's Jama't-e Islam.[19] The United States prevailed upon the interim government to disband the entire ministry, including the battalions trained to that point, and start again with a clean slate. Two years after the fall of the Taliban, Afghanistan again had no army.

Adaptation in the Afghan Army during this period was directed by its external sponsors in Western governments in response to strategic pressures, most prominently the desire to create a symbolic army distinct from existing warlord militias. There was little if any operational pressure for adaptation. Western governments believed that the Taliban had been defeated, with the remnants being pursued by Western special operations forces, and they did not foresee a need to engage in direct military competition with warlord militias. These pressures would change in the coming years.

Founding a New Model Army: 2004–6

The disbandment of the Afghan National Guard paved the way for a new model Afghan National Army. In 2004 the interim Afghan government and international donors agreed that the ANA would be all volunteer, professional, ethnically balanced, and free from warlord influence.[20] It would be larger than key warlord militias, at seventy thousand personnel, and unlike previous versions it would have a sizable institutional framework with a Ministry of Defense, General Staff, and an air corps. This signaled a shift to a more ambitious vision for the ANA that would take much longer to develop. Once developed, however, the more independent, professional, and capable ANA would presumably have more operational capacity and provide a more natural counterpart to the ambitious state-building project occurring in other sectors of the Afghan government.

The centerpiece of the new model ANA was a substantially expanded training and advisory effort overseen by a more capable US Combined Security Transition Command—Afghanistan (CSTC-A) in Kabul.[21] CSTC-A's main effort would be the reinvigoration and expansion of the Kabul Military Training Center (KMTC). The facilities at KMTC were improved in late 2003, and additional US trainers arrived in January 2004. Under the new model all ANA recruits received twelve weeks of individual training before being assigned to a *kandak*, which would subsequently receive additional collective training. The expanded facilities at KMTC allowed four *kandaks* to be trained simultaneously, creating a foundation for future growth.[22]

More important, all newly established *kandaks* would be supplemented by Embedded Training Teams (ETTs) or, later, NATO Operational Mentor and Liaison Teams (OMLTs) when they completed the program at KMTC.[23] The ETTs and OMLTs accompanied the new units to their duty stations and remained with them indefinitely to continue training and mentoring. Many advisory teams, including all ETTs, accompanied their partner units on operations, to include combat. Some OMLTs were prevented from doing this, with severe if predictable implications for the quality of advising.[24]

The expansion of KMTC and the advisory team programs provided the institutional foundation for faster growth and improved quality. The army reached six thousand nominal end-strength in 2004 and eighteen thousand in 2005.[25] However, desertion continued to be an enormous challenge, and hence actual present for duty numbers remained a fraction of nominal strength.[26] KMTC found itself fighting an uphill battle as army end-strength expanded and desertion rates held steady as a proportion of that force. A steadily increasing number of new recruits were required simply to refill existing units, slowing the rate at which the army could be expanded.

At this point, in mid-2005, the Taliban made its renewed presence felt in southern and eastern Afghanistan. There had been indications of limited Taliban infiltration beginning in late 2003, but Western officials remained comfortable that it presented only an annoyance. By the summer of 2005 the Taliban was poised to roll back the limited government presence in its traditional Pashtun strongholds. Fighting blossomed throughout the south and east during the warm months, which, coupled with a bumper crop of poppy, created a crisis that forced GIRoA (the Government of the Islamic Republic of Afghanistan) and its international supporters to re-examine the ANA project. It soon became evident that the Afghan government and its international supporters were producing too few soldiers and police, of too little individual and collective quality, to provide sufficient security for the overall state-building endeavor.[27] The combination produced a significant adaptive moment for NATO and the ANA.

The response came in the form of the 2006 Afghan Compact, which announced a massive expansion of Afghan National Security Forces (ANSF). The army was to grow from 18,000 to 70,000 soldiers in four years, and the police were to be expanded too, from roughly 20,000 to 62,000 during the same period. This represented both an increase in headline goal (from 50,000 to 70,000 for the army and from 20,000 to 60,000 for the police), as well as a radical increase in the rate of expansion. More important, the quality of ANA and ANP units was to be enhanced by additional training, newer and heavier weapons, vehicles, communications gear, and additional training and advisory teams embedded within Afghan units.[28]

The scale of the desertion problem within the ANA, and the seriousness of the new emphasis on quality, can be discerned in the fact that ANA end-strength did not increase substantially in 2006 despite growing concerns in Kabul and Western capitals.[29] In the face of firm political guidance to expand the army, ISAF's expanded efforts were consumed with backfilling ANA units with soldiers to replace those unofficially absent without leave and replacing those lost to attrition.[30] Meanwhile, violence continued to worsen, and the Taliban visibly strengthened vis-à-vis the government and ISAF.

Years of Expansion: 2007–11

In line with the Afghan Compact, ANA training infrastructure was expanded again in 2007 with the opening of regional training centers and specialist schoolhouses. As a result, ANA growth accelerated in 2007, reaching 35,000 by the end of that year.[31] In 2008 the authorized size of the ANA was expanded again, to 134,000, and nominal end-strength reached more than 75,000.[32] ANA end-strength reached 95,000 in 2009, 134,000 in 2010, and a whopping 171,000 in 2011.[33]

As it expanded, the ANA also began to embark on a series of tactical adaptations that would have sweeping implications for the force. These included the conversion of ANA infantry battalions from smallish, light units to Western-style motorized infantry equipped with Western small arms and support weapons, the creation of a number of specialist infantry *kandaks*, and the establishment of the Afghan National Army Air Corps.

Among these adaptations the most important was the conversion of ANA infantry *kandaks* to a Western model. The *kandaks* were significantly expanded to include a weapons company of more than two hundred personnel and medical, signal, and maintenance and support platoons. The total size of an infantry *kandak* rose from approximately five hundred personnel to nearly eight hundred.[34] Additionally, all infantry *kandaks* were converted from foot mobile to motorized units with the delivery of Light Tactical Vehicles (Ford Ranger light pickup trucks) and High Mobility Multipurpose Wheeled Vehicles (HMMWVs, or "Humvees") sufficient to move the entire unit under tactical circumstances.[35]

Moreover, the infantry *kandaks* were re-equipped with Western weapons beginning in 2008. ANA riflemen were issued M16A2 assault rifles from excess US Marine stocks, and machine gun teams were fielded M249, M240B, or M2 machine guns.[36] The mortar platoons in the new heavy weapons companies were equipped with M252 82mm mortars.[37] All infantry *kandaks* were issued US portable and vehicle radios, and an increasing number of infantry units received night vision devices and personal protective equipment.[38] With these changes the ANA infantry *kandaks* began to look very Western—large, motorized, and heavily armed.

At the same time that NATO was expanding and rearming line infantry *kandaks*, it also created a number of specialist units. The most important of these were the ANA Commandos. The commandos were established in Rish Kvor, northeast of Kabul, in early 2007.[39] Volunteers were solicited from infantry *kandaks* and subjected to a rigorous screening process and ten weeks of additional training that followed the general parameters of US Army Ranger tactics and techniques. Once operational, the commando *kandaks* operated as special operations forces in a direct action role, conducting raids against high-value targets in conjunction with coalition forces.[40] The first *kandak* was established in July 2007, followed quickly by others in October 2007, January 2008, and May 2008.[41] As of 2011 the commandos were widely respected by coalition forces, but they remain dependent on the coalition forces for key enabling capacity, including rotary-wing lift, intelligence, and air support.[42]

In 2008 NATO also established infantry *kandaks* for counternarcotics and highway security missions. The counternarcotics units were created to provide additional support to Western operations that the Afghan National Police could not, or would not, provide.[43] Several were activated but within a short time converted to line infantry when the broader Western counternarcotics effort took second priority to counterinsurgency. The bulk of the counternarcotics operations passed to specialist units within the Ministry of Interior, operating in a direct-action mode.

The highway *kandaks* were created as highly mobile motorized units to provide additional security to ground lines of communication.[44] As an adaptation, they resulted partly from NATO frustration with the ANA practice of securing roads by garrisoning them. Key NATO officers held the view that a motorized unit could secure lines of communication much more efficiently. Several highway *kandaks* were created, and at least one operated in the highway security mode in conjunction with an American Stryker brigade conducting the same mission. Over time, however, they have also been rerolled as line infantry *kandaks* when the experiment (and its American Stryker-based variant) proved unsuccessful.[45]

In 2008 the ANA also renewed efforts to establish heavy *kandaks*, including an armored *kandak* and a mechanized infantry *kandak*, both located in the Kabul area.[46] The ANA succeeded, despite NATO objections, in adding these units to its order of battle, but the tanks and armored personnel carriers remain inoperable because of a lack of funding and maintenance expertise. The two units operate as line infantry *kandaks* instead, attached the 111th Capital Division.[47]

Finally, 2007 also saw the re-emergence of the Afghan Air Force (AAF). Afghan leaders pushed hard for the creation of a new air arm on both practical and symbolic grounds. They expressed a strong preference for the AAF to be equipped with modern Western systems, including the CH-47 cargo helicopter, C-130 transport, and F-16 fighters.[48] NATO agreed to fund the re-establishment of the AAF, but with less expensive Russian and European aircraft. In May 2007, NATO established the Combined Airpower Transition Force (CAPTF) to train and advise a small element of helicopters and fixed-wing aircraft.[49] The AAF opened its first operational facility, at Kabul International Airport, in April 2008, and its first operational wing, the Kabul Air Wing, in late 2008.[50] The Kabul Air Wing is composed of a Presidential Air Unit responsible for transporting Very Important People (VIPs), a fixed-wing squadron of C-27A and AN-32 cargo aircraft, and a rotary-wing squadron of Mi-17 transports and Mi-35 attack helicopters. By late 2009 the AAF had established smaller operational detachments with each of the corps headquarters.[51]

The operational adaptations launched in 2007 required the better part of two years to implement fully. As this was occurring, NTM-A began to make plans for a series of further adaptations to address gaps in ANA capabilities.[52] The most important of these is the adaptation of the ANA logistics system. Between 2001 and 2009, NATO forces routinely provided supply and movement support to ANA units.[53] In 2009 the ANA began developing a Combat Service Support (CSS) system with the establishment of the Corps Logistics Battalions (CLBs).[54] The CLBs, assigned one per corps headquarters, are designed to provide long-haul trucking support between central ANA depots and the corps-level depots. They are large units (approximately 450 personnel) equipped with heavy trucks and the associated maintenance equipment.[55] The ANA also established CSS *kandaks* in each brigade during this period. The CSS *kandaks* are primarily responsible for moving supplies from the corps depots to the brigade-level supply points. They are equipped with medium-size trucks and the associated maintenance equipment. With the advent of the CLBs and the CSS *kandaks*, the ANA became independent of NATO's movement and storage system, though the West continues to pay for virtually everything in the depots.[56]

The 2009–10 period also saw the establishment of Combat Support (CS) *kandaks* in each ANA infantry brigade. The CS *kandaks* provide a variety of reinforcing capabilities to the brigade, including a reconnaissance company that often acts as brigade-level commandos, as well as a battery of D-30 howitzers. The ANA pushed NATO for many years to reintroduce artillery, and the establishment of the CS *kandaks* fulfills that desire.[57] The CS *kandaks* also contain a large engineer company equipped for, among other missions, explosive ordnance disposal. These units often work closely with another recent addition to the force structure, a Route Clearance Company within each brigade. The RCCs are equipped with armored vehicles, rollers, and various other systems to neutralize improvised explosive devices.[58]

During 2009–10 it was also decided that each corps would receive a Quick Reaction Force (QRF) *kandak*. The QRFs serve as the corps-level reserve force, postured to assist fellow ANA units that may be hard pressed by the Taliban. The QRFs are equipped with up-armored Humvees and heavy weapons to provide it the capability to tip the balance of an engagement quickly.[59] These units are still being established as of late 2011.

Finally, during this period the ANA and NATO also agreed to establish the ANA Special Forces (ANASF). The ANA Special Forces are drawn from volunteers from the commando *kandaks*. They receive three months of extra training and operate in support of the Afghan Local Police and Village Stability Operations initiatives. The primary mission of the ANASF is not direct action, but rather the "indirect approach" of mobilizing villages in support of the government and organizing nonstatutory forces. As of late 2011 several ANASF *kandaks* had been created, but it appears likely that they will be disestablished and their constituent companies will instead be reintegrated with the commando *kandaks* to create a mixed direct/indirect capability at the battalion level.[60]

The ANA in Late 2011

The ANA has undertaken an epic level of adaptation since its re-establishment in 2003. The result is a very impressive looking army. As of October 2011 the nominal end-strength of the army was approximately 170,000.[61] ANA field forces are organized into six corps, one division, 25 combat brigades, and approximately 120 maneuver battalions, in addition to combat support and combat service support detachments, companies, and battalions.[62] Each corps is authorized approximately 16,000 personnel, including three infantry brigades, a corps logistics battalion, and a commando *kandak*. Each infantry brigade is composed of four infantry *kandaks*, a combat service support *kandak*, a combat support *kandak*, and route clearance company for counter-IED missions. Brigades are typically assigned 4,500 personnel.[63]

ANA nominal end-strength has increased rapidly in recent years, though very high levels of absenteeism mean that only a fraction of the nominal number are typically present for duty. Reliable data on absenteeism is not collected by the ANA or NATO, but publicly reported estimates of 20 percent absent without leave are consistent with this author's observations.[64] The Ministry of Defense (MoD) is unquestionably struggling to maintain adequate quality in such a large army, and there are reports of the ANA inducting recruits who lack the required mental and physical capacity to be soldiers.[65] Many NATO officers cite an illiteracy rate of 90 percent.[66] There are similar reports regarding drug use.[67] Public estimates assert that as much as 85 percent of the force "would fail a drug test," though that is far higher than the author has observed.[68] There are also significant concerns regarding discipline and subversion; there have been at least twenty incidents of ANA soldiers attacking NATO advisors.[69]

ISAF and the MoD seek to manage these challenges by improving retention and training. As of mid-2011 the ANA operated more than a hundred training facilities with a total capacity of 27,000 soldiers. The ANA training base includes the flagship Kabul Military Training Center, six regional military training centers, a military high school, a very competitive military academy, twelve branch schools, and dozens of specialized courses for ranks. Every ANA recruit receives literacy instruction to at least the first grade level, with noncommissioned officers, commissioned officers, and specialists receiving more.[70]

ANA Combat Effectiveness in 2011

In spite of these adaptations, the ANA remains a force of very limited operational capability. NATO, as of 2011, assesses ANA units via the Commanders

Unit Assessment Tool (CUAT).⁷¹ The CUAT measures inputs to unit capability, including personnel, materiel, training, and processes. The assessment of inputs is supplemented by a subjective assessment of the unit's capability and their dependence on NATO for advice or assistance.⁷²

In early 2011 NATO assessed zero ANA units as capable of conducting independent operations. Approximately sixty units, or 30 percent of the force, were assessed as capable of operating with NATO advisory support. The other two-thirds of the force were assessed as requiring NATO units in the lead, or as unable to operate even under those conditions.⁷³

The situation had changed slightly by the end of 2011. NTM-A adjusted its rating system over the course of 2011 to, inter alia, replace the "Independent" label with the less ambitious "Independent with Advisors" for the top rating, marking a significant lowering of the bar.⁷⁴ Of the 204 ANA units assessed in late 2011, one received the top rating, though even that unit, 2nd *Kandak*, 2nd Brigade, 205th Corps, "remains dependent on ISAF for combat support and combat enablers."⁷⁵ Approximately one-third of ANA units were rated as "Effective with Advisors" in October 2011, meaning that they can conduct internal functions reliably but require ISAF to provide routine support that ought to originate from other ANA units, primarily combat support and enablers as well as occasional advisory assistance.⁷⁶ The other two-thirds of the ANA cannot function, even internally, without constant ISAF assistance.⁷⁷ At an aggregate level this paints a picture similar to that of the April 2011 report, implying that there has been little progress in that six-month interval.

The bottom line, from NATO's perspective, is that ANA operational effectiveness remains unsatisfactory. NATO explains this by citing three factors. The first is training. Despite the enormous expansion of the ANA training base, and enhanced partnership with NATO units in country, training Afghan forces to conduct effective operations requires longer timelines than those to which Western officers are accustomed, in part because most Afghan soldiers are not literate or numerate when they enter the service.⁷⁸ Western weapons, vehicles, and night vision devices are difficult to master for such personnel.⁷⁹ NATO has therefore concluded that additional individual and collective training would increase the capability of the ANA by enabling units to make use of appropriate tactics and exploit the full capabilities of Western materiel.⁸⁰

The second reason is materiel. ISAF notes that equipment shortages are still common in the ANA, and too many units retain obsolescent systems.⁸¹ Improving ANA capabilities therefore requires the procurement of thousands of additional weapons, vehicles, radios, and sustainment items. Moreover, it requires the establishment of national, regional, provincial, and local logistics systems that can process and deliver materiel throughout the country, a parallel system of national, regional, provincial, and local maintenance facilities and processes, and a line-haul and airlift capability resident in the armed forces. This implies an ISAF judgment that a complex and robust logistics and maintenance system will allow the ANA to sustain operations against the Taliban and other internal threats more effectively.

Finally, NATO emphasizes the role of corruption in limiting the effectiveness of the ANA.[82] There are widespread reports that key officer positions must be purchased from ministry officials who control personnel assignments.[83] Indeed, officers are reportedly required to continue paying on a fixed schedule as long as they hold the position. This creates enormous incentives to skim whatever resources are placed at the disposal of the officers, including pay for subordinates, foodstuffs, fuel, and disbursements for locally contracted consumables.[84] The result is corruption that is viewed as not just odious but also detrimental to the operational effectiveness of the ANA.[85] NATO's response has been to enhance leader development mentoring and training and to seek to inculcate an "ethos of stewardship" in the Afghan Army in hopes of stemming corruption, thereby enhancing the effectiveness of the ANA against the Taliban.[86]

In sum, ISAF's diagnosis and prescription for the ANA remain decidedly orthodox. The Afghan Army's weakness is explained in terms of shortfalls in inputs, primarily training, personnel, institutional structure, and materiel. The resulting prescription is for the allocation of more of precisely those inputs. The implicit rationale is that more of these inputs, and more time, will allow the ANA to become independent of NATO operational support and mentoring. This, in turn, will improve the security situation and allow GIRoA to begin resolving the governance vacuum.

Yet there is clearly something missing in the idea that doing more of the same will transform the ANA. The Afghan Army has benefited from years of focused large-scale assistance from ISAF. From a technical perspective, Afghan forces receive better individual and collective training, equipment, facilities, and logistics support than the security forces of comparable states in the region. Yet zero Afghan units are rated by ISAF as ready for independent operations, and there is no evidence that the situation is improving.

Hence the puzzle: With the evidence of adaptation plain to see, from new unit designs to Western equipment and the massive expansion of the overall force, why has there been so little improvement in operational effectiveness?

A CLOSER LOOK AT ADAPTIVE PRESSURES ON THE ANA

The answer to the puzzle is that the process of adaptation within the ANA has been distorted by interaction among different types of adaptive pressures. In particular, operational pressures from the battlefield have interacted with strategic pressures from the political realm to produce widespread adaptation without corresponding improvements to operational effectiveness. The conceptual framework developed for this volume, with its focus on shaping factors and adaptation at different levels, highlights this phenomenon nicely.

Operational Pressures and Extrinsic Adaptation

The ANA certainly confronts operational pressures. The Taliban remains a powerful adversary despite ISAF's recent surge, and much of the country suffers from acute insecurity. In the south and east, particularly, the levels of violence remain unacceptably high. ANA units suffer losses daily, sometimes in large numbers, and the Taliban and other antigovernment forces are able to carry off spectacular attacks even in the heart of Kabul.

The manner in which these operational pressures are exerted on the ANA is somewhat peculiar, however, because of NATO's presence in the country. NATO appears to distort those pressures in at least two ways. The first is that it creates a moral hazard. The presence of more than 100,000 NATO combat forces ensures that the Taliban could not overthrow the Afghan government even were the ANA to collapse completely. ANA commanders and Afghan political leaders thus know that when operational challenges arise, ISAF possesses the capacity to deal with them directly. Thus the ANA has strong incentive to preserve itself and allow NATO to do the fighting, which comports with ISAF's instinctive "get it done" mentality in any case. When situations arise in which ISAF, against its natural inclinations, wishes the ANA to take the lead in responding to an operational challenge, the ANA can safely fail to do so because it understands that ISAF will limit the resulting operational downside. Indeed, where and when the ANA demonstrates that it can cope with operational challenges it is "rewarded" with reduced NATO assistance and funding. So the ANA has very limited, if any, incentive to directly adapt to operational pressures.

The second distortion is rooted in NTM-A's character as an advisory command. When confronted with operational pressures, NTM-A's assigned mission is to help the ANA adapt to those pressures. However, NTM-A demonstrates a (perhaps unconscious) tendency to replicate its own Western approaches as it advises the ANA. In 2002 the immediate response of NATO officers and policy-makers to the security vacuum in Afghanistan was to create an all-volunteer professional army for Afghanistan. The natural response to the rise in Taliban operations in 2006 was to Westernize the ANA further by motorizing its maneuver units, equipping them with American weapons and expanding the force to achieve favorable force-to-population ratios derived from previous Western experience. In the latter part of the decade NTM-A has attempted to solve a perceived problem with the "sustainability" of ANA operations by replicating Western logistics, personnel, and management systems in the ANA. This tendency toward what might be labeled coercive emulation in NTM-A patron-client relationship with the ANA is certainly not unique among Western advisory efforts in recent decades.[87] Indeed it mirrors a dynamic that has been much commented upon, and lamented, in great power advisory operations in southeast Asia, Africa, the Middle East, and Latin America.[88]

The ANA and NATO are therefore engaged in what might be termed extrinsic adaptation, in which the requirement for adaptation and the form of the "appropriate" response are both located outside the organization that is meant to be adapted. The operational pressures that act upon the ANA are mediated through a NATO organization that naturally leans toward Western-style adaptive responses, whether or not they are entirely appropriate to the Afghan context. Moreover, the very presence of the NATO mission creates disincentives to adopt certain types of adaptations that are difficult or dangerous for the ANA. It might be argued that these distortions inherent in extrinsic adaptation alone are sufficient to explain the gap with combat effectiveness, but it is in fact the interaction between extrinsic adaptation and the strategic pressures of the Afghan political system that produce the full result.

Strategic Pressures and the ANA's Place in Afghan Political Order

The strategic-level drivers of ANA adaptation are derivative of the nature of the contemporary Afghan political system and the role played within that system by the Afghan Army. These are subtle pressures and unfamiliar to many Western security analysts, but they are central to understanding how the ANA has reacted to NATO pressures over the last ten years.

The Afghan political system is a "limited access order" of the sort that often exist where geographic circumstances have historically conspired against high-density patterns of human settlement, as they do in Afghanistan, leading to the development of decentralized social and political systems.[89] Limited access orders tend to be dominated by a handful of warlord elites who oversee sprawling patrimonial networks built around ethnicity or regional affiliation.[90] In limited access orders the patrimonial networks that compete for power have recourse to instruments of large-scale violence, often in the form of militias. The state thus lacks a classic Weberian monopoly of force.[91] Patrimonial elites compete with each other for political power and commercial opportunities, such as control of natural resources or trade routes, from which the elites extract exorbitant profit (spoils or "rents" in economic terms). The rents are then used to maintain the network through patronage. The patrimonial elites compete, or bargain, with each other but also have a common interest in preventing the emergence of new rivals, whether political or economic. They act together to prevent others from gaining access to the spoils of patronage, thus creating a "limited access order."[92]

Limited access orders need not be unstable, particularly where substantial rents can be extracted from natural resources or external actors such as development donors and/or great power patrons. In these cases the competition between patrimonial elites will tend to evolve toward a mutually agreeable bargain in which different networks control rents in proportion to the relative balance of power. In a stable limited access order, the rents accruing to each patrimonial network exceed those available if large-scale violence were to erupt. A form of stability results.[93] Competition between networks continues but in terms of targeted low-level violence and criminality.

In this context the state apparatus is enormously valuable as a source of influence and resources. Patrimonial elites will compete for control over the state apparatus, and a stable order often emerges from an arrangement in which different networks control different portions of that apparatus. This applies to all elements of the state but especially the security forces.[94] In this way the state apparatus represents a subsystem, or microcosm, of the overall character of the political order.[95] It cannot exist separately from the patrimonial bargain that underpins stability in the political system without profoundly destabilizing that system. It must be incorporated into the bargain.[96]

In a limited access order the security forces tend to be divided into internal patronage networks linked to external elites.[97] Parallel structures proliferate as elites compete and bargain with each other to maintain a balance of influence within the security forces. In an unstable limited access order this can lead the security forces to factionalize, and a civil war is an ever-present risk. Even in stable orders this arrangement tends to destroy the operational effectiveness of the security forces. Parallel structures create distrust and inhibit coordination. The capabilities of different military organizations are often developed in relationship to each other, rather than external threats or antigovernment forces, in order to maintain a balance of military power within the security forces. Officers gain appointments through links to external elites and the manipulation of internal patronage. Leadership posts are assigned to "loyal incompetents" to prevent the development of a new and destabilizing power base inside the security forces. Perhaps most important, the majority of the time and energy of key officers is devoted to managing the internal patronage networks for individual gain. The security forces are often preoccupied with this internal jockeying rather than the technocratic tasks of managing the armed forces or the professional tasks of preparing for future operations.[98]

There are numerous powerful patrimonial networks in the Afghan political order. These include Hamid Karzai's largely southern Popalzai network and connected figures such as Hazrat Ali and Pacha Khan Zadran. Other Pashtun networks include those led by Gul Agha Shirzai in Kandahar and Sher Mohammad Akhundzadeh in Helmand. The former Northern Alliance united several Tajik networks led by powerbrokers such as Burhannudin Rabbani (recently assassinated), Muhammad Qasim Fahim, and Atta Mohammad Noor in Balkh. Ismail Khan is a major Tajik powerbroker in Herat, while the most prominent Uzbek network, Junbesh, is led by Abdurrashid Dostum. Many others might be included in this list, and parsing the competition and cooperation among these patrimonial elites is far beyond the scope of this chapter.[99]

For our purposes it is sufficient to note that many of these patrimonial elites have links into patronage networks within the ANA officer corps.[100] The ANA networks are widely recognized to exert substantial influence on officer assignments and promotions, with well-connected officers receiving prominent and remunerative assignments and rapid promotions as powerful patrons manage the relative balance of plum assignments among their respective networks.[101] The patronage system also distorts the ANA's discipline and system of military justice. Many officers are wary of disciplining subordinates who are connected to other networks, and powerful patrons routinely prevent action against "their guys."[102] Finally, the ANA patronage networks control financial corruption at the wholesale level within the ANA, ranging from fuel smuggling to skimming pay and supplies to protection rackets.[103] As a result, many ANA officers devote enormous time and energy to managing the patronage networks, and the army as a whole appears deeply distracted by the machinations.[104] While ISAF has devoted substantial effort to ensuring that ANA units draw from a mix of nationalities and, partly as a result, there is little evidence to suggest that entire units are aligned with one or another patrimonial network, the links between ANA patronage networks and broader patrimonial networks are widely recognized.[105]

This is not to suggest that all, most, or even many ANA officers are not loyal to the government. It is, rather, that officers as individuals and the ANA as an institution exist within a broader political context that is characterized by patrimonial networks. The picture is much more complicated than an Afghan national government beset by insurgents. In the view of many, the Taliban and other antigovernment forces are simply another group of patrimonial networks in competition with the networks that happen to be interwoven with the Afghan government. Afghan unit commanders will fight the insurgents, but their most pressing task is often related to managing internal army politics.

To Western eyes this appears to be corruption, pure and simple. It is that, but as North et al. and Giustozzi observe, it is also necessary for the overall functioning of the political system. Were the Afghan Army to become purely professional, it would simply be an instrument with which the patrimonial elite currently atop the government structure could overthrow the current order. Too much power would accrue to a single faction. Given the broader political context, it actually benefits stability if the army remains a sprawling, divided, and distracted collection of patronage networks. This however has the downside, from NATO's perspective, of distorting how the ANA adapts to operational pressures.

The Interaction of Operational and Strategic Pressures

Operational and strategic pressures interact to shape adaptation in the ANA in at least three situations. The first situation is that where operational pressures, as mediated by NATO, are congruent with strategic pressures. In these situations the ANA adapts readily because all pressures point in the same direction. As an example, the expansion of ANA end-strength virtually every year from 2006 to 2010 was operationally desirable and also increased the scope and power of patronage networks within the officer corps. Likewise, the expansion of the upper echelons of the ANA into multiple layers of parallel organizations, including the MoD, General Staff, Ground, special operations forces (SOF), and Air Forces Commands, created additional management capacity but also multiplied opportunities for patronage and corruption.[106] The establishment of specialized units ranging from the commandos and Counter-Narcotics Infantry *Kandaks* to the Quick Reaction Forces, 111th Division in Kabul, and even the Afghan Air Force, responded to operational pressures but also created the opportunity to establish parallel layers of security forces to watch each other. Finally, the expansion of the ANA logistics capability, from MoD to brigade levels, created new capacity to support ANA operations while it also moved major flows of NATO resources from coalition channels into internal ANA channels.[107] In all of these cases NATO proposed adaptations in response to operational pressures, and the ANA readily cooperated while quietly adapting the adaptations to suit strategic level pressures. All were arguably beneficial from a combat effectiveness standpoint, but their ultimate impact has been limited by subtle subordination to higher pressures.

The second type of interaction occurs when operational and strategic pressures are incongruent, usually when NATO seeks to persuade the ANA to undertake adaptations in response to operational pressures that also run counter to strategic pressures. Examples include personnel and logistics reforms that would make these important functions more transparent and responsive to operational requirements yet would also hamper important forms of corruption and patronage. The ANA has resisted both.[108] The same can be said of proposals for rotating ANA units through the most violent areas of the country, which would undoubtedly reduce attrition and enhance combat effectiveness by committing fresh units to the fight yet would also uproot carefully constructed patronage networks where units are currently located. This too has been resisted. In circumstances where operational and strategic pressures collide the ANA often agrees in principle to adopt the proposed adaptation but in fact nullifies the decision by dragging its feet or otherwise undermining the proposed reform.

The third type of interaction occurs when NATO advocates for operational adaptations that are likely to be unsuccessful, or even counterproductive, yet comport with the strategic pressures on the ANA. The motorization of the ANA is an excellent example of this phenomenon, driven by mimetic impulses among NATO officers and actively harmful to the counterinsurgency campaign, which is not a motorized operation. Yet the ANA enthusiastically embraced motorization, likely because the fuel, spare parts, and replacement vehicles would vastly increase the financial value of the logistics pipeline. Much the same could be said for the adoption of Western infantry and support weapons, which many in the ANA and NTM-A recognize is likely to reduce the combat effectiveness of the ANA because of poor maintenance (in the memorable words of General Amlaqallah Patyani, chief of ANA training, "We have no clue how to operate the weapons that NATO gives us. And even if we did, will the weapons keep coming after 2014?").[109] Nevertheless, NATO advocated for the shift to NATO weapons and the ANA complied, filling warehouses with expensive and hard-to-track hardware.[110]

The result of these interactions is an Afghan National Army that shares much in common with the many other armies that have been developed by great power patrons, from Vietnam and Egypt to the ANA's communist predecessor. The ANA looks more capable that it behaves, particularly when the patron is committed on the ground to mitigate the risk of failure. It has gained a reputation for inefficiency, bloat, and inactivity except where opportunities arise for corruption, when it proves to be dynamic and opportunistic. It adapts readily, but the promised improvements to combat effectiveness never seem to materialize.

IMPLICATIONS

This assessment of adaptation in the ANA has several implications for Afghanistan and beyond. Where the ANA is concerned, it suggests than a major adaptive moment is approaching in the lead-up to transition in 2014. As NATO withdraws its combat forces and enablers, the ANA will become less insulated from operational pressures—the moral hazard will fade. At the same time NATO funding is also likely to decline substantially, and the ANA is far too large to be supported by Afghan society and its economy. Thus operational pressure is likely to increase as the ANA's capacity is decreasing. This will create a major test for the ANA and perhaps reveal its true operational capacity, concealed today by moral hazard.

The withdrawal of Western support is precisely the type of exogenous shock that can destabilize the bargain at the core of a limited access order. Moreover, because the ANA has become so large and expensive, it represents a wasting asset in terms of internal competition. It will degrade more or less as a function of the decline in Western financial support, likely by first withdrawing from the Afghan hinterlands (where the security return on investment has never been very high) and creeping inward toward the center as funding dries up. As the ANA degrades it will create additional points competition for control over how much it degrades in what areas, and its status as a wasting asset will give patrimonial elites powerful incentive to seize and employ the residual combat capability before it disappears. Thus there is a significant danger of fragmentation, escalation, and civil war of the sort witnessed in Afghanistan and Yugoslavia during the 1990s. That would be a catastrophe for Afghanistan and the West.

Such a result is not inevitable, however. From the perspective of the patrimonial elites who presently participate in the Afghan political system, the descent into another civil war is likely to be a worse result than the emergence of a new, if perhaps poorer and less powerful, bargain. Thus an adaptive moment also looms for the West, which has abandoned the comprehensive state-building project in Afghanistan without replacing it with a coherent alternative. The strategic interests that brought the United States and its allies to Afghanistan have not disappeared altogether. These interests are fundamentally aligned with those of the dominant coalition of patrimonial networks controlling the Afghan political system today. Neither wishes to see the current system implode, nor do they have any reason to welcome the networks, principally the Taliban, the Haqqanis, and Hekmatyar's network, that are currently excluded from the system. NATO has carried most of the burden for preventing this during the last decade. President Obama clearly wishes to shift that burden to the Afghan elites. The test for NATO will be to develop a strategy that will achieve this while drawing down the forces and resources that are its most persuasive assets.

The new strategy, unlike the present approach, will need to be adapted to the fundamental realities of the Afghan political system. The West has abandoned the long-term project of displacing patrimonialism in Afghanistan. If it is to continue to pursue its objectives in the country it will need to learn to work within the limited access order that exists. This will include working effectively with an Afghan Army that reflects the nature of the political system rather than the Western ideal, that is controlled more by patronage than the chain of command, and that responds to incentives that NATO has only just begun to understand.

NOTES

1. See discussion of NATO ISAF Operations Plan 38302, in *Report on Progress toward Security and Stability in Afghanistan* (Washington, DC: Department of Defense, October 2011), 7. Hereafter *October 2011 Progress toward Stability*.

2. *October 2011 Progress toward Stability*, 46. NTM-A's funding topped $11 billion in 2011, mostly from the US Afghan Security Forces Fund. Annual expenditures are projected to decline in coming years, though the ultimate numbers are not available at the time of this writing. NTM-A directly oversees forty-five hundred personnel. Thousands

more serve in embedded training teams, operational mentoring and liaison teams, and partner units. The total number of Western personnel involved in training, advising, or mentoring Afghan counterparts is well in excess of ten thousand. On NTM-A numbers, see, "NTM-A Update," ISAF, Kabul, September 26, 2011, 3.

3. *October 2011 Progress toward Stability*, passim. NTM-A has not existed for that entire period, but the vast majority of this expansion occurred under its watch.

4. Ibid., 43–44.

5. This is reportedly the conclusion of a draft National Intelligence Estimate discussed by Steve Coll in "Let's Hear from the Spies," *New Yorker*, Daily Comment Blog, November 24, 2011.

6. Agreement on Provision Arrangements in Afghanistan Pending the Re-Establishment of Permanent Government Institutions (Bonn Agreement), unpaginated.

7. Ibid.

8. Antonio Giustozzi, "Re-building the Afghan Army," London School of Economics Conference on State Reconstruction and International Engagement in Afghanistan, June 1, 2003, unpaginated.

9. Ibid.

10. Ibid.

11. Terrence Kelly, Nora Bensahel, and Olga Oliker, *Security Force Assistance in Afghanistan: Identifying Lessons for Future Efforts* (Santa Monica, CA: RAND, 2011), 23.

12. "Policing in Afghanistan: Still Searching for a Strategy," Asia Briefing No. 85, International Crisis Group, Brussels, BE, December 2008.

13. "A Blueprint for the Afghan Military," CJTF-180, November 2004.

14. Ibid.

15. Obaid Younossi et al., *Long March: Building an Afghan National Army* (Santa Monica, CA: RAND, 2009), 15.

16. "Afghanistan," in *The Military Balance 2002* (London: International Institute for Strategic Studies, 2002), 347; "Afghanistan," in *The Military Balance 2010* (London: International Institute for Strategic Studies, 2010), 337.

17. *Military Balance 2002*, 127.

18. Interview with Army Special Forces officer, US Army Special Operations Command, Fort Bragg, NC, October 2007.

19. International Crisis Group, *A Force in Fragments: Reconstituting the Afghan National Army* (Brussels: International Crisis Group, 2010), 8.

20. *United States Plan for Sustaining the Afghanistan National Security Forces* (Washington, DC: Department of Defense, June 2008), 4.

21. Kelly et al., *Security Force Assistance in Afghanistan*, 32.

22. Ibid.

23. The OMLT program was approved in 2005 and initiated by the British in May 2006. See Younossi et al., *Long March*, 35–36.

24. Kelly et al., *Security Force Assistance in Afghanistan*, 42.
25. "Afghanistan," in *The Military Balance 2004* (London: International Institute for Strategic Studies, 2004), 149.
26. Kelly et al., *Security Force Assistance in Afghanistan*, 41.
27. See, for example, General Barry McCaffrey memorandum, "Trip to Afghanistan and Pakistan," author files, dated June 2006.
28. Kelly et al., *Security Force Assistance in Afghanistan*, 33.
29. Ibid.
30. Conversations with NATO personnel, Bagram, March 2007.
31. Ibid.
32. *Military Balance 2010*, 347.
33. Desertion and "ghost personnel" remained a major issue for field units, however, and actual strength present for duty might have averaged two-thirds of this number. International Crisis Group, *A Force in Fragments*.
34. Interviews, ISAF, Kabul, January 2011.
35. An infantry *kandak* is assigned approximately three dozen of each vehicle. Ibid.
36. The M249 is a light machine gun (squad automatic weapon); the M240B is a medium machine gun; and the M2 is a heavy machine gun.
37. However, the weapons companies are still equipped with RPG-7s and SGP-9 recoilless rifles. Ibid.
38. Ibid.
39. *April 2010 Report on Progress toward Stability*, 108.
40. Ibid.
41. Ibid.
42. Ibid.
43. *June 2008 Report on Progress toward Stability*, 58.
44. Interviews with NATO and ANA officers, Lashkar Gah, December 2009.
45. Ibid.
46. Interviews, NTM-A personnel, Kabul, January 2011.
47. Ibid.
48. Interviews, NTM-A personnel, Tampa, FL, July 2010.
49. Michael Boera, *The Role of the Combined Air Power Transition Force (CAPTF) in Building Partner Capacity for Afghanistan* (Arlington, VA: Air Force Association, September 2010), 3.
50. Interviews, NTM-A personnel, Tampa, FL, July 2010.
51. Ibid.
52. *October 2011 Report on Progress toward Stability*, 30.
53. Interviews with RC-S personnel, Kandahar, December 2009.
54. Interviews with NTM-A personnel, Kabul, January 2011.
55. Ibid.
56. Ibid.

57. Ibid.
58. Ibid.
59. Ibid.
60. Sean Naylor, "No Easy Task: Making the Afghan Special Forces," *Army Times*, May 19, 2010.
61. *October 2011 Report on Progress toward Security and Stability in Afghanistan*, 13.
62. Special Inspector General for Afghanistan Reconstruction, "Quarterly Report to Congress: January 2011."
63. C. J. Radin, "Afghan National Security Forces Order of Battle," *Long War Journal*, 2010: available at http://www.longwarjournal.org/oob/afghanistan/index.php.
64. For public numbers, see Alim Remtulla, "In the (Afghan) Army Now," *American Interest*, October 2011.
65. See Kelly et al., *Security Force Assistance in Afghanistan*.
66. See, for example, "Department of Defense Bloggers Roundtable with Army Colonel John Ferrari, Deputy Commander for Programs, NATO Training Mission—Afghanistan," December 16, 2010.
67. See, for example, "Interview with Major Rich Lencz," Operational Leadership Experience Interview, Combat Studies Institute, Fort Leavenworth, KS, September 2008.
68. Remtulla, "In the (Afghan) Army Now." To be sure, the author has witnessed drug use by ANA soldiers, but not on the scale implied by some public sources.
69. Ibid. See also Kelly et al., *Security Force Assistance in Afghanistan*, 40–43.
70. *Report on Progress toward Security and Stability in Afghanistan* (Washington, DC: Department of Defense, April 2011), 12. Hereafter *April 2011 Progress toward Stability*.
71. Interview with ISAF Joint Command personnel, Washington, DC, June 2010.
72. Ibid.
73. *April 2011 Report on Progress toward Stability*, 38.
74. *October 2011 Report on Progress toward Stability*, 43.
75. Ibid.
76. Ibid.
77. Ibid.
78. Interview with NTM-A personnel, Kabul, December 2009.
79. Ibid.
80. Ibid.
81. *October 2011 Report on Progress toward Stability*, 29.
82. Ibid.
83. *April 2011 Report on Progress toward Stability*, 107.
84. *United States Plan for Sustaining the Afghanistan National Security Forces*, 6. See also "Interview with Major John Tabb," Operational Leadership Experience Interview, Combat Studies Institute, Fort Leavenworth, KS, May 2008.
85. International Crisis Group, *A Force in Fragments*, 11.

86. April 2011 Report on *Report on Progress toward Stability*, April 2011, 9.

87. I am indebted to Frans Osinga for introducing me to this term.

88. See, for example, William Mott, *Military Assistance: An Operational Perspective* (New York: Praeger, 1999), *Soviet Military Assistance: An Empirical Perspective* (New York: Praeger, 2001), and *United States Military Assistance: An Empirical Perspective* (New York: Praeger, 2002).

89. See, for example, Douglass C. North, John J. Wallis, Barry R. Weingast, *Violence and Social Orders: A Conceptual Framework for Interpreting Recorded Human History* (Cambridge University Press, 2009).

90. Thomas Barfield makes this argument about Afghanistan specifically in *Afghanistan: A Cultural and Political History* (Princeton: Princeton University Press, 2010).

91. Antonio Giustozzi echoes North on this point in *The Art of Coercion: The Primitive Accumulation and Management of Coercive Power* (London: Hurst, 2011) 69–70.

92. See North, Wallis, and Weingast, *Violence and Social Orders*.

93. Ibid.

94. See, for example, Kimberly Marten, "Patronage versus Professionalism in New Security Institutions," *PRISM* 2, no. 4 (2011): 83–98; and Noah Coburn, *Political Economy in the Wolesi Jirga* (Kabul: Afghan Research and Evaluation Unit, May 2011).

95. William Malley argues that this occurred at Bonn I, in *The Afghanistan Wars* (London: Palgrave, 2009), 236.

96. See Douglass North, John Wallis, and Barry Weingast, "The Natural State: The Political-Economy of Non-Development," in *Understanding the Process of Economic Change* (Princeton: Princeton University Press, 2005).

97. Giustozzi, *The Art of Coercion*, 60–70.

98. Ibid., 73.

99. Kenneth Katzman provides a concise summary in *Afghanistan: Politics, Elections, and Government Performance* (Washington, DC: Congressional Research Service, July 2011), 8–20.

100. In general, see International Crisis Group, *A Force in Fragments*. For a specific example, see Dipali Mukhopadhyay, "Disguised Warlordism and Combatanthood in Balkh: The Persistence of Informal Power in the Formal Afghan State," *Conflict, Security and Disarmament* 9, no. 4 (December 2009): 535–64.

101. Interviews with NATO and Afghan personnel in Regional Commands South and East, December 2009.

102. Ibid.

103. Kelly et al., *Security Force Assistance in Afghanistan*, 109.

104. For example, of the first ten ANA *kandaks* visited by the author in December 2009, only one commander was present in the garrison. Most of the absent commanders were in Kabul, and it was not clear when they would return.

105. International Crisis Group, *A Force in Fragments*, is effective on this point, as is

Eray Basar, *An Introduction to Corruption in Afghanistan* (NATO Civil-Military Fusion Center, November 2011).

106. Interviews, NTM-A personnel in Kabul, January 2009.

107. Interviews with NATO personnel in Regional Commands South and East, December 2009.

108. In both cases, this has taken the form of delayed implementation and backsliding.

109. "Cash Flows, but Can Afghan Training Legacy Last?" *Reuters Wire Report*, November 1, 2011.

110. Interviews with both NATO and ANA personnel, December 2009 and January 2011.

12 Conclusion: Military Adaptation and the War in Afghanistan

Frans Osinga and James A. Russell

In Bonn, on November 14, 2001, the foundation was laid for what has proven to be a severely challenging mission for all nations that have contributed forces to the International Security Assistance Force (ISAF). In Bonn, United Nations (UN) Security Council Resolution 1378 was agreed upon by the international community, thus expressing support for the creation of a new Afghan government. The new regime was to be broadly based, multiethnic, fully representative, and committed to peace and human rights regardless of gender, ethnicity, or religion. In Bonn, Afghan elites under the chairmanship of Hamid Karzai and an Interim Authority were to convene an Emergency *Loya Jirga* (Grand Council), hold parliamentary and presidential elections, draft a new constitution, and begin the transition to a democratic, market-oriented state. With that goal, the US, followed by the North Atlantic Treaty Organization (NATO) coalition, embarked on a hugely complex and increasingly ambitious state-building mission. State-building knows many definitions, but all involve the creation of new government institutions, rebuilding or the strengthening of existing ones so as to enable weak, failing, or failed states to (re-)gain the capacities to perform the core functions of modern states, and thus enabling governance capable of providing citizens with physical and economic security.[1]

While ambitious, the Bonn agreement flowed from a strategic logic framed by the UN's peacekeeping operations that went beyond the motive of counterterrorism, which informed the initial US attack on Afghanistan in 2001. Since the end of the Cold War, the UN has launched more than sixty missions in twenty-four countries. While the primary objective of all of these missions was to monitor, keep, enforce, or build peace, a second objective, which is intrinsically linked to the first, was to contribute directly or indirectly to the reestablishment of functioning statehood. Interventions were to facilitate the transition to liberal democratic polities, complete with elections and political parties, as the foundation for political legitimacy. Second, in cases of ethnic fragmentation these transitions were best initiated through consociational power-sharing arrangements.[2] Finally, peace-building projects would occur within the framework of a sovereign state, preferably centralized and institutionally well developed. Implicitly relying on modernization theory, this Westphalian and liberal model of state-building, sometimes termed the New York consensus, envisions a transition to a stable and peaceful society that is democratic, inclusive, and accompanied by institution-building to make possible effective government, the establishment of the rule of law, and the creation of a prosperous economy.[3] Together, these elements provide a blueprint for the rehabilitation of war-torn societies through international intervention that formed the basis for the NATO-ISAF approach to rebuilding Afghanistan after the US invasion.

The strategic logic contains both humanitarian motives as well as security political concerns and centers on the perception that failing and failed states constitute a risk both to individuals within states but also to the broader international community.[4] Thus it is now broadly assumed that one vital condition for sustainable peace is that the state-apparatus has the capacity to exercise the core functions of statehood in an efficient, nonviolent, and legitimate way. Consequently, peace-building is more and more seen as state-building, and this evolution is reflected in the development aid strategies of most nation states, in UN statements (for instance the doctrine of Responsibility to Protect—or R2P—of 2005) and NATO strategy documents (such as the 2010 Strategic Concept).[5] The ultimate goal of such efforts is to maintain or return to a stable, self-sustaining peace, a NATO information brochure states.[6] In short, state-building in Afghanistan served both the end of negating a sanctuary for the Taliban and Al Qaeda in the short term as well as the aim of improving regional and international security in the long run.

In hindsight, one can agree with Russell when he states that the US had no intention of mounting what turned into its most ambitious attempt at state-building since the Vietnam War. Nor was NATO or any of the NATO member states the other chapters suggest. What began as a relatively small-scale operation in the east of Afghanistan following the defeat of the Taliban, evolved into a systematic attempt to build an entirely new national political structure down to the district and subdistrict level, new government ministries, and security apparatus. For NATO, which became involved from 2002 onward, it meant a watershed in the sense of going "global" in an operation thousands of miles from Europe. It represented NATO's first mission at a strategic distance, and it ultimately involved taking responsibility for the security and reconstruction of an entire state.[7]

Not surprisingly, the impact of the mission has been significant for all participating militaries and their governments. The preceding chapters chronicle the dynamic interplay of drivers of adaptations (operational challenges and technological developments), the factors that shaped adaptation (domestic politics, alliance politics, strategic culture, and civil-military relations), and the dimension in which adaptation becomes manifest (strategy, force levels, resources, doctrine, training, plans, and operations). The case studies tell a story of frantic tactical and operational efforts by task forces and ISAF to adapt, experiment, and learn because of operational pressure (a mirror process can be seen within the Taliban). The mission, however, also tells a story of glaring strategic misalignment and of a glacial adaptation process at the strategic level where domestic politics, strategic culture, and alliance politics limited and shaped the ability of the militaries, ISAF, and NATO to respond effectively. Only in the case of the US (and by extension, NATO) does it appear that the risk of campaign failure produced sufficient adaptive pressure to innovate strategically.

OPERATIONAL ADAPTATION

The process of adaptation by units conducting tactical operations is not difficult to establish. At the lowest level, many units experimented with a variety of modes of operation, adapting to the time-space and infrastructural factors particular to their assigned province, the limited number of troops available, and the level of local Afghan support, as well as the mandate and constraints set by their respective national governments and MoDs (Ministries of Defense). All militaries had to come to grips with the vast task that Stabilization and Reconstruction entailed. A key determinant driving tactical adaptation was the intensity and sophistication of the opposition, which grew into a full-blown insurgency from mid-2005. The increased intensity of the war and the growing realization that the coalition faced strategic failure initially constituted the dominant adaptive pressure, especially in the contested southern provinces, but also, to the unpleasant surprise of the German Provincial Reconstruction Team (PRT), in the north. This tactical experience in the immediate period after initial deployment in all cases led to adjustment of plans.

The contributing militaries in the southern and eastern provinces discovered that they were manifestly unprepared for what awaited them. The US and Dutch task forces, the German PRT, the Canadians, and the UK brigades found that the Taliban had not been defeated and had no intention of abandoning their political objectives. All discovered that reconstruction and development could not commence until the security situation improved. However, that required substantial employment of troops and involved quite intense small unit combat, while diverting manpower away from the planned reconstruction activities thus delaying progress. The initial surprises and setbacks resulted in adjustment of plans and time-schedules, and experimentation at the tactical level.

The US entered Afghanistan with a conventionally structured land force overwhelmingly trained and equipped to conduct fire and maneuver operations, and thoroughly under-resourced for the job and without a clearly articulated link between the ends and means. In 2002, Operation Anaconda proved that the Taliban had not been defeated and that they returned to areas once US troops departed after their clearance operation. The attempt in 2003 by Lieutenant-General Barno, commander of Combined Forces Command Afghanistan (CFC-A), to discard the search and destroy strategy and apply a "clear-hold-build" approach lay dormant because of the limited available force of just 10,500 US troops. Entire provinces under US responsibility were in some cases policed by a single battalion.

The lack of troops compromised the American approach until the drawdown of troops in Iraq and the decision by the Obama administration to increase troop levels in 2009. Along with extra troops, the US flowed in critical supporting assets that previously had been deployed in Iraq, such as Intelligence, Surveillance and Reconnaissance (ISR) platforms and Special Forces. These forces helped establish a formalized Village Stabilization Program (VSP) in 2010 that had grown into a program involving Special Forces teams across 103 locations in Regional Command South (RC-S) and Regional Command East (RC-E). It made possible the execution of the "clear-hold-build" that Barno had envisaged six years earlier—particularly in the contested Taliban strongholds of Kandahar and Helmand provinces. Better intelligence now also enabled a dramatic increase in the number of night raids (reaching six hundred per month in 2010) against suspected insurgents as well as so-called High Value Targets (HVTs). The additional forces also made possible the training of Afghan police units, which in turn offered the option of bringing in Afghan security forces along with Afghan government officials in those areas cleared by US units in order to prevent the Taliban from returning.

Somewhat similar to the US, UK brigades (along with the Danish battlegroup), operating in Helmand, also tried a variety of approaches to address the Taliban's persistent strength. Wrongly believing that it was deploying to a quiet area, the first task force of the 3,150 men of 3 Para deployed in Lashkar Gar and Gerskh, postured for providing security and support development. Right away, however, it ran into fierce opposition, which delayed the deployment significantly. In part because of pressure from the provincial governor it was forced to spread its six hundred infantry thin among a number of other villages as well, thus implementing a "platoon house strategy." These small detachments of about forty to one hundred men came under almost constant attack and prevented the brigade from supporting development in Helmand.

When 3 Commando Brigade took over operations from 16 Air Assault Brigade in October 2006, while retaining the aim of supporting development efforts, the unit found itself focusing on the dire security situation. The new leadership gave up on the "platoon house strategy," and switched to maneuvers, setting up Mobile Operations Groups (MOGs) of 250 men and 40 vehicles each that were supposed to seek out and engage the Taliban. The Taliban responded with ambushes and tactics that denied the MOGs opportunities to do battle. The tactic also implied a reduced presence across the province, and hence the security situation did not improve sufficiently to allow the execution of development projects.

When 12 Mechanised Brigade took over Task Force Helmand in April 2007, it wanted to establish patrol bases in the population centers of the "green zone" along the Helmand River in order to demonstrate "persistent presence." It wanted to hold on to areas that had been purged of Taliban presence. The bases served as jumping off points for more logistically challenging mechanized operations. For guarding and defending those bases, it planned on reinforcements by ANA (Afghan National Army) and ANP (Afghan National Police) augmentees. That was not to happen, however, with the result that the Taliban could reinfiltrate territory that had been cleared by the brigade.

The 52 Infantry Brigade continued the approach of clear-hold-build when it took over the sector from 12 Brigade in October 2007. The unit, however, began to shift its emphasis to nonkinetic actions to win the consent of the population and de-emphasized direct action against the Taliban. This also informed clearance operations, in which the brigade would not forcefully enter a village and cause substantial destruction but instead would approach the town and rely on the local population to demand the Taliban to leave or seek safety in advance of the pending fight. Instead of rotating units through the forward operating bases (FOBs), it committed units to FOBs for the entire tour so as to provide assurance and enable more intense contact with the local population, which in turn could produce relevant intelligence. This approach was continued by 16 Brigade during its second tour in 2008. Expanding upon the theme of a population-centric approach, it also enhanced the integration with the British-led PRT in Helmand that saw a major increase in capacity during that same period. 3 Commando Brigade followed a similar approach during its second tour in the fall of 2008.

The Taliban found that engaging ISAF troops was prohibitively costly and had switched to the use of radio-controlled and pressure-plate Improvised Explosive Devices (IEDs). However, the focus on the north of Helmand also had left central Helmand exposed, a weakness the Taliban explored in a three-pronged attack by three hundred insurgents on Lashkar Gah. While 3 Commando repelled the attack, it forced the brigade to reassess the campaign plan. Central Helmand came back in focus with a series of large-scale clearing operations conducted by British and Danish troops as well as ANA elements. These were followed by setting up lodgment in villages. This scheme was adhered to by follow-on rotations of 19 and 11 Brigade, enhancing coherence and continuity.

A similar pattern repeated itself in Uruzgan Province, where the Dutch Task Force Uruzgan (TFU) had to adjust its mode of operations and expectations rapidly. The TFU started off in 2006 with a plan to establish Afghan Developement Zones (ADZs), have Special Forces disrupt and engage opposing military forces (OMF), and slowly expand outward from these once the security situation allowed it. This did not materialize. Instead of a permanent presence in a few limited areas, the TFU reverted to a roster of continuous platoon-size patrols and deliberately planned large-scale clearing operations. However, as in Helmand, the Taliban returned the moment Dutch troops left the cleared area. From 2007 onward these patrols escorted PRT personnel and assisted them in Stabilization and Reconstruction (S&R) work.

For the Dutch a real adaptive moment, akin to the Taliban attack on Laskar Gah in 2008, came with the Battle of Chora in June 2007, when a full battalion (along with dozens of airstrikes) was needed to defend that town against a well-organized Taliban attack of about five to fifteen hundred fighters. The swift, forceful response repelled the Taliban and sent a broader message that the

Dutch troops intended to protect the local population. It also demonstrated, however, that available resources precluded implementation of the 2006 plan. It forced the TFU to develop a new plan in 2008 that still used a Comprehensive Approach as the intellectual framework but was more Counterinsurgency (COIN) oriented by emphasizing security and concentrated efforts in just three ADZs. This emphasis on COIN reflected the dominant mindset within the units and with the work-up training program. By late 2008 and early 2009, as the civilian side of the task force gained in capabilities and as ANA forces became capable of conducting follow-up actions after large sweeping operations and maintain presence, TFU activities shifted back to the reconstruction mission. The changing environment in Uruzgan rippled back into the training program for follow-on Dutch units.

After their first casualties and under political pressure, German units opted not to venture outside their compound unless in case of militarily necessary. They did analyze ambushes and IED-patterns and changed tactics accordingly. Only after several years, from 2009 onward, did several developments combine to impart real change in the German approach. Mounting Taliban pressure in 2009 could have resulted in further entrenchment of the PRT, effectively conceding the area to the Taliban. However, the operational environment of the PRT had changed with a more assertive ISAF presence, and importantly, with the influx of two US "surge" brigades, led by an energetic and experienced US Army brigadier with a distinct COIN mindset. The US also augmented the PRT with aviation assets, to give them medical evacuation (Medevac) and Close Air Support (CAS). At the same time Germany incurred mounting losses. This time the US reinforcement stimulated and enabled a more robust offensive approach and higher operational tempo, without necessarily increasing operational risk for German troops. Another important factor was a new minister of defense, the then up-and-coming young and energetic media-savvy minister Guttenberg, who deliberately changed the political rhetoric surrounding the mission. When a Taliban ambush once again caused 3 German casualties in April 2010, the Germans responded with the deployment of Panzerhowitzers to bolster protection of their forces. In addition to the PRT, German units subsequently set up posts outside the wire so as to live among the Afghan population.

Over time, US, UK, and Dutch units within their provinces discovered the merits of cultural awareness, strategic communications, building up knowledge of the social and political infrastructure, and conducting key leadership engagement based upon that intelligence. US brigade headquarters (HQs) expanded to include all kinds of civilian expertise required to support the widening requirements of state-building. From 2008–9 onward, military units found themselves providing security to development-related activities, training Afghan soldiers and policemen, working on improving agriculture and irrigation, and setting up a Ministry of Defence.

As demonstrated by Giustozzi, these efforts directly competed with the Taliban, which were similarly undertaking efforts to provide communal services to villages such as fighting crime, setting up courts, and punishing local officials for corruption. In addition, Taliban re-established shadow government structures in a rapidly expanding number of provinces. Concern for its image resulted in a softer approach concerning adherence by Afghans to the strict social edicts the Taliban had enforced since the 1990s but that had proven to alienate the population. A similar deliberate softening of attitude was visible vis-à-vis nongovernmental organizations (NGOs) and development aid.

The Dutch military made significant adaptation in their approach to intelligence over their experience in Uruzgan, despite the framework of reconstruction and development and the fact that Dutch soldiers rapidly recognized that they were conducting counterinsurgency operations. The intelligence focus initially remained with the enemy. In Uruzgan, however, the Taliban was not the main source of insecurity. The divide-and-rule politics of former governor Jan Mohammed had caused violent contention, and the true problem was to realign marginalized societal fragments with the provincial government and prevent the Taliban insurgents from exploiting local feuds. To understand this required the population-centric intelligence typical for counterinsurgency campaigns. Under the influence of civilian expertise within the PRT that only slowly evolved over two years, a gradual shift occurred in the intelligence process. Through the use of a tribal adviser, and later cultural advisers, the units developed a nuanced picture of local society and convinced military intelligence officers of the necessity for population-centric intelligence. This process took until mid-2007 and came into focus following the battle for Chora—an adaptive moment. Thus, after gradually shedding its enemy-centric orientation, the emphasis in intelligence shifted from the Taliban to the local population.

CIVIL-MILITARY RELATIONS ALLOWED THE MILITARY LEEWAY

Most countries permitted their tactical commanders considerable leeway that, it could be argued, developed into strategic adaptation as myriad commanders across ISAF struggled to solve the misalignment between available forces, the opposition, and the size of the area of responsibility. Behind this attitude of noninterference may have been a political lack of interest in getting involved in detailed military planning (in the case of the UK government under Brown, for instance) or even sound civil-military relations. The hands-off approach allowed the military to make changes that in hindsight were of a strategic nature in the sense that the aggregation of changes resulted in a marked change in approach and nature of the mission.

The UK brigades are a case in point. Each subsequent brigade commander during the first two years in particular, sanctioned by superior national command, came up with his own approach to the problems in Helmand, which deviated from the official UK Joint Helmand Plan formulated in 2006. Similarly, the 2007 Helmand Road Map was the result of a bottom-up development process within 52 Brigade that then received approval by Whitehall.

Another example is Uruzgan Province. During the entire campaign, strategy for operations was formulated by the Dutch TFU staff in the province. This bottom-up process allowed for the initial strategy to be further developed, based on the reality of the operational environment. De facto, it resulted in a plan that—by including a focus on security—incorporated as many tenets of COIN doctrine as it did of the 3D approach (Defense, Development and Diplomacy). The 2008 Focal Paper adjusted the initial strategy into a campaign plan for counterinsurgency in Uruzgan, thus deviating significantly from the original political mandate and narrative.

In the case of Canada, a generational change in military leadership was able to institute the drastic changes and gain acceptance of the military nature of the mission and the risks that might involve. The appointment in 2006 of General Hillier, a former ISAF commander, as Chief of Defence Staff (CDS) meant a good-bye to a generation of senior officers whose risk aversion could be traced back to their Somalia peacekeeping experience in the 1990s, and heralded a cultural change in the Canadian Army.

US units experimented individually with approaches, often determined by the different social, political, and military situations in the various provinces. From 2002 through 2009, US units demonstrated immense variation in their approaches to the operational environment. As Russell shows, each valley constituted its own area of operations, with specific features dictating its own peculiar form of operations by the unit operating in it. Each battalion in its parent brigade in turn structured its operations in ways to support top-down guidance, but also showed significant differences in structuring their organizational outputs. The Commanding Officer of 4/25 brigade, Colonel Michael Howard, for instance, established a civilian-military board of directors to coordinate and synchronize all US government organizations in the province, and each battalion replicated this brigade structure, but designed its tactical operations in different ways.

DOMESTIC POLITICS SOMETIMES PROVIDED ADAPTIVE PRESSURE

While most militaries enjoyed considerable leeway in structuring their operations during the initial phases of the war, over time, deliberate changes in the political mandate eventually pushed military activities in another direction. Such a form of strategic adaptation was inspired by operational demands, but one profoundly filtered and shaped by domestic politics can be found in the Canadian case study. The changing political climate in Canada translated into changes in the place where Canadians soldiers served in Afghanistan, in their equipment, and in changes in the command set-up. The inability of the conservatives and liberals to form a majority government explains why Canadian forces first deployed as a combat unit as part of Operation Enduring Freedom, then as a peace-keeping and S&R deployment in Kabul. From 2006 onward, its forces served as a combat force engaged in conventional combat and then in counterinsurgency. After a change in government in 2011, Canadian forces became a training unit with explicit caveats that limited combat activities.

In Denmark, domestic politics did not shape military operations. Instead, once the Danish battlegroup was committed, the political discourse concerning employment was dominated, or indeed hijacked, by the Danish Army, which considered itself as a professional—*militær faglich*—community with unique expertise. Domestic politics started to have a strategic impact when the public began to question the rationale of the national contribution in light of the mounting number of casualties. This resulted in a move toward an exit strategy. From 2007 onward, the gap between ambition and achieved level of progress inspired the opposition to demand a clear end state and benchmarks that could indicate progress. It resulted in two plans that for the first time defined some sort of goal to be achieved by its national contingent in Helmand (that is, training of Afghan security forces that could, over time, take over from the Danes).

A similar dynamic can be seen in the Netherlands during the national debate in 2008 over whether to extend the mission for another two years. The decision to extend came attached with specific conditions that affected plans developed in Uruzgan (the Focal Paper and the 2009 Uruzgan Campaign Plan) by insisting that those plans provide a guideline for the transfer of authority in 2010 and subsequent exit from Uruzgan. While TFU strategy certainly adapted to meet the requirements of a plan for counterinsurgency in the specific operational environment of Uruzgan, domestic politics and government policy were the additional main drivers of this adaptation by changing the nature of the mission at a certain stage, much like that which occurred in the Canadian case. Dutch domestic politics actually dramatically informed the decision to terminate the mission because the debate over another extension in 2010, and lack of consensus within the coalition parties that made up the cabinet, led to the fall of the government.

The United States, by contrast, had no real national domestic political debate over the Afghan War until 2009, when the Obama administration was presented with a request for more forces by General Stanley McChrystal. While it is clear that national-level politics and civil-military relations shaped the environment that resulted in the increase in troops starting in 2009, public opinion per se and/or widespread opposition to the war had little impact on the conduct of the war. For the United States, its military involvement had been outsourced to the volunteer force, with little public engagement and interest in the conduct of the war. As in the other cases noted here, authority to structure and conduct military operations was delegated to military commanders.

ADJUSTMENTS TO EQUIPMENT IN RESPONSE TO TALIBAN TACTICS: THE IED THREAT

Also in one other respect most nations did adapt strategically, but often it was not perceived as such. All of the case studies, apart from Germany, discuss the development and application of an Urgent Operational Requirement (UOR) acquisition process in order to rapidly provide capabilities that the tactical units in the provinces required after their first encounters with Taliban and the increasing threat of IEDs.

The Taliban had profited from the relative absence of military pressure that existed in most parts of Afghanistan well into 2006. Giustozzi shows how during that period the Taliban was able to regroup, form a network based on tribal alliances and patrimonial patterns, and gain coherence despite being widely dispersed geographically. The arrival of NATO Special Forces and substantial task forces in the southern and eastern provinces surprised them, and caused an increasing number of casualties, which resulted in adaptive pressure. The Taliban subsequently invested in its command structure (a painful process of centralization), in recruitment, and in the quality of training. It also integrated foreign fighters in Taliban-led units (and by seeking assistance of Iran and Pakistan). Avoiding pitched battles with the stronger ISAF units, and also in response to effective ISAF key leadership targeting, from 2006 the Taliban shifted to an increasing emphasis on targeted assassinations of ever more junior officials, but also of officials close to Karzai. Finally, it resulted in a deliberately planned suicide bombing campaign and, probably with the aid of Iranian, Pakistani, and Iraqi experts, a sophisticated and widespread use of IEDs. Networks of specialists were developed for manufacturing, handling, and deploying IEDs. Placement tactics (for instance, using children) and IED technology evolved in response to ISAF countermeasures.

The evolution of the IED threat in particular inspired technological adaptation among ISAF task forces. Denmark lost forty-two soldiers by November 2011, the majority to IEDs, and suffered forty-seven wounded soldiers just in 2009. IEDs forced the Dutch TFU to change patrolling tactics. Inspired by the hearts and minds mantra, initially it had emphasized patrolling dismounted or in open patrol vehicles, and not in closed armored vehicles, as that would preclude making contact with the population and would intimidate instead of building trust—a point of critique leveled against the very cautious German approach. For similar reasons troops would not wear wraparound sunglasses, as this was considered culturally insensitive. The increasing IED threat, together with the rising number of incidents with Troops in Contact (TICs), forced a more robust stance upon the TFU and also resulted in the rapid acquisition of twenty-five Bushmaster armored vehicles. F-16 reconnaissance equipment was brought to bear to detect the placement of roadside bombs, and investments were made in datalink equipment to improve close cooperation between Forward Air Controllers (FACs) and Apache and F-16 CAS assets. The United States funded a multi-billion-dollar program to find solutions to the IED threat and subsequently built and deployed the MRAP vehicle to protect its troops.

For similar reasons—experiencing a fourfold increase in IED attacks from 2006 to 2008 and another doubling from 2008 to 2010—in 2006 the UK force acquired 100 Mastiffs for Iraq and 150 Ridgebacks for Afghanistan. Using the UOR scheme, both types of Mine-Resistant Ambush Protected vehicles (MRAPs) were to replace the lightly armored Land Rovers used on patrols and obtain "protected mobility" instead. In 2008, 700 of such MRAPS were set to be purchased. The UOR scheme was also applied to enhance UK soldier equipment such as body armor and night vision equipment, and legacy systems such as the Warrior were upgraded. In addition, as in Canada, investments were made in ISTAR (Intelligence, Surveillance, Target Acquisition, and Reconnaissance) capabilities, also to counter the IED threat. The five ASTOR (Airborne Stand-Off Radar)-manned Air Ground Surveillance systems were added in 2008 to three new types of UAVs (Unmanned Aerial Vehicles) that were introduced in 2007. Ground-based ISTAR also received a boost in order to improve FOB security. Most of this equipment—at a cost of 4.2 billion pounds—was fielded within a year.

The Canadians, interestingly, reversed a previous government decision to abolish tanks, and reintroduced Leopard tanks into the inventory by purchasing surplus equipment from Germany and the Netherlands. Denmark also deployed such heavy armor material. In both cases operational requirements (operation Medusa was underway in the case of the Canadians) were identified, grounded in Taliban tactics as well as in risk reduction for their own troops in the face of the threat of IEDs that lighter vehicles had proven to be very vulnerable to. Yet in Denmark it was less a case of adaptation than of positive experience in peacekeeping in the Balkans, and adhering to existing mechanized warfare doctrine. In Canada the lack of effective political opposition to the government was instrumental in getting this controversial decision implemented.

A shortage of helicopters—attack as well as transport—proved persistent across Afghanistan, and various NATO force generation meetings never solved this problem. On a national level, this led to difficulties. The UK MOD responded to UK Army request by deploying additional Sea King and Lynx helicopters, which were not quite suited for the physical environment and the climate. The introduction of seven Merlin helicopters helped out in that respect. Still, this proved insufficient, and despite a public complaint by Chief of the Army Dannatt in 2009, the UK Army was unable to get approval for an increase in capacity. Germany found out that, also in order to reduce vulnerability, it required more than the six CH53 transport helicopters. Canada on the other hand managed to boost its fleet of Chinooks relatively rapidly, through a mix of leasing and purchasing, again also on the grounds of reducing risk for its troops. Here the interests of the military—risk reduction and enhancing operational effectiveness by decreasing the reliance on vulnerable ground transport convoys—coincided once again with those of politicians who needed to minimize casualties in order to maintain enough support from the opposition.

STRATEGIC CULTURE, REPUTATION, ORGANIZATIONAL IDENTITY, AND ALLIANCE POLITICS

All four of the above factors played visible roles in the ways and trajectories of adaptation, often intertwined and sometimes with opposite effects, sometimes with interlocking influence, but often not conducive to strategic adaptation or innovation. In the wake of the 9/11 attacks and the transatlantic rift over Iraq, all the NATO members evinced a strong interest in participating in Afghanistan, if for no other reason than to preserve alliance solidarity. In Denmark, Germany, and the Netherlands in particular, strategic culture fostered an image of a military that was employed for humanitarian motives, supported by a narrative of reconstruction and development. Counterterrorism carried some weight in all of the countries (preventing a resurgence of the Taliban and thus negating a sanctuary for Al Qaeda), but in European countries distinctly less than in the US, for obvious reasons. This emphasis in European countries downplayed the use of force and offensive actions, and offered a problematic causal logic that made the presence of national contingents and casualties hard to explain and justify on strategic grounds. Apart from the US, in other countries the case for participation was not self-evident. The United States and its military institutions held no such commitment to the European humanitarian motives, but found itself dragged into the mission over the course of the war.

The UK Brigades, and the British government, were keen to restore their sense of pride and international reputation (in particular with their US partners) in the wake of the less than stellar performance in subduing the insurgency in Basra. Alliance politics too played a role here, as the UK wanted to bolster NATO unity. A forceful, robust presence fit that posture. Canada, in turn, was keen to contribute, being a smaller power neighboring the US, and Canadian units were keen to excise their reputation as blue helmet peacekeepers that would not engage in actual combat. The Canadian decision to buy tanks and deploy them not quite incidentally coincided with the organizational identity and interest of the armor branch of the Canadian Army.

The militaries of European nations meanwhile were mostly driven by different concerns. While the Danish MoD and the Ministry of Foreign Affairs stressed the comprehensive approach and the reconstruction side of the mission, the Danish military wanted to demonstrate commitment to the NATO Transformation agenda and conduct expeditionary missions that included combat. This reflected recent official MoD policy changes which stipulated that the Danish Armed Forces transform from a conscript force oriented on territorial defense to an intervention force. To that end they were keen to demonstrate that they could deploy and employ effectively and stomach the realities of war, leave nation-building to civilians, and were satisfied to have their 650-man battlegroup operate according to the plans of the UK brigades in Helmand. There operations were superficially in tune with COIN, but as Rasmussen states, the Danish mechanized battlegroup actually operated on a schema of mechanized warfare and quite deliberately ignored the Civil-Military Cooperation (CIMIC) side of the mission. This created an ever-increasing distance between political rhetoric at home and reality in Helmand, especially since it was Denmark that put the concept of the Comprehensive Approach on the NATO agenda, an idea that stressed increasing cooperation of civilians and the military.

Somewhat similar alliance political motives, combined with a strategic culture that fostered the transatlantic relationship, were at play in Dutch political discussions concerning the contribution of the Netherlands. Regarding itself as the big one among the smaller NATO nations and with a defense policy that included an ambition to operate also in complex expeditionary operations, the Netherlands took on a very significant task when it opted to take responsibility for Uruzgan as a lead nation. The Dutch Army units in turn carried with them a reputation tarnished by the Srebrenica massacre, a heavy load that hurt their pride in their professionalism, and one they were keen to shed. This informed in part a willingness to take operational risks when confronted by Taliban offensives. From 2003 onward Dutch Special Forces had been involved in various ISAF operations. Indeed, reputation played an explicit role in the decision in 2007 to protect Chora against the Taliban's attack.

In its own peculiar way, the story of the development of the ANA also vividly illustrates the impact of culture and domestic politics. NATO discovered that creating a national army and adapting it to the evolving security situation cannot occur in isolation from the particular social, cultural, and political patterns and dynamics of the host society. The ANA has undertaken an epic level of adaptation since its (re-)establishment in 2003. Based on the architecture of Western defense organizations, complete with corps-level headquarters, brigades, battalions (*kandaks*), companies, and platoons as basic organizational units, supported by first the Kabul Military Training Center, later by Combined Security Transition Command—Afghanistan (CSTC-A)—and NATO Operational Mentor and Liaison Teams (OMLT), and deliberately created to act as a symbol for national unity, the ANA became just another arena in which common Afghan societal patterns of favoritism and patronage were played out, and where the allegiance of officers did not necessarily go to the Karzai government. Officers as individuals and the ANA as an institution exist within a broader political context that is characterized by patrimonial networks. As Grissom observes, NATO never managed to erase the legacy of the "limited access order" formed by the warlord elites who oversee patrimonial networks built around ethnicity or regional affiliation.

When NATO initiatives concerning the ANA coincided with interests of patronage networks, adaptation generally occurred in line with NATO intent. For instance, the gradual expansion of the ANA between 2006 and 2010 to a nominal strength of 170,000 was necessary to deal with the increasing Taliban threat. The expansion of the army, however, in parallel increased the scope and power of patronage networks within the officer corps. Operational rationality likewise drove the development of management and C2 (Command and Control) structures, the expansion of ANA logistics and engineering capabilities, and the creation of specialized units such as Special Forces, the Air Force, Counter-Narcotics units, and Quick Reaction Forces. But these structures also provided major resource flows for corruption, as well as parallel layers of security forces that could watch each other and opportunities for patronage and favoritism that compromised the institution ISAF was trying to create. Indeed, when the ANA was forced to embrace motorization and start using modern Western military infantry equipment, it readily did so, not because it closed a performance gap vis-à-vis the insurgency. Instead, the fuel, spare parts, arms, and transport vehicles offered valuable and hard to track hardware, representing material, financial, and thus political capital for those Afghan officers who were "dual-hatted," in that they served the ANA on behalf of the central government, but also on behalf of the patronage network.

On the other hand, when NATO attempted to persuade ANA to undertake adaptations that were incongruent with patronage network interest, adaptation generally stalled—not through overt resistance but by foot-dragging and rising absenteeism and desertion. Grissom notes how efforts to introduce personnel and logistics reform so as to improve transparency, operational responsiveness, and accountability stalled, as these would hamper corruption and patronage. Similarly, proposals to rotate ANA units, and regularly insert fresh units into the most violent areas, while operationally sound, stranded because such a scheme would disrupt the patronage networks that had been constructed. The assurance that ISAF would compensate for ANA deficiencies also inspired risk averse behavior—indeed, demonstrating operational effectiveness and adaptability would only reduce ISAF support. Testifying to the impact of cultural dynamics within the Afghan society, Grissom concludes that the ANA adapted readily, but the promised improvements to combat effectiveness never seemed to materialize. Low levels of education—90 percent illiteracy rates—lack of discipline and subversion, and subsequent low training standards are some reasons for this, as are the deficiencies in logistical and administrative organization at the national, regional, and local levels. But underlying these issues, cultural and political dynamics explain best why, as of late 2011, after years of training and advice, and tens of billions in Western investment, not a single ANA unit was deemed capable of operating independently of NATO support, despite adaptive pressure applied by ISAF and the Taliban.

Germany offers perhaps the clearest example of the way strategic culture can produce inertia and stasis, but also how over time operational experiences and generational change among senior officers and political leadership can inspire a gradual shift in strategic culture. A culture of military restraint and an identity as a *Zivilmacht* formed the background for the German discussions concerning deployment and employment of German troops. This culture also structures civil-military relations in that a dominant and detailed civilian oversight is exerted by the Bundestag over all military affairs, and by civilian courts over incidents involving civilian casualties caused by German military actions. The antimilitaristic culture also explains why military expertise is not ranked highly, why it is in short supply within parliament, and why public interest in military affairs is very limited, despite the fact that with fifty-one hundred soldiers in Afghanistan it is the third largest troop contributor and it had lost fifty-three soldiers in battle by the end of 2011.

Both tactical and strategic adaptation were significantly hindered by the self-imposed limits concerning the use of force. On the one hand the political lack of expertise and indeed interest offered leeway for German troops to improvise tactically. But is also had the consequence that adaptation was never driven by the government or Bundestag, even when significant changes were called for—for instance, in the form of more liberal Rules of Engagement (ROEs) concerning the employment of weapons. Facing up to the reality that its troops were engaged in intense fighting if not war was politically unacceptable, as the mission was framed as a stabilization mission, and tolerated because of that—a fate that initially also befell the Dutch military. Indeed, only after nine years into the mission was the Law of Armed Conflict deemed applicable instead of German criminal law.

Unsurprisingly, this general attitude resulted in risk-aversion and caution in the fabric of the German Army and indeed it shaped the mode of operations of the PRT. That was also the very aim of instructions handed down from Berlin to commanders that warned them to avoid casualties at all costs and to issue extraordinarily restrictive ROEs that even precluded engaging identified Taliban elements unless German troops were actually being shot at. Another illustration of this was Minister Jung's insistence that movement outside of the fortified bases was to be conducted only in armored vehicles, which deliberately also limited the potential of troops having contact with the local populations. Similarly restrictive and dangerously unresponsive was the procedure for requesting fire support, a request that required approval from Berlin, seven time zones removed from the tactical environment. Moreover, Tornado reconnaissance planes were prohibited from assisting combat operations, even German ones. Such an attitude percolated downward within the army as micromanagement with PRT commanders requesting lieutenant-general-level approval in Berlin, for tactical decisions in Kunduz. It would require a combination of mounting casualties and Taliban pressure, peer pressure by US units under the command of the German PRT, the Kunduz incident (the bombing of a fuel truck captured by the Taliban which killed dozens of civilians), an ensuing generational change in leadership at the MoD with actual experience in Afghanistan and a line of energetic ministers of defense to enforce alignment between Berlin rhetoric and Kunduz reality and to effect a shift in the German strategic culture.

LIMITED INSTITUTIONALIZATION

When German units did try to adapt to the challenges in Afghanistan from 2003 onward, the tactical improvisations never rose to the level of organizational adaptation. The military culture of the Bundeswehr itself is partly to blame for this, as it does not inspire open debate and honest intellectual analysis, out of fear for political repercussions (again, the Dutch officers ran a similar risk). Rid and Zapfe thus doubt whether the German Armed Forces are institutionally ready to absorb, retain, process, develop, and deploy the lessons it learned. Indeed, there is only limited evidence of institutionalization among the various militaries studied here.

It took most militaries several years before best and worst practices were shared and turned into changes in tactics, doctrine, or training programs. Farrell notes that it took the British two years to adapt a new approach to operations, to change training and doctrine. Yet he also observes that 16 Brigade on its second tour made a point of not repeating the concept of operations applied during the first tour but instead adhered to the concept put in place by 52 Brigade that appeared successful. From 2009 onward the approach ventured by 19 Brigade in central Helmand became the model for subsequent rotations, suggesting tactical institutionalization. Doctrine development, however, did not follow suit. When the UK did bring out a manual on COIN doctrine as well as on stabilization operations in 2010, it had taken six years since its first encounter with COIN in Iraq and four years after the US had published COIN manual FM 3–24.

Size may matter here, and somewhat negate the necessity of having a formal institutionalization process. For instance, in the case of the Netherlands, the adaptation of the predeployment training program points at an interesting feature of a small military like the Netherlands Army. Despite the lack of a formal organization to institutionalize lessons learned, informal as well as professional personal contacts led to a relatively quick adaptation of the training program. This allowed the use of experiences of previous rotations in order to establish a tailor-made program that prepared the troops and civilians for the conduct of an integrated counterinsurgency campaign in Uruzgan's demanding operational environment. Informal learning also occurred through various so-called informal doctrine notes and articles authored by TFU, PRT, and battlegroup commanders who had just returned from their deployment (something that was notably absent in Germany). The drawback is that such bottom-up learning may not result in improved doctrine or a formalization of new training standards. Indeed, training was structurally altered only with the fourth TFU rotation onward. This change included the addition of more cultural awareness lectures and more attention to CIMIC issues and lectures on state-building. Doctrine took more time and a formal evaluation, started only when the last TFU returned home. Indeed, the lack of an institutionalized lessons learned process was identified as one of the key issues by a former RC-South commandant who was appointed Commandant of the Army in 2011.

Similarly, within the small Danish Army all commanding officers know each other personally, get together frequently, and hold operational seminars that aim to exchange lessons learned rapidly. However, in both countries such learning often remains limited to the tactical level. Only recently have the Dutch Armed Forces established a new policy that aims to install a formalized lessons learned process that also couples the tactical to the strategic level. While Danish troops adjusted their Tactics, Techniques and Procedures (TTPs) concerning patrolling, the use of tanks, and avoiding IEDs, these adjustments were not considered to constitute a significant deviation from accepted doctrine, the Field Manual. In the Danish case, a somewhat dogmatic adherence to the Field Manual actually seems to have prevented genuine adaptation to the Helmand conditions. While changing TTPs enabled the battlegroup to operate its equipment and units with a certain level of security, this did not make the strategic situation more favorable because the changes did not address the appropriateness of the structure and approach of the battlegroup vis-à-vis the task and opposition in Helmand.

The US Army and Marines Corps provide the most obvious example of deliberate institutionalization. As Russell noted, the change in institutional focus of the US military happened through deliberate bureaucratic efforts in the rear echelon in combination with learning and innovation in the field. It was a two-way traffic. In 2004 and 2005 important concepts and directives were promulgated by the Joint Forces Command and Department of Defense (DoD) that emphasized the importance of stabilization operations and laid out doctrinal tenets that reflected lessons learned in Iraq and Afghanistan. In December 2006 FM-3-24 was published after an intense development process involving experienced commanders and academic experts, to be followed in 2007 by an Irregular Warfare concept.

EMULATION

Considering the very public debate in the US but also in Europe on the experiences in Iraq and the attention in the Western media and military journals in particular for the subsequent publication of FM 3-24, the case studies curiously suggest a relative scarcity of emulation. The one clear case in which (coercive) emulation did take place—the creation of the Afghan National Army—it might actually have had a somewhat counterproductive effect. Literature would suggest that among the various units within the theater one would observe instances in which a unit would learn from the experience, doctrine, organization, and tactical practices of a senior contributing nation such as the UK or the US, in particular when such nations had units attached to units of smaller nations, in adjacent provinces or when combined operations were conducted. Having a mature ISAF command structure in place after a while, with strong senior leadership, also would suggest that units would adopt practices that such leadership would support.

Indirectly, US COIN doctrine, and/or the debate in open literature, seems to have had some influence in the sense that it informed training of units—for instance, Dutch TFU units—and offered legitimacy to tactics that a unit was considering. In contrast, the Danish Army rejected US COIN developments. Similarly, Germany has not developed a COIN doctrine of any sort, and most operational changes have remained at the tactical level. Indeed, its culture mitigated against adopting the very term "counterinsurgency."

Germany is one example in which the peer pressure exerted by the influx of US troops, along with the introduction of heavy US equipment within the German province, helped German units overcome reluctance to operate robustly. Another example is the Danish case, where the fact that the Danish contingent was entirely absorbed and encapsulated within the UK-led taskforce in Helmand made it evident that Danish troops could operate effectively and with any measure of security only if they intimately indeed symbiotically aligned their planning and operations with the UK brigade, and later also with US Marines when they entered the Helmand Province in 2009. When the brigade adapted and changed tack, the Danish unit adapted too, not necessarily to changes in operational demands but rather to the adaptations of the UK Brigade to Taliban tactics; hence Rasmussen's assertion that this was a case of second-order adaptation.

STRATEGIC ADAPTATION AND INNOVATION

Strategic adaptation throughout the NATO alliance was limited. At the tactical level the distinction between COIN and the Comprehensive Approach proved irrelevant, as several commanders indicated in interviews right from the start. Most units in the field either conducted an S&R mission with a COIN flavor or vice versa, and strategic adaptation in practice involved only a change in degree on that limited spectrum of options. This tactical and operational adaptation by task forces (apart from the US) hardly ever transcended the confines of their province, and from a campaign perspective cannot be considered strategic.

Strategic adaptation in most European countries occurred in the sense that the narrative (only) gradually shifted from a purely S&R mission toward one that incorporated the reality of a violent insurgency conducted by a determined adversary. There often existed a gap between reality at the tactical and perceptions at the strategic level. Indeed, as the German and Dutch cases discussed, several countries very deliberately had politically framed their commitment in terms that made clear that their approach differed from the counterterrorism and kinetic action oriented COIN approach. Such (naive) expectations among all countries, apart from the US, concerning the nature of the mission supported by a narrative that was informed by the limited mandate set by the fragile political bargain between government and parliament persisted until the first visits by Chiefs of Defence Staffs, defense ministers, and/or members of parliament to the respective province of their national contingent. Only when the insurgency gained strength and visibility, and after mounting casualties and reports of combat incidents, the evidence that there was a COIN dimension to the mission could not be ignored within the European capitals. Still, domestic politics often hindered adaptation when this required relaxing the national ROEs or increasing the level of commitment.

In Canada a minority government meant only delicate changes could be considered that always required bargaining. During 2006 and 2007 in the UK, Blair and Brown were focused on domestic political challenges and their own rivalry from 2008 onward. Brown, who never had a deep interest in defense, was preoccupied with his own fragile power position. On top of that, four different UK defense ministers cycled through the ministry in five years, which prevented continuity and resulted in fragmented leadership. A fragile ruling political coalition and waning public support for the mission in the Netherlands put strict limits on the nature and scope of the contribution. The Dutch TFU organization was initially tailored for a reconstruction and development mission with the battlegroup and the PRT as its two main components. Uruzgan's contentious environment, however, forced an extension of battlegroup assets, and thereby initially more emphasis was placed on the kinetic parts of the organization. Troop levels were increased up to a permanent contribution of fourteen hundred soldiers (a limit imposed by the political level in The Hague) and as many as six hundred temporarily deployed additional troops. Domestic politics, not operational considerations, however, had put a stringent cap on the size of the national commitment to ISAF. Overall, most countries—in particular Germany, Denmark, the Netherlands, and Canada—proved very reluctant to step up and achieve alignment between requirements in terms of troop levels, ROEs, and resources on the one hand and the tactical situation in the assigned province combined with the scope of the mission on the other.

The UK is among the few examples, along with the obvious case of the US, of significant, albeit reluctant, strategic adaptation in terms of manpower. It increased its force levels within two years to more than 8,000 troops. The commitment in Iraq precluded a larger increase till 2009, when the number of UK troops in Afghanistan rose to 9,000 and eventually to 9,500 in 2010. Yet this was still considered insufficient by the UK Army, in light of their assessment of the scope of the mission in Helmand. The requested additional 2,000 troops were declined, though, despite the Army Chief Dannatt's going public with his concerns. The British, like the Germans in Kunduz, benefited greatly from the addition of substantial US contingents in the fall of 2009 (totaling eventually almost 20,000). Those forces released British forces from southern Helmand, allowing them to focus on central Helmand.

Indeed, the United States stands out as the only country that made a determined step to make an impact at the campaign level by surging its force by an additional 30,000 troops, reaching a total of about 100,000 as well as several thousand civilians by mid-2011. The increase in additional troops, however, did not take place until the Obama administration started to decrease its commitment to Iraq—which was always the US strategic priority. The US military commitment in Afghanistan was undeniably and systemically affected by the priority of the Iraq War on the US military. Afghanistan was neglected by the United States from 2002 through 2009. Despite the increase in forces after 2009, the coalition could never boast more than 130,000 men: while a sizable number in absolute terms, it was still limited relative to the size of the theater and scope of the mission. Moreover, the force did not materialize until after precious momentum had been lost in the preceding years.

In general one can observe that, after the decision to commit national units to ISAF, very little suggests that European governments seriously re-evaluated that decision from a strategic perspective. They never addressed the issue of whether S&R (or COIN for that matter) could achieve strategic success. Russell argues that the US approach developed not through a strategically driven process that articulated objectives at the war's outset and matched ends to means. Instead the approach evolved iteratively on an ad hoc basis, dictated by strategic circumstance framed by the war in Iraq as well as deteriorating circumstances on the ground in Afghanistan. Subsequently, operational adaptation in the field in the period from 2002 to 2009 was ad hoc and often lacked strategic focus, which systematically compromised the application of US military force.

As Farrell concluded, British strategy did not really adapt, and he argues that there was an absence of a clear and coherent strategy. In Denmark, the Netherlands, and Germany, the Comprehensive Approach became the mantra and was mistaken for strategy, and in the UK too, strategic principles (that laid out the case for a comprehensive approach along the familiar lines of operations that that approach entails) took the place of strategy proper. The smaller nations in particular limited their discussion to questions of whether their units had sufficient resources to operate tactically more or less effectively within their assigned areas. Taking part, or not, and with what sort of effort, became the key strategic decision. Why and how the mission exactly was to contribute to national objectives and whether the forces were sufficient for the task were analytical questions that appeared more in critical journal articles and after-action parliamentary hearings than in official governmental documents. Certainly in the case of Denmark, the Netherlands, and Canada, but also Germany and the UK, whether that commitment was to an endeavor that had any realistic chance of being successful strategically was not considered an issue to be addressed on a national level, but one that NATO was supposed to address. Nations tacitly deferred to NATO and the US when it came to developing a coherent overall strategy.

NATO'S LIMITS AS A STRATEGIC ACTOR

NATO as an organization showed a very limited ability to develop a coherent campaign strategy and enforce it upon the various contributing forces. In an irony of the wartime experience, the Afghanistan experience brought major changes to the organization even as the organization proved itself unable to develop strategic coherence on the ground in Afghanistan. The Afghan mission has shaped the reform of the integrated command structure, drastically reducing the static structure, shortening the lines of responsibility between theater HQs and SHAPE (Supreme Headquarters Allied Powers Europe), and introducing a deployable Joint Forces HQ (JFHQ). Reflecting the Afghan experience, NATO developed a crisis-management capacity as part of the 2010 Strategic Concept. NATO showed a distinct inability to support and integrate the nonmilitary side of the mission in Afghanistan. In addition it will develop the capacity for cooperating with partner nations, reflecting the fact that ISAF included for instance Australian units. Indeed, Afghanistan by 2010 had turned into an engine of change for NATO, as argued by Rynning.

ISAF presented NATO with an entirely new type of mission, and one at an unprecedented distance from the transatlantic region. No planning document before 2001 had included a scenario of such wide geopolitical reach, and very few countries had made the switch yet to a professional military with an expeditionary orientation.[8] Whether NATO should have such a strategic—almost global—ambition was a contentious issue that divided the alliance in a transatlantic camp, a status quo camp, and a Eurocentric camp.[9] Similarly, whether NATO was to adopt comprehensive peacekeeping missions or maintain a purely military focus remained inconclusive, as was the discussion to what extent global terrorism was to be a major concern for NATO, beyond mere rhetoric. The exact nature of the Afghan mission itself proved a hot political issue too, dividing the allies with one group in favor of a traditional peacekeeping mission and another in favor of a more forceful approach (COIN and counterterrorism). The War on Terror and the rift over Iraq had divided the allies with long-lasting echoes within NATO.

Adaptation and innovation did take place, but predominantly at the campaign level in the form of a slowly maturing command structure and increasing coherence among the national task forces. From 2002 to 2008 innovation proceed at an erratic, glacial pace. PRTs were the main new vehicle introduced by NATO and were an answer to the worsening situation that NATO experienced after the expansion of the mission into the southern provinces. The PRTs would look after the reconstruction and development side of the mission in the provinces, releasing troops that had been preoccupied with creating a secure environment. The PRTs were intended as interagency hubs where diplomats and aid workers would get together to help improve local governance. However, ISAF forces were simply unable to create the required level of security, and within the provinces there were no governance structures in place that connected to Kabul. Instead, the PRTs dealt with local strongmen involved in drug trafficking who had connections to the Taliban. The tougher campaign instigated by US Commander of ISAF (COMISAF) McNeill in 2007 did not produce the results expected—a prime factor being the inability of NATO to resource the campaign adequately in both military and diplomatic terms. The US was preoccupied with the civil war in Iraq, and the European nations were uninspired to make up for US failure to finish the mission in Afghanistan after 2001.

Innovation occurred only from 2009 onward, as a result of a strategic review initiated by the Obama administration, and also because of the realization among NATO member states that NATO faced possible failure, and that its survival was at stake. NATO's survival had already become part of the agenda of various NATO summits since 2002 but became more acute as the war dragged on. In 2008, NATO had agreed upon a Comprehensive Strategic Political-Military Plan (CSPMP) for ISAF that was new in the sense that it meant NATO was in the business of state-building, and an actor that had the remit to incorporate nonmilitary lines of operations in its activities too. Debate on this topic among the member states dated to 2004. The political endorsement in 2009 in Bratislava of a COIN template for its operations, and agreeing on a transition strategy, were additional important decisions of a strategic nature. Never before had NATO known such a political openness to COIN thinking. Other innovations included creating the NATO Training Mission Afghanistan (akin to the one in Iraq), which folded the US-led Operation Enduring Freedom CSTC-A mission into it. In actuality, NATO Training Mission Afghanistan (NTM-A) merely took over what CSTC-A had created, and NATO made good use of the double-hatting solution by appointing the US CSTC-A commander also the NTM-A commander. It also took a change in president, a US military surge, and forceful new US generals to take command of ISAF to introduce changes within ISAF. Indeed, much came down to US political and military initiatives and experiences.

THE CENTRALITY OF THE US

The US case vividly demonstrates how strategic choices of elected leaders exerted adaptive pressure at various stages of the campaign; campaign progress of the coalition in Afghanistan must be understood against this backdrop of US politics. Initially, the war in Afghanistan became one element in the War on Terror, which became linked to the war in Iraq, a war that relegated Afghanistan to secondary importance until the election of Obama in 2009. As early as 2002 US troop levels in Afghanistan were limited as the Bush administration prepared for the mission. Initial assumptions held by Pentagon and White House leadership in both wars, fueled by a belief in the superiority of high-tech warfare, were that force could be applied quickly, easily, and with a light footprint of ground forces. Preemption and preventive war characterized the National Security Strategy of 2002, a document that inspired much critique across the Atlantic. It aired the neoconservative agenda, emphasized an aggressive military posture, and downplayed the role of the military as a nation-building force and the responsibility for the aftermath of wars. Subsequently, in both wars military victory was confused with political victory, and theater commanders in Iraq and Afghanistan initially were not provided with sufficient forces for their mission—which, starting in 2001 with the Bonn conference, gradually included an expanding array of tasks and objectives. As Russell observes, an informal cap on troop levels was established of just 7,000 military personnel for the US Combined Joint Task Force-180 in 2002, which was raised to 10,500 by the end of 2003.

By 2008 the message that the security situation had dramatically deteriorated since 2005–6 had become accepted in Congress and the Pentagon. With the risk of strategic failure looming, the new presidency of Obama provided the political opportunity for a significant shift in priority: a quick withdrawal from Iraq and a refocus on the very reasons Afghanistan was invaded in the first place—Al Qaeda and the Taliban. This shift was also fueled by the intent not to be regarded as the president who lost the war in Afghanistan. This shift came attached with specific timelines and goals in terms of achieving sufficient progress so as to enable a withdrawal of US troops from Afghanistan following the transition of responsibility for security to Afghan security forces. This shift materialized in the decision to surge the US presence, a surge that was requested by the newly appointed COMISAF, General McChrystal, based upon his own assessment report of August 2009.

As Farrell and Rynning note in their chapters, the ISAF chain of command remained weak until the arrival of McChrystal in 2009. McChrystal, along with US Ambassador Eikenberry, established a proper campaign plan with prioritized objectives (the Integrated Civil Military Campaign Plan), based on COIN principles, and promulgated his Counterinsurgency Guidance that stressed population-centric operations, followed up by a Counterinsurgency Training Guidance that directed commanders to master COIN theory. He also created a corps-level headquarters to take charge of the running of the campaign through the Regional Commands down to the Task Forces and create more coherence between actions in the provinces. US experience in Iraq inspired the division of labor between a four-star COMISAF for political-strategic issues and a (US) three-star Joint Command in charge of operational level issues. McChrystal's successor, General Petraeus, adhered to McChrystal's key directives, reinforcing them and elaborating upon them in the 2010 COIN Guidance Tactical Directive, which stressed the reduction of civilian casualties, and the 2011 Integrated US Government Civil-Military Campaign Plan, which aimed to stimulate coherence among the activities of the host of US military and non-military organizations in Afghanistan. NATO was not central to these changes, nor was it central to the development of the COIN plan endorsed by NATO in 2009. The US was the leading actor here, and both McChrystal and Petraeus enjoyed considerable leeway in defining the campaign plan.

Although slow in coming, and only after coming close to alliance disintegration and campaign failure, those changes gave the ISAF campaign coherence for the first time in the war. For instance, judging by the similarity in reconstruction and development lines of operations between the two documents, the new Afghan National Development Strategy seems to have informed the Dutch Uruzgan Campaign Plan in 2009. Across the various contingents, a similarity in the graphic illustrations by which they communicated their plans also suggests that units gleaned from each other and/or from ISAF briefings. The enhanced ISAF capacity also induced change within NATO in the sense that in 2010 it resulted in the decision to strengthen the nonmilitary side of the mission. This included the reinforcement of the position and staff of the Senior Civilian Representative (SCR) in Afghanistan and bringing it up to the four-star level. The politically empowered SCR was now to help the Afghan government develop capacity so as to take over from ISAF in 2014. Similarly, the move toward the so-called population-centric approach, which included an increased awareness of the adverse impact of civilian casualties, as opposed to the kinetic approach, explicitly emphasized by US ISAF commanders since McChrystal in light of their experience in Iraq, also seems to suggest a positive and direct influence. Civilian casualty figures showed a decline after 2009 in response to more restrictive rules of engagement promulgated by ISAF.

THE PROBLEMS OF CREATING COALITION COHERENCE

However, while for US units this shift may have been significant, for various European units this policy merely brought their existing national approach in line with ISAF guidelines. Constraining national ROEs, strict interpretation of the goal of the mission (reconstruction), and motive for the national contribution (humanitarian concerns and alliance obligations), experience, and/or limited resources never allowed Germany, Canada, and the Netherlands to adopt an overtly kinetic approach. Indeed, it took quite a number of casualties for the political level in those countries to convince them that the S&R mission might involve intense combat, and to loosen those constraints.

Already before the end of 2009, other ISAF units—German, British, and Dutch in particular—also increasingly incorporated Afghan units in their plans and operations. OMLTs were introduced to train ANA troops. Steadily the focus shifted to providing security to the Afghan population. Units in the field had learned how to cooperate with PRTs, NGOs, and OGOs (Other Governmental Organizations) at the tactical level, trying to breathe life into the idea that was alternatively known as Effects Based Approach to Operations (EBAO), the Comprehensive Approach, or the 3D approach. A gradual increase of civilian personnel within the tactical units was key to improvements. From 2009 onward the German PRT saw enhanced civil involvement. Although the Netherlands TFU strategy resulted from a comprehensive conceptual framework, involvement of civilian actors was minimal at first. The largest adaptation of the organizational aspect, therefore, was the influx of civilian personnel. The year 2008 especially saw a surge in civilian personnel and a radical change in the overall command structure as the senior civil representative became the commander of the PRT and the co-commander of the TFU. Moreover, the gradual increase of NGOs contributing to the mission (from six NGOs in 2006 to fifty in 2010) and the establishment of the Dutch Consortium for Uruzgan further augmented the civilian capacity needed for the execution of a comprehensive campaign.

Also at the tactical level, a more population-centric approach was practiced well before McChrystal put out his assessment and guidance. The European militaries had shifted to a nonkinetic approach by this point in the war. It did not require ISAF guidance in 2009 for most units to change this aspect of the mission. In fact, it had required substantial effort to convince some governments that the S&R approach might have to be adjusted to the extent that, while not adopting the term "COIN," some of the tenets had to be incorporated in the national plans. Several countries had deliberately avoided referring to the "COIN" word, as it suggested that combat activities against insurgents would take priority over reconstruction efforts. 3 Commando Brigade in Helmand, for instance, deliberately intended to tread softly and escalate to violence only if necessary and with due care, to minimize collateral damage, and 52 Infantry Brigade in 2007 also harped on the theme of population-centric operations. In 2009, 19 Brigade informed civilians of impending attacks so they could evade fighting and/or persuade Taliban elements to leave the village, a practice that was applied with success by other UK units during subsequent operations. Such "messaging" came attached with the reassurance that locals would not be abandoned after the assault.

Overall, as Rynning observed, despite the campaign-level directives of ISAF commanders since McChrystal, ISAF has struggled to make an imprint on the established campaign organization that was characterized by disunity. The Comprehensive Approach and the coalition character of ISAF made any ambition to unity of command an illusion.[10] The result has been an ad hoc approach to security and development.[11] NATO as an institution—the Brussels bureaucracy and command structure—also struggled in genuine strategic thinking. NATO had no strategy: it assumed control of ISAF without a firm understanding of the demands, level of commitment, or level of resources such a mission would entail.[12] The North Atlantic Council (NAC), the Military Committee, and SHAPE were drawn into discussions of operational details, not of strategy. The default policy of NATO when confronted with setbacks and disappointment has been to add more of the same: more troops, more financial support, and better coordination among the various actors involved in state-building.[13] As before, in the Balkans, development of overarching guidance in the form of a campaign plan has proven elusive, any collective plans generally appearing months or years after initial deployment (if at all), predominantly developed by one participant (the US).[14]

ISAF AND THE STRUCTURAL PROBLEMS OF STATE-BUILDING

Despite the demonstrations of tactical and operational adaptability and sacrifices by ISAF and its the task forces, it has turned out to be a good war gone wrong, as one of the editors recently concluded after reviewing a number of scholarly studies.[15] A Sisyphean mission, another called it.[16] Indeed, after a decade of foreign intervention, various assessments paint a bleak picture of the effectiveness of NATO's mission in Afghanistan. Afghanistan remains a "phantom state," exercising only limited sovereignty outside of major cities, and even there authority is rarely ubiquitous. A dysfunctional government and endemic corruption undermine the legitimacy of Hamid Karzai's regime, and the inability of the central state to provide public goods and security pushes popular loyalties away from Kabul toward local strongmen, warlords, and the Taliban. Instead of being rebuilt as a secure, forward-looking, and democratic regional ally to the West, Afghanistan has languished as a broken, ineffective, and externally dependent "rentier" state facing a well-organized insurgency, an uncontrolled and politically pervasive opium trade, and continued penetration by regional criminal networks.[17] After stumbling into Afghanistan, NATO might well shun taking on another such difficult mission.

Perhaps one should not be surprised. The NATO experience seems to underline Mandel's set of fallacies concerning strategic interventions. As in the past, in the case of Afghanistan, governments of the contributing nations overestimated their understanding of the host society and the obstacles to accomplishing their objectives. They have also overestimated the capabilities of armed forces and their technologies in subduing local resistance, and their ability to pursue both stability and justice. In addition they seem once again to have overestimated the ease and speed of transforming the political system, the ease of transfer of power to local authorities, and the readiness of the local population to adopt external values and norms. At the same time they have, as in previous missions of similar complexity and scope, underestimated the costs and time involved in economic reconstruction, the severity of local social turmoil in the posthostilities phase, and the problems of obtaining and maintaining external legitimacy accorded to postwar arrangements.[18]

Indeed, the mission in Afghanistan has been an encounter with the verities concerning counterinsurgencies and state-building missions. State-building is one of the most complex undertakings a state can attempt, involving facilitating three interrelated and very disruptive societal transitions. First, it entails a *social transition*, from internecine fighting to peace, often in the midst of an extensively damaged and fragmented social environment conditioned by long periods of war and conflict. Second, interveners must encourage an *economic transition*, from war-warped accumulation and distribution to equitable, transparent postwar development, which in turn reinforces the peace transition. Third, interveners must arrange for a *political transition* from wartime government (or the absence of government) to postwar government, often contending with entrenched ethnic and regional divisions, and the possibility of limited experience among domestic groups with alternative forms of governance.[19]

The coalition experience has all the marks of the structural problems of state-building identified by Paris and Sisk. The first tension, they note, is that outside intervention is used to foster self-government and international control of the region, and the local institutions aim to establish local ownership. Moreover, short-term imperatives often conflict with longer-term objectives: post-conflict operations may require bargaining with existing ruling elites whose continued power is undesirable and counterproductive for building indigenous legitimate and effective institutions in the long term. In addition, state-building is a long-term enterprise, but prolonged presence can also result in passivity of the local population and overdependence on external assistance, and erode domestic political support in the donor countries. Finally, universal values are promoted as a remedy for local problems, but those universal—liberal—values may be incompatible with local customs, cultural expectations, and political traditions of the host society.[20]

That contradiction seems clearly present in the NATO mission. The liberal internationalist and democratic reconstruction model adopted by the West seems to have been part of the problem. This model overlooked the difficult questions of political legitimacy, insecurity, and clientelism.[21] Several of the underlying assumptions of the Bonn agreement, of the Comprehensive Approach, and of the revamped Western COIN doctrines proved false or indeed counterproductive in Afghanistan. For instance, the model of democratic statehood, which assumes a willingness for power-sharing, now seems unsuited to the realities of post-Taliban Afghanistan.[22] Power-sharing, as Migdal argued in 1988, does not come naturally in weak, socially fragmented societies such as Afghanistan.[23] It depends on leaders moderating their own demands. The essential problem of nation-building in a socially fragmented society like Afghanistan is that the effort to create a central government takes place within a context of divided legitimacies. The newly created government must vie with all the rest for both control and citizen support. Existing authorities do not cede their authority but hold on to it all the more fiercely.[24] To overcome such resistance, incentives are created such as ministerial titles and/or territorial authority, effectually empowering them with the means to back their demands.

To wit, in Bonn, state offices and ministerial positions were allocated to different groups to ensure representation of all the main factions. Few Afghan commandants however, as Martin observed, shared the Western vision of a small, effective state, regulating a market economy, instead regarding the new state as an exploitable instrument, funded by foreign donors, with which to maintain power and status through patronage networks. Unable to remove military power from the hands of these strongmen, Karzai's government bestowed grand but empty titles upon them, nominally incorporating them into the state apparatus but exercising little authority over their activities. Inclusion and exclusion from power were based not on legitimate leadership capabilities compatible with a Weberian model of legal-rational institutions, but on personal authority over ethnic networks of clientelism and the ability to threaten instability in exchange for concessions—the dynamic described by Grissom. The necessity of conceding autonomy and resources to regional power-brokers thus introduced incentives to abuse the system. Organs of the state, such as the ANA, as Grissom has elegantly demonstrated, thus became "fiefdoms" under the control of particular factions, whose interest in maintaining a desirable status quo overwhelmed their desire to advance the state-building process.

Elections, and an early emphasis on developing a centralized government too, may stimulate such a counterproductive dynamic. Without elections, foreign interveners find themselves attempting to empower a regime that has difficulty establishing itself as a legitimate authority. The early focus on establishing a Western model of central government in the immediate postconflict phase may actually have contributed to ISAFs problems, in particular in light of the limited ISAF presence during the first two to three years after the Bonn agreement, the "lost years" according to Grissom. It deepens societal division and generates conflict, and importantly is often at odds with peripheral interest groups and does not cater to the immediate concerns of the local population; they care about security, food, and water.[25]

In that respect, the initial failure to expand ISAF into the provinces, the very slow materialization of the promised financial support (almost $20 billion), and the reliance on the inadequate resources of the Afghan National Army were not helpful, as they left most of Afghanistan in a security vacuum and created an expectation gap. Such a space has been filled quickly by local militias, and in the case of the south and east, a resurgent Taliban. As in any postconflict phase, demonstrating momentum and making good on promises is key to persuading local fence-sitting power-holders that the new regime is the party to choose sides with. The failure to re-establish a legal economy has forced many Afghans to resort to illegal forms of income—in particular narcotics. Another factor eroding support was the policy to fund and arm local warlords. While necessary in the short term to fight the Taliban, in the long run it undermined the legitimacy of the central state that the coalition aimed to strengthen.

Thus, against initially high expectations following the fall of the Taliban regime, and in the absence of manifest progress in effective official governance, these dynamics undermined the embryonic government and fostered fragmentation. Alternative local governance structures stepped in that were built around kinship, trust, clientelism, mutual interests at local level, with deep cultural roots, and with no real interest in solving the anarchic situation that this may produce at the national level. Indeed, as Talentino concluded, in particular in identity-based conflicts in which societal fragmentation characterizes the country, such as in Afghanistan, negotiated settlements—power sharing, elections, effective centralized government—fall apart and state-building is least likely to be successful.[26] Or, as Moran noted, mastery of the complex human terrain of Afghanistan is possible only for those who wish to live and rule there—that is to say, for the Afghans themselves.

Some have called into doubt the very possibility at all, regardless of the intensity of the efforts of the interveners. They argue that the track record of state-building in territories without a preexisting modern state structure has been poor, pointing at studies that indicate that interventions serve to extend the time that combatants fight; that biased interventions are more likely to lead to an escalation of a civil war than its termination and increase the time until the onset of negotiations, and do nothing to decrease the duration of a civil war. External interventions also seem not to promote democratic transitions; actually, they tend to restrict rather than facilitate democratic transitions.[27]

While intrusive missions have fared a tick better than nonintrusive, neither has been very successful at boosting progress in policy fields other than security, such as economic growth, rule of law, and effective government—that is, in policy fields that cannot easily be outsourced to third parties but require domestic reform and domestic capacity-building. Zürcher thus laments that the problem that needs to be solved is not whether intrusive missions are better or worse than nonintrusive missions. Rather, we need to ask why even intrusive missions seem poorly equipped to induce change in three policy fields crucial for state-building: democracy, institutional capacity, and economic capacity.[28]

Similar doubts exist concerning the chances of success in fighting an insurgency. More often than not COIN campaigns fail. In their classic studies on the problems that strong states have with small wars, Gil Merom and Andrew Mack have come up with a number of generic explanations.[29] As Merom states, democracies lose small wars because they find it extremely difficult to escalate the level of violence and brutality to that which can secure victory. There is an impossible trade-off in protracted wars between expediency and moral dicta that arise from an intricate interplay between forces in the battlefield and at home. Andrew Mack adds that in every case success for the insurgents arose not from a military victory on the ground but from progressive attrition of their opponent's political capability to wage war. The asymmetry of stakes involved produces a commitment differential. Superior strength of commitment compensates for military inferiority. The stronger side has lower political tolerance for loss of blood and treasure—indeed, the insurgent's superiority is not his ability to harm but in his greater willingness to be harmed. Another factor explaining the difficulties is that one can identify no fewer than twenty-seven variables that may contribute to success and failure. Such a high number of variables imply weak predictive value for any approach taken. What did become evident was that Afghanistan scores particularly bad on several accounts.[30]

THE LEGACY OF THE AFGHANISTAN MISSION

Against that sobering empirical background, and the assessment offered by Daniel Moran in Chapter 2 concerning the experience of four armies in Afghanistan that preceded ISAF, the foundation of the ISAF mission, its goals, and strategy as agreed upon in Bonn may in hindsight be criticized as being overly ambitious, lacking a strategy for implementation, based upon flawed assumptions, and as manifestly under-resourced. On the other hand it suggests that one should be careful in passing judgment on the efforts of the militaries to achieve success in a very challenging, complex, and novel mission they have had to conduct under very difficult circumstances. Despite the best efforts of the militaries and governments involved, there is an argument to be made that several structural problems particular to the Afghan environment have severely put limits on the potential to be effective, almost regardless of the specific actions taken at the local level by tactical units. In Afghanistan the West may have stumbled upon a society that was least amenable to the state-building approach.

Unsurprisingly, the strategic logic of state-building missions is now being challenged, and in early 2012 most countries were preparing for an exit out of Afghanistan. A decade of Afghan experience has come to suggest that the West will have reservations in undertaking another similarly ambitious mission. They seem inversely related to the security interests of most Western countries. Afghanistan is seen in a similar light as Somalia, where, according to Fareed Zakaria, the odds of success are low and the risk of unintended consequences very high. It has cost the contributing nations dearly in money and lives (2,526 military casualties among coalition troops by October 2011). Yet for all the hundreds of billions spent (US$1,200 billion for Iraq and Afghanistan combined, just for the US), Afghanistan still relies for 97 percent of its Gross Domestic Product (GDP) on foreign funding, it is rife with corruption, and its state apparatus is failing. More important, perhaps, is that the evident failings of the West have hurt the West's credibility and legitimacy in the eyes of large parts of the world, including the West itself. It has challenged the (indeed ill-founded) assumption that Western economic, cultural, political, and military power on a global scale equals transformative power on a local scale. Officials question whether state failure is related to the threat of terrorism, and hence whether state-building is related to national security interests.[31] Indeed, US Secretary of Defence Gates indicated as much just before his retirement, and his successor has repeated such concerns. A similar fading interest in, and skeptical attitude toward, external state-building was voiced by President Obama when he told US citizens that he was actually more interested in state-building in the US itself.

As of 2012 countries are showing a fading interest in the mission and the fate of Afghanistan. The reasons for staying in Afghanistan have become tenuous, and NATO Secretary General Rasmussen has urgently called upon the member states not to neglect the mission in Afghanistan. The political reasons for the gradual retreat from Afghanistan have been varied, but most combined an increasing lack of interest in Afghanistan, a problem in explaining the mounting casualties to their own citizens in light of the lack of evident progress in developing the country, a failing narrative concerning the fight against terrorism, in particular now that bin Laden is gone, a sense that alliance obligations have been sufficiently honored, a growing military problem in sustaining such an operation at such a distance, and the high costs involved in a time of ever tightening government budgets. Moreover, in early 2012 newspapers cited ISAF reports which argued that Afghan troops increasingly cooperated with the Taliban once ISAF troops had vacated an area, and another one stating that ANA soldiers had killed several ISAF soldiers.[32]

Regardless of the exit, a decade of war in Afghanistan has left a deep imprint on the militaries that were involved in that complex NATO mission. It has educated a generation of officers. Whether it has affected the culture of the organizations is an open question, as is the question of whether lessons concerning civil-military cooperation will be shared and institutionalized not only by the militaries but also in particular by other ministries and other governmental organizations. There is a distinct risk that most organizations will revert to dominant organizational schemas—that is, back to normalcy. That would be a shame and a risk. While NATO may not consider taking on extreme challenges such as Afghanistan, there is no doubt that there will be cases of state failure in which the strategic logic of state-building will be asserted. For such missions NATO forces will need to retain the valuable knowledge obtained in developing and applying COIN and Stabilization doctrines, including concepts such as PRTs and OMLTs. The rediscovery and modernization of COIN theory has had the beneficial effect of sensitizing Western militaries to the fact that the dominant canon of military thought that focuses primarily on interstate warfare was too limited. Afghanistan and Iraq were pedagogic events in that sense.[33] Finally, Afghanistan is markedly different from Vietnam. One should not overlook the fact that despite the significant challenges, NATO has not suffered a costly defeat comparable to previous COIN missions—a fact in no small part the result of the impressive adaptability of NATO's militaries.

NOTES

1. This definition borrows from Francis Fukuyama, *State Building: Governance and World Order in the Twenty-First Century* (Ithaca, NY: Cornell University Press, 2004), ix.

2. For a critique on the consociational approach in Afghanistan, see, in particular, the excellent analysis by Philip Martin, "Intervening for Peace? Dilemmas of Liberal Internationalism and Democratic Reconstruction in Afghanistan," *Journal of Military and Strategic Studies* 13, no. 3 (Spring 2011): 1–24.

3. See Miles Kahler, "Statebuilding after Afghanistan," in Ronald Paris and Timothy Sisk, eds., *The Dilemmas of Statebuilding* (Abingdon: Routledge, 2009), 287–303.

4. For this, see, for instance, Christopher Coker, *War in the Age of Risk* (Cambridge: Polity Press, 2009); and Mikkel Rasmussen, *The Risk Society at War* (Cambridge: Cambridge University Press, 2007).

5. "Larger Freedom: Towards Development, Security and Human Rights for All," United Nations, March 2005; NATO Strategic Concept, Brussels, 2010.

6. See, for instance, the information brochure "Political Guidance on Ways to Improve NATO's Involvement in Stabilization and Reconstruction," NATO, Brussels, 2011, 2.

7. Andrew Hoehn and Sarah Harting, *Risking NATO: Testing the Limits of the Alliance in Afghanistan* (RAND: Santa Monica, 2010), 3.

8. For that problematic transformation process, see Theo Farrell, Terry Terriff, and Frans Osinga, *A Transformation Gap?* (Stanford: Stanford University Press, 2010).

9. Timo Noetzel and Benjamin Schreer, "Does a Multi-tier NATO Matter?," *International Affairs* 85, no. 2 (2009): 211–26.

10. Jaïr van der Lijn, *3D "The Next Generation"; Lessons Learned from Uruzgan for Future Operations* (The Hague: The Clingendael Institute and Cordaid, November 30, 2011), 6.

11. Cyrus Hodes and Mark Sedra, *The Search for Security in Post-Taliban Afghanistan*, IISS Adelphi paper 391 (Routledge: Abingdon, 2007), 48.

12. Hoehn and Harting, *Risking NATO*, x.

13. Astri Suhrke, "The Dangers of a Tight Embrace: Externally Assisted Statebuilding in Afghanistan," in Paris and Sisk, *The Dilemmas of Statebuilding*, 227–28.

14. Russell Glenn, *Band of Brothers or Dysfunctional Family? A Military Perspective on Coalition Challenges during Stability Operations* (RAND: Santa Monica, 2011), xii.

15. Theo Farrell, "A Good War Gone Wrong? Review Essay," *RUSI Journal* 156, no. 5 (October/November 2011): 60–64.

16. Bing West, "Groundhog War," *Foreign Affairs*, September/October 2011; See the ensuing debate by Christopher Sims, Fernando Luján, and Bing West, "Both Sides of the COIN: Defining War after Afghanistan," *Foreign Affairs*, January/February 2012, 178–85.

17. For such a bleak assessment, see, for instance, Suhrke, "The Dangers of a Tight Embrace, 227–51; and Martin, "Intervening for Peace?"

18. Robert Mandel, *The Meaning of Military Victory* (Boulder, CO: Lynne Rienner Publishers, 2006).

19. Paris and Sisk, *The Dilemmas of Statebuilding*; see also Paul Miller, "The Case for Nation-Building," *Prism* 3, no. 1 (March 2011): 70.

20. Robert Paris and Timothy Sisk, "Conclusion: Confronting the Contradictions," in Paris and Sisk, *The Dilemmas of Statebuilding*, 305–9.

21. David Roberts, "Post-conflict Peacebuilding, Liberal Irrelevance and the Locus of Legitimacy," *International Peacekeeping* 18, no. 4 (August 2011): 410–24; Sonia Grimm, "External Democratization after War: Success and Failure," *Democratization* 15, no. 3 (2008): 525–49; and Astri Surhke, "Democratizing a Dependent State: The Case of Afghanistan," *Democratization* 15, no. 3 (2008): 630–48.

22. Martin, "Intervening for Peace?"; see also Edward D. Mansfield and Jack Snyder, *Electing to Fight: Why Emerging Democracies Go to War* (London: MIT Press, 2005).

23. Joel S. Migdal, *Strong Societies and Weak States: State-Society Relations and State Capabilities in the Third World* (Princeton: Princeton University Press, 1988).

24. Andrea K. Talentino, "Nation-building or Nation-Spitting? Political Transitions and the Dangers of Violence," *Terrorism and Political Violence* 21, no. 3 (2009): 378–400.

25. Mats Berdal, "Consolidating Peace in the Aftermath of War—Reflections on 'Post-Conflict Peace-Building' from Bosnia to Iraq," in John Andreas Olsen, ed., *On New Wars* (Oslo: Norwegian Institute for Defense Studies, 2007).

26. Talentino, "Nation-Building or Nation-Spitting?"; see also Grimm, "External Democratization after War."

27. Patrick M. Regan, "Interventions into Civil Wars: A Retrospective Survey with Prospective Ideas," *Civil Wars* 12, no. 4 (2010): 456–76.

28. Christoph Zürcher, Nora Roehner, and Sarah Riese, "External Democracy Promotion in Post-Conflict Zones: A Comparative-Analytical Framework," *Taiwan Journal of Democracy* 5, no. 1: 1–26; Christopher Zürcher, *Is More Better? Evaluating External-Led State Building after 1989*, CDDRL Working Papers (Stanford: Center on Democracy, Development, and the Rule of Law, Stanford Institute on International Studies, 2006).

29. For a concise summary of those arguments, see Jeffrey Record, *Beating Goliath: Why Insurgencies Win* (Washington, DC: Potomac Books, 2007), ch. 1.

30. Christopher Paul, *Counterinsurgency Scorecard: Afghanistan in Early 2011 relative to the Insurgencies of the Past 30 Years* (Santa Monica, CA: RAND, 2011).

31. Anna Simons and David Tucker, "The Misleading Problem of Failed States: A 'Socio-geography' of Terrorism in the Post 9/11 Era," *Third World Quarterly* 28, no. 2 (2007): 387–401; Aidan Hehir, "The Myth of the Failed State and the War on Terror: A Challenge to the Conventional Wisdom," *Journal of Intervention and Statebuilding* 1, no. 3 (2007): 307–32.

32. As reported by Elisabeth Miller, "Panetta Says US to End Afghan Combat Role as Soon as 2013," *New York Times*, February 1, 2012.

33. See David Ucko, "Counterinsurgency after Afghanistan," *Prism* 3, no. 1 (2011): 3–20, for this argument. Similarly, see the debate in Sims, Luján, and West, "Both Sides of the COIN"; and Frans Osinga and Julian Lindley-French, "Leading Military Organizations in the Risk Society: Mapping the New Strategic Complexity," in Joseph Soeters, Paul C. van Fenema, and Robert Beeres, eds., *Managing Military Organizations: Theory and Practice* (Abingdon: Routledge, 2010).

Index

Index

AAF, *see* Afghan Air Force
Abbott, Augustus, 30
Accommodation, 139–40
ACO, *see* Atlantic Command Operations
ACT, *see* Allied Command Transformation
Adaptation: analytical framework, 3, 8; avoidance, 25–27; definition, 2; dialectical nature, 76; drivers, 3, 8 (table), 8–10; emulation, 280, 307–8; extrinsic, 276; failures, 4–5, 26–28; importance, 3–5; leadership, 18; Piaget on, 139–40; pressures, 2, 26–27; relationship to innovation, 6–7, 7 (fig.); schemas, 139–40, 142; second-order, 136, 138–39, 140, 145, 151, 155, 308; shaping factors, 8 (table), 10–18; by small states, 137–39, 140; during wars, 3–4. *See also* Learning; Operational adaptation; Strategic adaptation
Adaptive moments, 8
ADZs, *see* Afghan Development Zones
Afghan Air Force (AAF), 270
Afghan Compact, 267–68
Afghan Development Zones (ADZs): definition, 169; in Helmand, 91, 92, 110, 111; in Uruzgan, 169, 170, 173, 178, 293
Afghan government: Bonn Agreement, 60–61, 66, 97, 264, 288, 319; Defense Ministry, 85, 264, 265, 266, 272; elections, 61, 74, 319; foreign funding, 322; Interior Ministry, 269; officials assassinated by Taliban, 248–49, 261n19, 299; weakness, 317–18. *See also* Karzai government
Afghanistan: complexities, 51, 54; literacy, 28, 48n7, 272, 304; modernization, 39; societal fragmentation, 164–65, 174–75, 179–80, 185, 320; state-building, 288, 289–90, 312–13, 321–22. *See also* Anglo-Afghan Wars
Afghanistan War: Bush administration policies, 52, 57, 60–61, 62–63, 310, 312, 313; effects on NATO, 83–84, 97–104, 311–12; failure, 317–18; link to Iraq War, 57–58, 310, 313; planning, 64–65; public support, 11–12, 56–57, 230–31, 298. *See also* International Security Assistance Force; Operation Enduring Freedom
Afghan National Army (ANA): adaptation, 264–72, 303–4; adaptive moments, 267, 281; adaptive pressures, 275–81; air corps, 268; battalions, 265, 272; battles with Taliban, 250, 275; casualties, 275; combat effectiveness, 263–64, 272–74, 275, 280–81, 304; commandos, 269, 271, 272; corruption, 274, 278, 281, 303; counterinsurgency operations, 74; desertions, 267, 268, 272, 284n33; distorted incentives, 275,

329

276; equipment and vehicles, 269, 270, 273–74, 280, 303; establishment, 266; expansion, 263, 267–71, 279, 303; future challenges, 281–82; in Helmand, 112, 116, 292, 293; highway *kandaks*, 269–70; infantry *kandaks*, 268–69, 270, 272; logistics system, 270–71, 274, 280; NATO advisors, 86, 263, 267, 272, 275–76, 323; NATO goals, 264–65, 266; NATO influence on, 275–76, 280–81; patronage networks, 277, 278–79, 280, 303; political context, 276–79, 303; recruitment, 81n61, 267, 272; shaping factors in adaptation, 275; size, 61, 263, 264–65, 266, 267–68, 272; social context, 302–3; specialist units, 269–71, 280, 303; tanks, 270; training, 62, 263, 265, 266–67, 268, 272, 273, 274; in Uruzgan, 294; weapons, 269, 271, 273–74, 280–81, 284nn36–37
Afghan National Army Special Forces (ANASF), 271
Afghan National Development Strategy, 171, 315
Afghan National Guard, 265–66
Afghan National Police (ANP), 72, 112, 265, 267–68, 269, 292
Afghan National Security Forces (ANSF): battles with Taliban, 116; building up, 55, 59, 69, 89, 167, 263; counterinsurgency operations, 72, 74; expansion, 267–68, 291–92; in Helmand, 116, 119–20; joint operations with Germans, 206; shortcomings, 112; suicide bomb attacks on, 254; training, 85–86, 105n6, 137, 219, 228, 235, 238n8, 263; transition to, 59, 61, 89, 281–82, 314; in Uruzgan, 167, 169, 173, 180, 185
Aircraft, 179, 185, 203–4, 270. *See also* Helicopters
Akhonzada, Sher Mohammad (SMA), 91, 97, 109, 278
Allen, John, 86, 89
Alliance politics: adaptation shaped by, 12–14; in world wars, 12–13
Alliance politics, NATO: Britain and, 121–22; burden-sharing issue, 13–14,

92, 104; divisions, 92–93, 97, 311–12; domestic politics and, 14; ISAF and, 13–14, 83–84, 88–89, 97, 311–12; Kosovo War and, 13; solidarity, 162, 212–13, 301; US-European relations, 92–93, 95
Allied Command Transformation (ACT), 99
Al Qaeda, 51, 66–67, 90, 212. *See also* Bin Laden, Osama
Amanullah Khan, 38, 39
ANA, *see* Afghan National Army
ANASF, *see* Afghan National Army Special Forces
Anglo-Afghan Wars: Afghan resistance, 27, 32; aims, 28, 31, 33–34, 46; first (1839), 29, 35; peaceful periods, 34; second (1878–80), 31–33, 37, 108; third (1919), 33, 37–38, 49n14; Treaty of Gandamak, 32, 33
ANP, *see* Afghan National Police
ANSF, *see* Afghan National Security Forces
Assimilation, 139, 147
Atlantic Command Operations (ACO), 98

Balkans: Bosnia, 85, 146, 232, 233; Dutch forces, 167, 176, 302; Kosovo War, 13, 85, 96, 98; peacekeeping operations, 300; wars, 11
Bandit gangs, 245
Barno, David, 67–68, 291
Basra, Iraq, 4, 121–22, 301
Bech, Gitte Lillelund, 144
Bergdhal, Bowe, 74
Berlijn, Dick, 167
Bin Laden, Osama, 51, 66, 323. *See also* Al Qaeda
Bisserup, Bjørn, 152, 153
Bjerga, Kjell Inga, 140
Blair, Tony, 121, 309
Bonn Agreement, 60–61, 66, 97, 264, 288, 319
Bosnia, 85, 146, 232, 233. *See also* Balkans
Boyd, John, 140
Brezhnev, Leonid, 39, 43, 47
Brezhnev Doctrine, 39

INDEX 331

Britain: civil-military relations, 123–26, 296; competition with Russia in Central Asia, 24, 25, 28, 29, 30–32, 34; Crimean War, 31, 36; domestic politics, 118, 309; Ministry of Defense, 124, 125, 127, 300, 309; Parliament, 121, 125; public support for Afghanistan war, 11; relations with Denmark, 14, 153; special relationship with United States, 14, 15; strategic culture, 15, 301

British Army: Cardwell Reforms, 36–37; counterinsurgency field manual, 120; helicopters, 300; in Iraq, 16; organizational culture, 15–16; training, 120; Waterloo, 35; in World War I, 4, 12–13, 37–38; in World War II, 12

British Army in Afghanistan: Army of the Indus, 29; casualties, 49n14; failure to adapt, 25, 27, 34–36, 38; Kabul garrison massacre, 29–30, 35, 36; rescue of prisoners, 30–31, 36; retreat (1842), 29–30; strategic success, 46; tactics, 34, 37, 38. *See also* Anglo-Afghan Wars

British military: counterinsurgency doctrine, 120, 129, 306; Iraq War, 4, 16, 121–22, 301; reputation, 121–22; strategic adaptability, 7; strategic culture, 15, 301; in World War I, 4; in World War II, 12

British military in Afghanistan (2002–present): adaptive pressures, 121–26; Army units, 109; constraints, 108; counterinsurgency operations, 72, 91–92; deployment, 121; domestic political influences, 118; equipment, 126, 127; helicopters, 124–25, 300; ISAF command and, 119–20; local governors and, 91; operational adaptation, 108, 117–21, 127, 128–29, 138, 292–93, 299–300, 305–6; Operation HERRICK, 108, 114; Royal Marines, 110–11, 115–16; strategic adaptation, 108, 121–26, 129, 309–10; troop levels, 11, 122–24, 123 (table), 125, 309–10. *See also* Helmand campaign

Brooks, Risa A., 17
Brown, Gordon, 125, 126, 309

Browne, Des, 122, 124, 125
Brydon, William, 30
Bucknall, James, 213
Bundeswehr, *see* German military
Bureaucracies, 53, 63–64, 211, 224, 236
Bush administration: Afghanistan War planning, 64–65; democracy promotion, 95; financing of wars, 57; military priorities, 63; mistakes, 313; NATO expansion and, 95; objectives in Afghanistan, 60–61, 313; relations with allies, 89–90; strategic neglect of Afghanistan, 52, 57, 62–63, 310, 312; use of force policies, 60, 62, 63. *See also* Afghanistan War; Iraq War; War on terror

Bush Doctrine, 60, 63, 313
Butler, Ed, 122, 123
Butler, Elizabeth Thompson, Lady, *The Remnants of an Army*, 30, 48n9

Caldwell, William, IV, 86
Cameron, David, 153
Canada: civil-military relations, 226–28, 232–35, 236, 237; Department of Foreign Affairs and International Trade, 221, 234; Department of National Defence, 225, 232–33; development projects in Afghanistan, 234; minority governments, 219–20, 228–30, 234, 236, 297, 300, 309; public support for Afghanistan war, 11, 230–31, 231 (fig.)

Canadian Army: Land Staff, 223, 224; tanks, 223, 301

Canadian Expeditionary Forces Command (CEFCOM), 225, 232–33, 239n16

Canadian Forces (CF): Can'tbat nickname, 232; Chief of Defence Staff, 225, 227, 231, 232, 235, 296; command structure, 225, 239n19; generational change, 231–34, 236, 296; NATO missions, 220; peacekeeping missions, 222, 229, 231–32, 233; prime ministers and, 230, 234–36; reputation, 301

Canadian Forces in Afghanistan:

adaptation, 222–28, 235–36; battles with Taliban, 222, 235, 255–56; casualties, 219, 223, 229–30, 231 (fig.), 236; challenges, 221–22, 229–30, 235, 237; counterinsurgency operations, 221; deployments, 219, 220, 222, 222 (fig.); discretion of field commanders, 226–28, 232–34, 236, 240n34, 241n55; domestic politics and, 219–20, 228–31, 234–36, 237, 297, 300; helicopters, 224–25, 230, 236–37, 300; Kabul-centric training mission, 219, 228, 235, 238n8; Kandahar mission, 219, 220–21, 227, 233–34; Operation Medusa, 178, 222, 223, 227, 236–37; shaping factors in adaptation, 220, 236–37; special forces, 222; strategic adaptation, 219, 296; strategic motives, 220–21; tanks, 146, 223–25, 236–37, 300; weapons systems, 223–24, 228, 230, 236–37
CAPTF, *see* Combined Airpower Transition Force
Cardwell Reforms, 36–37
Carleton-Smith, Mark, 114
CEFCOM, *see* Canadian Expeditionary Forces Command
CENTCOM, *see* Central Command
Center for Naval Analyses (US), 54
Central Command (CENTCOM), US, 62, 65
Central Intelligence Agency (CIA), 64–65, 66
CF, *see* Canadian Forces
CFC-A, *see* Combined Forces Command Afghanistan
Change, *see* Adaptation; Innovation; Technological change
Cheney, Richard, 60
Chora, battle of, 175–76, 180, 293–94, 302
Chretien, Jean, 221, 229
CIA, *see* Central Intelligence Agency
C-IED, *see* Counter-Improvised Explosive Device measures
CIMIC, *see* Civil-military cooperation
Civilian casualties: in counterinsurgency operations, 70, 71; decline in, 315; German criminal law and, 192, 200–201; from IEDs, 248; Kunduz airstrike, 209–10, 214, 305. *See also* Population-centric warfare
Civilians: innovation and, 17; Red Army advisers, 41; in Task Force Uruzgan, 173, 175, 176, 180, 184, 294, 316; with US military, 55–56, 59, 61, 72, 73, 295, 297. *See also* Nongovernmental organizations; Provincial Reconstruction Teams; Senior Civilian Representative
Civil-military cooperation (CIMIC): in Afghanistan, 71, 73, 87–88, 297, 315–16; Danish views, 142–44, 147, 154–55, 302; in Task Force Uruzgan, 173–74, 175, 176; training courses, 177. *See also* Comprehensive approach
Civil-military relations: adaptation shaped by, 17–18, 295–97; in Britain, 123–26, 296; in Canada, 226–28, 232–35, 236, 237; in democracies, 17, 18; in Germany, 195–96, 205, 304, 305; noninterference, 295–96; in Provincial Reconstruction Teams, 114, 173, 315–16; in Soviet Union, 17–18, 28, 43; in United States, 58–59, 63. *See also* Domestic politics
CJTF, *see* Combined Joint Task Force
Clark, Wesley, 96
Clausewitz, Carl von, 26, 138, 198
Clinton administration: cruise missile attacks on Afghanistan, 51; military operations, 62
COIN, *see* Counterinsurgency
Cold War, end of, 11, 140
Combined Airpower Transition Force (CAPTF), 270
Combined Forces Command Afghanistan (CFC-A), 67–68, 291
Combined Joint Task Force (CJTF), 62–63, 99, 313
Combined Security Transition Command—Afghanistan (CSTC-A), 85–86, 266, 313
COMIJC, *see* Commander ISAF Joint Command
Commander International Security

Assistance Force (COMISAF), 85, 86, 89, 91–92, 226–27, 233, 312, 314. *See also* McChrystal, Stanley
Commander ISAF Joint Command (COMIJC), 86
Comprehensive approach: Danish views, 137, 142–44, 147, 151–52, 154–55, 302; of ISAF, 87, 95, 315; as substitute for strategy, 310–11. *See also* Civil-military cooperation; Provincial Reconstruction Teams
Comprehensive Strategic Political-Military Plan (CSPMP), 94–95, 96–97, 312–13
Conolly, Arthur, 24
Corruption, 58, 274, 278, 281, 303, 322
Counter-Improvised Explosive Device (C-IED) measures: British, 125, 126–27; effects, 251; investment in, 77n11, 253, 299–300; technologies, 126–27, 179, 298, 299–300. *See also* Improvised Explosive Devices
Counterinsurgency (COIN): absence of Danish doctrine, 149–50; adapting to, 16; Afghan forces, 72, 74; British doctrine, 120, 129, 306; British operations, 72, 91–92; Canadian operations, 221; civilian casualties, 70, 71; clear, hold, build strategy, 71–72, 112–13, 221, 291; failures, 321; Field Manual 3-24, 18, 64, 199, 204, 307; German operations, 193; influence operations, 113, 114; intelligence collection, 174, 295; in Iraq, 7, 18, 71, 315; ISAF operations, 72, 85, 87, 89, 90, 94, 314–15; ISAF strategy, 68–69, 70, 71, 72–73, 75, 89, 314; legacy of Afghanistan, 323; NATO doctrine, 103, 105n12, 313; oil-spot strategy, 91, 96–97, 178; operational adaptation, 72–75; population protection, 69, 71, 112–14, 119–20, 207; Task Force Uruzgan, 161, 170, 172, 178, 184, 294, 295, 296, 306; training, 71; US doctrine, 307–8; use of term, 316; US operations, 53, 67–68, 70–75, 291. *See also* Population-centric warfare

Counterterrorism, 86, 96. *See also* War on terror
Cowper-Coles, Sherard, 97, 106n22
Crimean War, 31, 36
CSPMP, *see* Comprehensive Strategic Political-Military Plan
CSTC-A, *see* Combined Security Transition Command—Afghanistan
Cultural sensitivity, 177–78, 179, 294, 295, 299
Culture, strategic, *see* Strategic culture

Dadullah, Mullah, 254, 255, 256
Danish Armed Forces: in Balkans, 141, 142, 146, 300; comprehensive approach, 137, 142–44, 147, 151–52, 154–55, 302; field manuals, 149–50, 307; in Iraq, 137; military metier, 147–48, 149–50, 151, 154, 297; officers' networks, 148–49, 306; tanks, 146–47; training, 149; transformation, 141–42, 154–55, 301–2
Danish Armed Forces in Helmand: adaptation, 136–37, 293; arrival, 141; battles with Taliban, 138, 150, 154–55; casualties, 150–51, 299; challenges, 154; domestic politics and, 151, 152–53, 297; Husar companies, 137, 138, 141, 144, 146; lack of civil-military cooperation, 142–44, 147, 302; learning, 148–49, 306–7, 308; mechanized units, 146–47, 149, 150; mission, 137, 142–43, 152–53, 301–2; operational adaptation, 145, 147, 148, 149, 150, 151, 154, 306–7; operational environment, 136, 138, 144–50; Operation Panchai Palang, 116; relations with British, 123, 138–39, 145–46; Royal Life Guards, 138, 146, 150; second-order adaptation, 136, 138–39, 145, 151, 155, 308; tanks, 300; training activities, 137, 154; troop levels, 137; withdrawal, 137, 154, 155. *See also* Task Force Helmand
Danish Army, transformation to expeditionary force, 136–37, 141–42, 154–55
Danish Development Agency, 143

334 INDEX

Dannatt, Richard, 125, 300, 310
Daoud, Mohammad, 109–10, 122
Daud Mohamed, 39, 49n20
Davis, Bill, 75
DCU, *see* Dutch Consortium for Uruzgan
Defense Department, US: budget increases, 59; bureaucratic reorientation, 53, 63–64; directives, 63–64; Joint Improvised Explosive Device Defense Organization, 77n11, 249; lessons learned, 307; Revolution in Military Affairs, 10, 61–62, 63, 142; war on terror strategy, 63
Defense Ministry, *see* Ministry of Defense
Democracies: civil-military relations, 17, 18; wars, 10–11
Democratization: in Afghanistan, 60–61, 66, 162, 288; Bush administration promotion, 95; state-building, 288–89, 318–19
Denmark: defense budget, 141–42; domestic politics, 151, 152–53, 297; foreign policy, 140, 152; Ministry of Defence, 137, 141, 143–44, 152–53, 301–2; Ministry of Foreign Affairs, 137, 141, 143, 301; Parliament, 152, 154; public opinion, 11, 152, 297; relations with Britain, 14, 153. *See also* Danish Armed Forces
Department of National Defence (DND), Canada, 225, 232–33
Dixon, Carl, 127
DND, *see* Department of National Defence
Doctrine, Concepts, and Development Centre (DCDC; UK), 120
DoD, *see* Defense Department, US
Domestic politics: adaptation shaped by, 10–12, 36, 297–98, 309; in Afghanistan, 276–79; alliance politics and, 14; in Britain, 118, 309; in Canada, 219–20, 228–31, 234–36, 237, 297, 300; in Denmark, 151, 152–53, 297; in Germany, 193, 194, 196, 197, 205, 210, 212, 213; NATO and, 10–11; in Netherlands, 160, 162–63, 165–66, 167, 297–98, 309; public support for wars, 11–12, 230–31, 298; in

United States, 56–59, 313–14. *See also* Civil-military relations
Dost Muhammad, 29, 30–31
Drones, *see* Unmanned aerial vehicles
Drugs: trafficking, 91, 245, 269, 312, 320; use in ANA, 272, 285n68
Durand, Henry, 33
Durrani tribes, 164, 165, 175
Dutch Approach, 166–67, 176
Dutch Armed Forces, *see* Netherlands Armed Forces
Dutch Consortium for Uruzgan (DCU), 174, 316

Eastern Europe, Soviet occupation, 40
EBAO, *see* Effects-based approach to operations
Education: literacy, 28, 48n7, 272, 304; schools, 61, 212, 248; Taliban policies, 248
Effects-based approach to operations (EBAO), 65, 315
Eikenberry, Karl, 71, 314
Elections, 61, 74, 116, 319
Ellenborough, Edward, 30
Elphinstone, William, 30, 35, 36, 37
Embedded Training Teams (ETT), 86, 267
Emulation, 280, 307–8
Enduring Freedom, *see* Operation Enduring Freedom
ESC, *see* Executive Steering Committee
Estonian military, 123
Ethnic groups: in Kandahar, 221; Soviets and, 40–41; in Uruzgan, 164. *See also* Networks; Pashtun tribes
Ethnographic intelligence, 174–75, 176, 179–80, 185
ETT, *see* Embedded Training Teams
EU, *see* European Union
Europe: demilitarization, 11; Eastern, 40; public support for Afghanistan war, 11; strategic cultures, 14–15. *See also individual countries*
European Union (EU), 92–93, 100, 101
Executive Steering Committee (ESC), 87, 105n9
Exum, Andrew, 202

INDEX   335

Fahim, Mohammad Qasim, 264, 265, 266, 278
FAO, *see* Food and Agriculture Organization
Farrell, Theo, 6, 7, 139
Feith, Doug, 60
Field Manual 3-24 (FM 3-24; US Marine Corps Counterinsurgency Doctrine Publication), 18, 64, 199, 204, 307
First Afghan War (1839), 29, 35
FM 3-24, *see* Field Manual 3-24
Food and Agriculture Organization (FAO), 174
France: Napoleonic wars, 35; Normandy invasion, 18; public support for Afghanistan war, 11; strategic culture, 15; in World War I, 12–13; in World War II, 4–5
Franks, Tommy, 62
Fraser, David, 227
Fraser, Sheila, 223
Freers, Werner, 211

Gaddafi, Muammar, 194–95. See also Libya
Gass, Simon, 89
Gates, Robert, 105n11, 322
Gentilini, Fernando, 88
German military: in Balkans, 211; defensive purpose, 195; expeditionary forces, 195; future challenges, 214–15; lack of counterinsurgency doctrine, 199; lack of lessons learned process, 193–94; officers, 198–99, 210–11; operational adaptation, 193; organizational culture, 203; parliamentary control, 195–96, 204; strategic culture, 14–15, 214, 304; in World War I, 12–13, 15; in World War II, 4–5
German military in Afghanistan: adaptive pressures, 204–5, 207, 214, 305; Afghan forces and, 206; aircraft, 203–4; armored vehicles, 202; battles, 192, 207–8, 209, 294; casualties, 197, 198, 202, 205, 294, 304; changing perceptions, 209–10; civilian casualties, 192, 200–201, 210, 214; civilians serving with, 205; constraints, 193, 202–4, 305; counterinsurgency operations, 193; domestic politics and, 205, 213; fire support, 203–4, 205; helicopters, 203; Kunduz airstrike, 209–10, 214, 305; legal environment, 192, 200–201, 304; lessons learned, 214–15; medals, 192, 207, 208–9; operational adaptation, 193, 201, 203, 204–7, 214, 294, 304; population-centric warfare, 199–200; Provincial Reconstruction Teams, 201, 209, 294, 305, 315–16; resources, 202–3; rules of engagement, 203, 304, 305; special forces, 193, 196, 206; stabilization mission, 197, 201, 212; strategic culture, 15, 194–201, 207, 208–9, 304–5, 308; strategy, 201–2, 211–13; troop levels, 196–97, 304; US forces and, 204, 206–7, 294, 308
German Military Police (*Feldjäger*), 192
Germany: basic law, 195; Bundestag, 195–96, 204, 210, 304; bureaucracy, 211; civil-military relations, 195–96, 205, 304, 305; domestic politics, 193, 194, 196, 197, 205, 210, 212, 213; Islamic extremism, 212; Ministry of Defense, 197–98, 200, 204, 205, 210–11, 215, 294; strategic culture, 14–15, 193, 194–201, 207, 208–9, 214, 304; view of NATO, 14
Ghilzai tribes, 164, 165, 175, 179–80
Gorbachev, Mikhail, 43, 44, 47
Government of the Islamic Republic of Afghanistan, *see* Afghan government
Graham, Bill, 225
Grant, Tim, 227
Great Game, 24, 25, 31, 34, 38, 46
Grissom, Adam, 6, 7
Gromyko, Andrei, 44
Guttenberg, Karl-Theodor zu, 197–98, 210, 294

Haaland, Torunn Laugen, 140
Halberstam, David, 24
Haqqani network, 243, 256, 281–82
Harper, Stephen, 228–29, 230, 234, 235–36
Helicopters, 124–25, 203, 224–25, 230, 236–37, 300

Helmand campaign (British): battles with Taliban, 109–12, 114–16, 118, 128, 293; challenges, 108–9, 117–18, 128; counterinsurgency operations, 91–92; initial deployment, 109, 292; Mobile Operations Groups, 110, 111, 147, 292; operational adaptation, 117–21, 128–29, 138, 292–93, 299–300, 305–6; Operation MOSHTARAK, 72, 120; Operation PANCHAI PALANG, 116, 117, 120, 125; Operation SOND CHARA, 116; Operation TOR SHPAH, 116–17; plans, 118, 119, 296; platoon house strategy, 110, 122, 144, 292; population protection, 112–14, 119–20, 293, 316; resources, 123–25, 126, 310; securing central Helmand, 115–17, 129, 293; stabilization activities, 113–14, 115; training, 120; US Marines and, 117, 129, 138. *See also* Danish Armed Forces in Helmand

Helmand province: Afghan National Army in, 112, 116, 292, 293; elections, 116; governors, 91, 97, 109–10, 278; Green Zone, 111, 130n13, 146, 147, 149, 292; ISAF expansion, 90; Provincial Reconstruction Team, 114, 116; Taliban insurgency, 109–12, 114–16, 118, 128, 244, 245, 250, 259, 292, 293

Helsø, Jesper, 141, 142–43, 146, 147, 151, 153, 154

Henault, Ray, 232

Hillier, Rick, 223, 224, 225, 226–27, 232–33, 235, 296

Hizb-i Islami, 256, 258

Howard, Michael, 73, 297

Humanitarian wars, 11

Huntington, Samuel, 147–48, 235

Hutton, Neil, 114

IEDs, *see* Improvised Explosive Devices
IJC, *see* ISAF Joint Command
Improvised Explosive Devices (IEDs): casualties, 125, 127, 150, 229–30, 248, 249, 299; deployment, 251, 252; effectiveness, 250, 252–53, 253 (fig.); evolution, 9, 251, 299; in Iraq, 252; neutralized, 253 (fig.), 271; number of attacks, 126, 127, 249, 253, 253 (fig.), 259, 299; production, 251–52; specialists, 250, 251, 252; Taliban use, 115, 250–53, 298–99; in Uruzgan, 179, 299. *See also* Counter-Improvised Explosive Device measures

IMU, *see* Islamic Movement of Uzbekistan

India, 28, 29, 31, 33, 36

Innovation: causes, 89–90; civilian leadership, 17; definition, 6; by ISAF, 84–95, 97, 103; military failure and, 9; by NATO, 83; relationship to adaptation, 6–7, 7 (fig.); shaping factors, 90–95, 97; technological change and, 10; in twentieth century, 10. *See also* Technological change

Insurgencies: in Iraq, 4, 250, 252, 254; successful, 321. *See also* Counterinsurgency; Taliban insurgency

Intelligence, Surveillance, and Reconnaissance (ISR) platforms, 69, 291, 300

Intelligence collection: adaptation, 295; in counterinsurgency, 174, 295; ethnographic intelligence, 174–75, 176, 179–80, 185; population-centric warfare, 174–75, 176, 179–80, 185, 190n61; by Taliban, 254; in Uruzgan, 174–75, 176, 179–80, 185, 295; by US military, 70

Intelligence services: Afghan, 265; CIA, 64–65, 66; Dutch, 175, 176; KGB, 41, 49n20; Pakistani, 246

International Security Assistance Force (ISAF): adaptation, 84; alliance politics and, 13–14, 83–84, 88–89, 97, 311–12; casualties, 90, 249, 322; challenges, 91–94, 95–97, 104, 311, 320, 321–22; coalition coherence, 315–17; command structure, 84, 85, 86–87, 87 (fig.), 97–98, 99, 103, 314; Comprehensive Strategic Political-Military Plan, 94–95, 96–97, 312–13; coordination with

civilian organizations, 87–88, 315;
counterinsurgency operations,
72, 85, 87, 89, 90, 94, 314–15;
counterinsurgency policies, 71, 204;
counterterrorism, 86; expansion, 1,
90–91, 95, 121, 201; force levels, 93;
innovation, 84–95, 97, 103; legacy,
322–23; mission, 153–54; motives
of participating countries, 301–2;
NATO alliance politics and, 13–14;
non-NATO partners, 102; planning,
94–95, 119–20; Special Forces, 86,
100; transition strategy, 89, 323; troop
levels, 96; withdrawal plans, 281–82.
See also NATO Training Mission
Afghanistan; *and individual national
forces*
Iran: Pasdaran, 247, 257; support of
Taliban insurgency, 252
Iraq: NATO Training Mission, 85; social
and institutional engineering, 77n9
Iraq War: British military, 4, 16, 121–22,
301; Bush administration policies, 61,
63, 313; counterinsurgency operations,
7, 18, 71, 315; ending, 314; European
assistance, 92–93, 160, 166; Improvised
Explosive Devices, 252; insurgents,
4, 250, 252, 254; link to Afghanistan
War, 57–58, 310, 313; resources, 57, 59;
targeting process, 70; troop levels, 57,
58 (table); US command structure, 86
Irregular warfare, 27, 53, 54, 63, 64, 307.
See also Counterinsurgency
ISAF, *see* International Security Assistance
Force
ISAF Joint Command (IJC), 86, 87, 89,
119, 204
Islamic Movement of Uzbekistan (IMU),
258
ISR, *see* Intelligence, Surveillance, and
Reconnaissance platforms
Italy, 265

Jakobsen, Peter Viggo, 145, 148
Jama't-e Islam, 247, 265, 266
Japan: mission in Afghanistan, 265;
strategic culture, 14

JFCOM, *see* Joint Forces Command
JHP, *see* Joint Helmand Plan
JIEDDO, *see* Joint Improvised Explosive
Device Defense Organization
Jihadists, 79n36, 242, 246, 247, 258. *See
also Mujahedeen*
Joint Force Headquarters (JFHQ), 84,
98–99, 102, 311
Joint Forces Command (JFCOM), US,
64, 307
Joint Helmand Plan (JHP), 118, 119, 296
Joint Improvised Explosive Device
Defense Organization (JIEDDO),
77n11, 249
Jung, Franz Josef, 197, 202, 208, 210, 212,
305

Kabul: airport, 270; British garrison,
29–30, 35, 36; Dutch embassy, 174–75;
US embassy, 68, 71. *See also* Afghan
government
Kabul-centric training mission, 219, 228,
235, 238n8
Kabul Military Training Center (KMTC),
266–67, 272
Kamp, Henk, 163, 170
Kandahar airfield, 67, 124
Kandahar province: battles, 255–56;
Canadian forces, 146, 219, 220–21,
227, 233–34; geography, 221; ISAF
expansion, 90, 123; local interest in
joining Afghan army, 81n61; Operation
Hamkari, 72; Operation Medusa, 178,
222, 223, 227, 236–37; Pashtun tribes,
221; Taliban insurgency, 69, 221–22,
244, 252, 255; Village Stabilization
Program, 69–70
Karzai, Hamid: associates and relatives
assassinated by Taliban, 249, 299;
Bonn process and, 288; Helmand
campaign and, 122; Popalzai network,
165, 278; provincial governors and, 91,
97, 165; Senior Civilian Representative
and, 88
Karzai government: incompetence
and corruption, 58; legitimacy,
317; negotiations with Taliban, 97;

patronage networks and, 319. *See also* Afghan government
Kean, John, 29, 35, 36
KGB (Komitet Gosudarstvennoy Bezopasnosti; Committee for State Security), 41, 49n20
Khan, Jan Mohammed, 165, 167, 174–75, 179–80, 185, 187n20, 295
King, Anthony, 146
Kipling, Rudyard, 24
Klein, Georg, 209–10
KMTC, *see* Kabul Military Training Center
Köhler, Horst, 198, 208
Kosovo War, 13, 85, 96, 98
Kratzer, David, 62
Krause, Joachim, 200
Kreps, Sarah, 11–12
Krulak, Charles, 16
Kunduz: German airstrike, 209–10, 214, 305; German casualties, 205; Provincial Reconstruction Team, 201, 209

Lacroix, Jocelyn, 226
Laroche, Guy, 227
Leadership: adaptation, 18; military, 37; of Taliban insurgency, 70, 242–44, 246, 256–57
Learning: cycles, 54; institutionalization, 159–60, 181–83, 305–7. *See also* Adaptation
Learning organizations, 15, 74
Leslie, Andrew, 226, 239n19
The Liaison Office (TLO), 167
Libya, 104, 194–95
Limited access orders, 276–78, 281
Lorimer, John, 111–12
Luhmann, Niklas, 136, 139

Machiavelli, Niccolò, 83
Mack, Andrew, 321
Mackay, Andrew, 113
Maddison, Greg, 232
Maizière, Thomas de, 208
Mandel, Robert, 317
Manley, John, 229, 234
Manley Panel, 224, 229–30, 235, 236–37

Marine Corps, *see* US Marines
Marine Corps Warfighting Laboratory (MCWL), 16
Marshall, Andy, 61–62
Martin, Paul, 221, 225, 229
MC, *see* Military Committee
McChrystal, Stanley: campaign planning, 71, 89, 96, 119–20, 138, 150, 314; counterinsurgency strategy, 68–69, 70, 71, 72–73, 75, 89, 314; Obama administration and, 68, 96; request for additional forces, 68, 298, 314; rules of engagement, 203; Senior Civilian Representative and, 88
McInnes, Colin, 11
McKiernan, David, 68, 86
McNeill, Dan K., 91–92, 312
MCWL, *see* Marine Corps Warfighting Laboratory
Meigs, Montgomery, 232
Merkel, Angela, 210, 212, 213
Merom, Gil, 321
Messenger, Gordon, 116
Michel, Alexandra, 148, 153
Middelkoop, Eimert van, 163
Migdal, Joel S., 319
Military adaptation, *see* Adaptation
Military Committee (MC), North Atlantic Council, 94, 317
Military innovation, *see* Innovation
Militias: drug trafficking, 91; incorporation in Afghan National Army, 264–66; local, 109, 320; tribal, 176, 277
Mine-Resistant Ambush Protected (MRAP) vehicles, 126, 127, 299–300
Ministry of Defence, Denmark, 137, 141, 143–44, 152–53, 301–2
Ministry of Defence, Netherlands, 159–60, 166, 167, 183
Ministry of Defense (MOD), UK, 124, 125, 127, 300, 309
Ministry of Defense, Afghanistan, 85, 264, 265, 266, 272
Ministry of Defense, Germany, 197–98, 200, 204, 205, 210–11, 215, 294

Ministry of Foreign Affairs, Denmark, 137, 141, 143, 301
Ministry of Foreign Affairs, Netherlands, 166, 167, 173, 175
MOD, *see* Ministry of Defense
Møller, Lars, 149, 150
Moltke, Helmuth von, 3–4, 198
Moscow Institute of Oriental Studies, 41
MRAP, *see* Mine-Resistant Ambush Protected vehicles
Mujahedeen: Pakistani support, 256; resistance to Soviets, 28, 42, 47; in Uruzgan, 164–65; US support, 44, 45, 50n28. *See also* Jihadists
Mulholland, Sean, 204
Mullen, Michael, 58
Mulroney, David, 230

NAC, *see* North Atlantic Council
Nagl, John A., 15–16, 200
Narcotics, *see* Drugs
National Directorate for Security (NDS), Afghanistan, 265
Nation-building, *see* State-building
NATO (North Atlantic Treaty Organization): adaptation, 83–84, 97–104, 312–13; Bonn process, 60–61, 66, 97, 264, 288; collective defense mission, 98–100; command structure, 98–99, 311; cooperative security, 101–2; counterinsurgency doctrine, 103, 105n12, 313; crisis management, 98, 100–101, 107n32, 311; dialogues with non-member states, 101–2; domestic political influences, 10–11; effects of Afghanistan War, 83–84, 97–104, 311–12; expansion, 95; expeditionary force planning, 99–100; future threats, 103; institutional network, 93, 97; Joint Force Headquarters, 84, 98–99, 102, 311; Kosovo War, 13, 85, 96, 98; lack of strategy, 316–17; military transformation, 142; political role, 88; Senior Civilian Representative, 87, 88, 89, 100, 103, 105n10, 315; as strategic actor, 290, 311–13; strategic adaptation, 84, 308; Strategic Concept (2010), 98, 100, 101, 104, 289, 311; summits, 93–95, 100, 121, 142, 153, 312. *See also* Alliance politics; International Security Assistance Force
NATO advisors: attacks on, 272, 323; Embedded Training Teams, 86, 267; embedded with ANA, 86, 263, 267, 275–76; operational mentoring and liaison teams, 86, 105n6, 173, 206, 267, 283n23, 315
NATO Training Mission Afghanistan (NTM-A): ANA adaptation and, 270, 275–76, 280–81; assessments of ANA capabilities, 272–73; budget, 263, 282n2; commanders, 86, 313; establishment, 85–86, 87, 313; personnel, 263, 282–83n2; training centers, 263
NATO Training Mission Iraq (NTM-I), 85
Natynczyk, Walt, 228, 233, 235, 240n34
NDS, *see* National Directorate for Security
Netherlands: domestic politics, 160, 162–63, 165–66, 167, 297–98, 309; embassy in Kabul, 174–75; foreign policy, 166; Ministry of Defence, 159–60, 166, 167, 183; Ministry of Foreign Affairs, 166, 167, 173, 175; National Audit Office, 159; NATO membership, 162; Parliament, 169, 173; public support for Afghanistan war, 11; strategic culture, 162; strategic motives for Afghanistan deployment, 162–63
Netherlands Armed Forces: Battle Groups, 173, 177, 184; effects of Srebrenica massacre, 167, 176, 302; as expeditionary force, 159, 160; funding, 185; intelligence service, 175, 176; in Iraq, 160, 166; lack of lessons learned process, 159–60, 161, 182–83, 306; operational adaptation, 159, 160; organization, 160; peacekeeping operations, 160; size, 160; special forces, 302; stability projection, 166, 167; technology, 161; 3D approach, 159, 162, 166, 181, 296; training, 177–78,

181–82, 185, 306; weaknesses, 159–60. See also Task Force Uruzgan

Networks: in Afghan National Army (ANA), 277, 278–79, 280, 303; in Afghan society, 277–79, 281–82, 303, 319; in Danish Armed Forces, 148–49, 306; Popalzai, 165, 175, 179–80, 278; in Taliban insurgency, 243–44, 256–57, 279

Newman, Dean, 65–66

Nongovernmental organizations (NGOs): military cooperation with, 144, 315; Taliban and, 248; in Uruzgan, 167, 173–74, 184–85, 316

Noonan, Steve, 227

Norms, 14, 16–17, 177–78

Norris, J. A., 36

North Atlantic Council (NAC), 13, 87, 92, 94, 96, 102, 317

North Atlantic Treaty Organization, see NATO

Northern Alliance, 264, 278

Nott, William, 36

NTM-A, see NATO Training Mission Afghanistan

NTM-I, see NATO Training Mission Iraq

Obama administration: Afghanistan withdrawal plans, 282, 314; counterterrorism, 96; international reactions to election, 89–90; NATO policy, 95–96, 104; state-building in US, 322; strategic reviews, 59, 69, 96, 312; strategy in Afghanistan, 52, 68–69, 75, 90, 95–96, 314; troop build-up in Afghanistan, 58–59, 68–69, 75, 78n16, 291, 298, 310

Odierno, Raymond, 86

OEF, see Operation Enduring Freedom

Omar, Mullah, 164, 243, 247, 248

OMLTs, see Operational mentoring and liaison teams

Operational adaptation: by British military, 108, 117–21, 127, 128–29, 138, 292–93, 299–300, 305–6; challenges, 5; in counterinsurgency, 72–75; by Danish military, 145, 147, 148, 149, 150, 151, 154, 306–7; definition, 2–3; by Dutch military, 159, 160, 176–81, 185–86, 293–94, 295, 299; by German military, 193, 201, 203, 204–7, 214, 294, 304; institutionalizing, 159–60, 181–83, 305–7; by NATO, 84; pressures for, 290–91; by Red Army, 42, 43; by Taliban, 250–58, 293; by US military, 51, 53, 54–55, 68, 72–76, 291–92, 294–95, 296–97, 310. See also Adaptation

Operational challenges, as adaptation driver, 3, 8–9, 275–76, 279–80

Operational mentoring and liaison teams (OMLTs), 86, 105n6, 173, 206, 267, 283n23, 315

Operation Enduring Freedom (OEF): Canadian forces, 219, 220; defeat of Taliban, 52, 65–66, 242, 266; in eastern provinces, 90; German special forces, 196; invasion, 51, 52, 63, 64–66; ISAF and, 85; training mission, 85–86, 266. See also Afghanistan War; US military in Afghanistan

Pakistan: allies of Taliban, 242, 244; Directorate for Inter-services Intelligence, 246; Durand Line, 33; support of Taliban insurgency, 247, 250, 252, 256

Panjpai Durrani, 164, 165, 175

Paris, Ronald, 318

Pasdaran, 247, 257

Pashmul, battle of, 255–56

Pashtun tribes: in Kandahar, 221; networks, 165, 175, 179–80, 278; Popalzai, 165, 175, 179–80, 278; in Uruzgan, 164–65, 175, 179–80; warriors, 259

Patrimonial networks, see Networks

Patyani, Amlaqallah, 281

Petraeus, David, 69, 70–71, 75, 86, 314

Piaget, Jean, 139, 142

Police: Afghan national, 72, 112, 265, 267–68, 269, 292; local, 69–70, 271; training, 69–70, 291

Politics: adaptation pressures, 2;

bureaucratic, 224, 236. *See also* Alliance politics; Domestic politics
Pollock, George, 36
Popalzai network, 165, 175, 179–80, 278
Population-centric warfare: by British, 293; by Dutch, 161; by European militaries, 316; German policies, 199–200; intelligence collection, 174–75, 176, 179–80, 185, 190n61; ISAF emphasis, 315, 316; in Task Force Uruzgan, 174–75, 176, 179–80, 185; by US military, 68–72, 119, 315
Posen, Barry R., 9
Powell, Hugh, 116
Provincial Reconstruction Teams (PRTs): British, 105n15, 114, 116, 293; challenges, 312; civil-military relations, 114, 173, 315–16; Dutch, 166, 173, 174; Executive Steering Committee, 87, 105n9; German, 201, 209, 294, 305, 315–16; in Helmand, 114, 116; institutionalization, 100; interagency cooperation, 73; introduction, 67, 91, 105n15, 312; in Kunduz, 201, 209; lack of explicit guidance, 53; role, 87, 91, 100; training, 177; transition to local governance, 88; in Uruzgan, 173, 174, 176, 177, 180, 184, 293, 295, 316
Public opinion: in Canada, 11, 230–31, 231 (fig.); in Denmark, 11, 152, 297; in Germany, 196, 208, 210, 212, 213; pressure for adaptation, 36; in United States, 56–57; on wars, 11–12, 230–31, 298

Qaeda, *see* Al Qaeda
Quagmires, 24–25
Quick Reaction Forces (QRF), 192, 196, 207, 271

Rasmussen, Anders Fogh, 96, 323
Rawlinson, Henry, 30
RC, *see* Regional Command
Red Army in Afghanistan: adaptive capacity, 40, 41–43; adaptive pressures, 28, 42; Afghan resistance, 27–28, 42, 44, 45, 47, 255; casualties, 45–46; Central Asian soldiers, 40–41; civilian advisers, 41; competing missions, 41; failure to adapt, 4–5, 28, 47; invasion, 24, 28, 40; leadership, 42; number of troops, 43; tactical adaptation, 42, 43; weapons, 42, 44; withdrawal, 43–44, 45. *See also* Soviet war in Afghanistan
Regional Command East (RC-E), 90, 291
Regional Command North (RC-N), 196, 201, 203, 207. *See also* German military in Afghanistan
Regional Command South (RC-S), 119, 123, 178, 181, 227, 291. *See also* Kandahar province; Task Force Uruzgan
Rentenaar, Michel, 176
Revolution in Military Affairs (RMA), 10, 61–62, 63, 142
Richards, David J., 91
Ringsmose, Jens, 143–44
RMA, *see* Revolution in Military Affairs
Roberts, Frederick, 32–33, 37
Rocca, Christina, 60
Royal Marines, 110–11, 115–16
Rumsfeld, Donald, 60, 61–62, 64, 65, 67
Russell, James A., 6, 7, 18
Russia: competition with British in Central Asia, 24, 25, 28, 29, 30–32, 34; Crimean War, 31, 36; revolution, 38. *See also* Soviet Union
Rynning, Sten, 143–44

SACEUR, *see* Supreme Allied Commander Europe
Scheffer, Jaap de Hoop, 93–94, 163
Schemas, 139–40, 142
Schneiderhan, Wolfgang, 210, 211
SCR, *see* Senior Civilian Representative
Second Afghan War (1878–80), 31–33, 37, 108
Second-order adaptation, 136, 138–39, 140, 145, 151, 155, 308
Sedwill, Mark, 88, 89
Sembritzki, Jason, 207–8
Senior Civilian Representative (SCR), 87, 88, 89, 100, 103, 105n10, 315
Serbia, Kosovo War, 13, 85, 96, 98
SHAPE, *see* Supreme Headquarters Allied Powers Europe

Shuja Shah Durrani, 29–30
Sisk, Timothy, 318
SMA, *see* Akhonzada, Sher Mohammad
Small states, 137–39, 140, 151–52, 310–11.
    See also Denmark; Netherlands
SOF, *see* Special Operations Forces
Somalia, 231, 232, 322
Soviet Union: Brezhnev Doctrine, 39; civil-military relations, 17–18, 28, 43; collapse, 46; establishment, 38; KGB, 41, 49n20; relations with Afghanistan, 39. *See also* Russia
Soviet war in Afghanistan: civil-military relations, 28; consequences, 43–44; goals, 28; international context, 28; *mujahedeen*, 28, 42, 44, 45, 47, 50n28; as quagmire, 24–25, 43, 45, 47, 48n3; Soviet goals, 39–40, 41, 43, 46–47; US response, 44, 45, 47. *See also* Red Army in Afghanistan
Special Operations Forces (SOF): of Afghan National Army, 271; Canadian, 222; Dutch, 302; German, 193, 196, 206; role in NATO, 100; Soviet, 42; US, 55, 63, 64–66, 69–70, 72, 86, 206, 265–66
Stability, Security, Transition, and Reconstruction (SSTR), 63–64
Stabilization activities: by British military, 113–14, 115; by German military, 197, 201, 212; by US military, 67; Village Stability Operations, 271; Village Stabilization Program, 69–70, 291
Stabilization and Reconstruction (S&R), 162–63, 166, 167, 170, 177, 181, 293, 316
State-building: in Afghanistan, 288, 289–90, 312–13, 321–22; avoidance, 62; challenges, 317–22; definition, 288; democratization, 288–89, 318–19; economic transition, 318, 320; failures, 320–21; future missions, 323; goals, 289; peacekeeping operations and, 288–89; political transition, 318, 319–20; skepticism, 60, 62, 322; social transition, 318
State Department, US, 68, 71, 73
Stinger anti-aircraft missiles, 45, 50n28
Stogran, Pat, 226, 232

Strategic adaptation: in Afghanistan, 308–11; by British military, 7, 108, 121–26, 129, 309–10; by Canadian Forces, 219, 296; challenges, 5, 309; definition, 2; by Dutch military, 169–72, 184, 296; by NATO, 84, 308; by Taliban, 245–49, 259, 299; by US military, 53, 64–72, 75, 307. *See also* Adaptation
Strategic culture: adaptation and, 14–17, 301–5; British, 15, 301; changes, 15, 16–17; definition, 14; Dutch, 162; French, 15; German, 14–15, 193, 194–201, 207, 208–9, 214, 304, 308; Japanese, 14; US, 15
Struck, Peter, 212
Suicide bombings, 254
Supreme Allied Commander Europe (SACEUR), 13, 96
Supreme Headquarters Allied Powers Europe (SHAPE), 94, 98, 99, 100, 311, 317

Tactical Combat Advisory Planning Framework (TCAPF), 74
Tactics, techniques, and procedures (TTPs), 7, 149, 150, 151, 214, 306–7
Talentino, Andrea K., 320
Taliban: defeat by US, 52, 65–66, 242, 266; high value targets, 70; leaders captured or killed, 206; link to Al Qaeda, 212; negotiations with, 97; in Pakistan, 66, 67, 242, 247, 252
Taliban insurgency: battles, 66–67, 109–12, 114–16, 118, 128, 250, 255–56, 267, 275; casualties, 115, 244–45; Code of Conduct, 248; command structure, 65; communications, 254, 258; foreign fighters, 79n36, 242, 246, 250, 258, 299; in Helmand, 109–12, 114–16, 118, 128, 244, 245, 250, 259, 292, 293; IED teams, 251, 252; image concerns, 248; intelligence collection, 254; in Kandahar, 69, 221–22, 244, 252, 255; leadership, 70, 242–44, 246, 256–57; looting, 245; Military Commission, 256–58; mobilization, 242–43; motives, 244–45; networks, 243–44, 256–57,

279; number of attacks, 247, 249, 249 (fig.); operational adaptation, 250–58, 293; planning and coordination, 254–58, 259–60; recruitment, 246–48; relations with populace, 244, 248, 258, 295; resilience, 56, 58, 243; rules of engagement, 248; shaping factors in adaptation, 244, 259, 298–99; strategic adaptation, 245–49, 259, 299; strategic interaction with ISAF, 52; strategy, 243, 245–46; suicide bombings, 254; surveillance and targeting, 70; tactics, 114–15, 250, 254, 258, 292; targeted assassinations, 248–49, 261n19, 299; training, 247–48, 250; in Uruzgan, 164, 170, 175–76, 293–94. *See also* Improvised Explosive Devices

Task Force Helmand (TFH), 112, 113, 116, 123, 153–54, 292–93, 310. *See also* Danish Armed Forces in Helmand; Helmand campaign

Task Force Uruzgan (TFU): aircraft, 179; battle of Chora, 175–76, 180, 293–94, 302; campaign plans, 167–69, 168 (fig.), 170, 171–72, 296, 298, 315; challenges, 161, 176; civilian advisers, 175, 176, 180, 184, 294, 316; costs, 183; counterinsurgency operations, 161, 170, 172, 178, 184, 294, 295, 296, 306; countries contributing to, 173, 175; domestic politics and, 160, 162–63, 165–66, 167, 297–98, 309; end of, 160–61; ethnographic intelligence, 174–75, 176, 179–80, 185; extension in 2008, 160, 163, 169, 184, 297–98; intelligence collection, 295; large-scale operations, 180–81, 185; learning, 159, 177, 181–83, 306; nonkinetic activities, 176, 179, 180, 182, 185; operational adaptation, 159, 176–81, 185–86, 293–94, 295, 299; operational environment, 164–65, 167, 170; organization, 172–74, 184–85; patrols, 178–80, 185, 299; population-centric warfare, 174–75, 176, 179–80, 185; Provincial Reconstruction Team, 173, 174, 176, 177, 180, 184, 293, 295, 316; reconstruction and development mission, 171, 173, 183, 184; size, 172–73, 184, 309; stabilization and reconstruction goals, 162–63, 166, 170; strategic adaptation, 169–72, 184, 296; strategic motives, 162–63; strategy, 165–67, 169–71, 172, 183–84, 293, 296, 298; training, 177–78, 185; vehicles, 179, 299

TCAPF, *see* Tactical Combat Advisory Planning Framework

Technological change: as adaptation driver, 3, 9–10, 50n23; counter-IED, 126–27, 179, 298, 299–300; in Netherlands Armed Forces, 161; in nineteenth century, 37; in twentieth century, 50n23

Terriff, Terry, 6, 7, 16

Terrorism, *see* Al Qaeda; Counterterrorism; War on terror

TFH, *see* Task Force Helmand

TFU, *see* Task Force Uruzgan

Third Afghan War (1919), 33, 37–38, 49n14

Thomas, Jerry, 110

Thruelsen, Peter Dahl, 145

TLO, *see* Liaison Office

Tora Bora, 66

Training: of Afghan National Army, 62, 263, 265, 266–67, 268, 272, 273, 274; of Afghan National Security Forces, 85–86, 105n6, 137, 219, 228, 235, 238n8, 263; of British Army, 120; by British forces, 120; by Canadian Forces, 219, 228, 235, 238n8; in counterinsurgency, 71; in cultural awareness, 177–78; by Danish Armed Forces, 137, 154; of Danish Armed Forces, 149; Embedded Training Teams, 86, 267; Joint Force Training Center, 99; Kabul Military Training Center, 266–67, 272; of Netherlands Armed Forces, 177–78, 181–82, 185, 306; of police, 291; of Provincial Reconstruction Teams, 177; of Taliban, 247–48, 250. *See also* NATO Training Mission Afghanistan

TTPs, *see* Tactics, techniques, and procedures

UAVs, *see* Unmanned aerial vehicles
UCP, *see* Uruzgan Campaign Plan
UN, *see* United Nations
UNAMA, *see* United Nations Assistance Mission in Afghanistan
United Kingdom, *see* Britain
United Nations (UN): crisis management, 101; Food and Agriculture Organization, 174; peacekeeping operations, 11, 101, 222, 288–89; Security Council Resolution 1378, 288
United Nations Assistance Mission in Afghanistan (UNAMA), 87, 88, 101, 174
United States: centrality in ISAF, 313–15; civil-military relations, 58–59, 63; domestic politics, 56–59, 313–14; embassy in Afghanistan, 68, 71; objectives in Afghanistan, 55, 59, 60–61; public support for Afghanistan war, 11, 56–57, 298; relations with Afghanistan, 39; relations with Europe, 92–93, 95; State Department, 68, 71, 73; strategic culture, 15; support of *mujahedeen*, 44, 45, 50n28; Vietnam War, 16, 24, 74. *See also* Bush administration; Defense Department; Iraq War; Obama administration
United States Agency for International Development (USAID), 73
US Air Force, 209–10
US Army: adaptation in World War II, 18; Field Manual 3-24, 18, 64, 199, 204, 307
US Marines: battles with Taliban, 115; in Europe, 16; Field Manual 3-24, 18, 64, 199, 204, 307; in Helmand, 117, 129, 138; organizational culture, 16; Warfighting Laboratory, 16
US military in Afghanistan: adaptive pressures, 55–64, 75, 310; capacity-building mission, 55–56; challenges, 51–52, 54, 66; civilians serving with, 55–56, 59, 61, 71, 72, 73, 295, 297; counterinsurgency operations, 53, 67–68, 70–75, 291; defeat of Taliban, 52, 65–66, 242, 266; domestic political influences, 56–59, 313–14; humanitarian role, 65; intelligence collection, 70; learning, 64, 294–95; mission creep, 55–56; objectives, 55, 59, 60–61, 62–63, 75; operational adaptation, 51, 53, 54–55, 68, 72–76, 291–92, 294–95, 296–97, 310; Operation Anaconda, 66–67, 291; phases of war, 52, 66; planning, 64–65; population-centric warfare, 68–72, 119; shaping factors in adaptation, 52, 63–64; Special Forces, 55, 63, 64–66, 69–70, 72, 86, 206, 265–66; stabilization missions, 67; strategic adaptation, 53, 64–72, 75, 307; strategies, 54, 59–60; Tora Bora battle, 66; troop build-up, 55, 58–59, 68–69, 75, 78n16, 291, 298, 310; troop level cap, 62, 68, 313; troop levels, 57, 58 (table), 68, 75; withdrawal plans, 59, 282, 314. *See also* Operation Enduring Freedom
Unmanned aerial vehicles (UAVs): British, 124, 126; Canadian, 230, 236–37; ISAF use, 70, 251
Urgent Operational Requirements (UORs), 126, 127, 298, 299–300
Uruzgan Campaign Plan (UCP), 171, 172, 172 (fig.)
Uruzgan province: Afghan Development Zones, 169, 170, 173, 178, 293; ethnographic research, 174–75; governors, 165, 167, 174–75, 295; ISAF expansion, 90; Pashtun tribes and subtribes, 164–65, 175, 179–80; social fragmentation, 164–65, 174–75, 179–80, 185; Taliban insurgency, 164, 170, 175–76, 293–94. *See also* Task Force Uruzgan

Vance, Jonathan, 227, 233–34
Verhagen, Maxime, 162
Vietnam War, 16, 24, 74
Village Stability Operations (VSO), 271
Village Stabilization Program (VSP), 69–70, 291
Vleugels, Theo, 167, 170, 175
VSO, *see* Village Stability Operations
VSP, *see* Village Stabilization Program

Warlords, 179, 264, 265, 276–77, 278, 317, 319
War on terror, 59–60, 61–62, 63, 89–90, 312, 313
Wars: adaptation during, 3–4; of democracies, 10–11; humanitarian, 11; military challenges, 3–4
Westerwelle, Guido, 194, 199
Wichert, Peter, 210
Wolfowitz, Paul, 60

World War I, 4, 12–13, 15, 37–38
World War II, 4–5, 12, 18, 41–42

Yugoslavia, *see* Balkans

Zakaria, Fareed, 322
Zirak Durrani, 164, 165, 175
Zisk, Kimberly Marten, 17–18
Zürcher, Christopher, 321